Experimentalist Regional Governance

Jens Matthias Sorg

# Experimentalist Regional Governance

## Policy Analysis in Geographic-Functional Regions

PETER LANG

**Bibliographic Information published by the Deutsche Nationalbibliothek**
The Deutsche Nationalbibliothek lists this publication in the Deutsche Nationalbibliografie; detailed bibliographic data is available online at http://dnb.d-nb.de.

**Library of Congress Cataloging-in-Publication Data**
A CIP catalog record for this book has been applied for at the Library of Congress.



ISBN 978-3-631-83383-4 (Print)
E-ISBN 978-3-631-83635-4 (E-PDF)
E-ISBN 978-3-631-83636-1 (EPUB)
E-ISBN 978-3-631-83637-8 (MOBI)
DOI 10.3726/b17629



Peter Lang – Berlin · Bern · Bruxelles · New York · Oxford · Warszawa · Wien



This publication has been peer reviewed.

www.peterlang.com

*"Solutions are not 'optimal' – the word is irrelevant since no one can know how well any solution will work until it is tried, and no one can know how the alternatives would work – the untried alternatives and even those not as yet conceived."*

*(Neale 1990: 146).*

# Contents

# Abstract

Following their predicted death in the 1980s, cities and geographic-functional regions such as metropolitan regions like FrankfurtRhineMain (FRM) becoming increasingly popular again (Newton 2012) – economical and space-wise, but especially regarding their regulatory structure of governance and political decision-making processes, because political boundaries in which involved institutions operate do not necessarily follow geographic borders. The regulatory structure of governance, the decision-making processes, and the institutional arrangement in geographic-functional regions are the focus of the following research. Current approaches of policy analysis seem to lack a sound analytical framework addressing the functional character of metropolitan regions and linking the structural perspective of political science and the procedural character of policy-making, which Mayntz already proclaimed in the 1980s (Mayntz 1983). The presented research intends to help closing the identified gaps in current governance research.

Coming from the field of jurisprudence and focusing on the European Union (EU), a promising approach seems to be Sabel and Zeitlin's idea of experimentalist governance. It focuses on the regulatory framework of governance described by five consecutive elements – discussion of shared perceptions of joint challenges; definition of goals and measurement systems; implementation; report and review; review of the measurement system, the decision-making process, and the stakeholders (Sabel and Zeitlin 2012b; Sabel and Zeitlin 2010a).

By applying Sabel and Zeitlin's approach to geographic-functional regions below EU-level, the presented research sets the center of attention on regional aspects. Focusing on the regulatory framework of governance, the decision-making process, and the institutional arrangements in the geographic-functional regions of FRM, the concept of experimentalist governance is analyzed in political science for the first time. That enhances the original idea and leads to a focus on the region. With that, experimentalist regional governance (ERG) is derived as a theoretical-analytical framework focusing on the examination of the regulatory framework of governance, the decision-making process, and the institutional arrangement in geographic-functional regions.

The presented research will analyze the assumptions and assertion of Sabel and Zeitlin by evaluating the governance processes of the airport extension in the geographic-functional regions like metropolitan FRM.

Based on a sound scientific approach more than 400 documents with over 31,000 pages are analyzed by content analysis – including decisions, memoranda, minutes, reports, and transcripts. With that, the regulatory structure of governance, the decision-making processes, and the institutional arrangments during the airport extensions of Frankfurt Airport are traced by applying the theoretical derived concept of ERG.

The findings of the presented research have three major aspects. First, the assumptions and assertion lying in the five aspects of the theoretical derived concept of ERG are tested against the social reality in the metropolitan region of FRM, namely the regulatory structure of governance and the decision-making process of the official proceedings of the extension of Frankfurt Airport since the 1960s – the regional planning procedure (RPP) and the road scheme evaluation (RSE) – and the mediation process since 1997. Second, the institutional arrangements including the involved stakeholder of the extension processes are presented. Finally, the research will conclude on governance learning.

Overall, the presented research contributes to a better understanding of governance in geographic-functional regions, like FRM.

# Kurzfassung

Nach ihrem prognostizierten Tod in den 1980er Jahren erfreuen sich Städte und geographisch-funktionale Regionen, wie Metropolregionen, zum Beispiel FrankfurtRheinMain (FRM) wieder zunehmender Aufmerksamkeit (Newton 2012) – wirtschaftlich und räumlich, aber vor allem aber auch hinsichtlich ihrer ordnungspolitischen Strukturen von Governance und politischen Entscheidungsprozessen, denn politische Grenzen, in denen die beteiligten Institutionen agieren, folgen nicht notwendigerweise geographisch-funktionalen Grenzen. Die Regulierungsstrukturen von Regierungsführung, die Entscheidungsprozesse und die Anordnung beteiligert Institutionen in geographisch-funktionalen Regionen stehen im Mittelpunkt der folgenden Forschung. Gegenwärtigen Ansätzen in der Governanceforschung, die sich mit dem funktionalen Charakter von Metropolregionen befassen und die strukturelle Perspektive der Politikwissenschaft mit dem prozeduralen Charakter der Politikgestaltung verbindet, den Mayntz bereits in den 1980er Jahren proklamierte (Mayntz 1983), scheint ein fundierter analytischer Rahmen zu fehlen. Die vorligende Forschung soll dazu beitragen, die identifizierten Lücken in der aktuellen Governanceforschung zu schließen.

Aus dem Bereich der Rechtswissenschaft kommend und auf die Europäische Union (EU) fokussiert, scheint die Idee von Sabel und Zeitlin zu Experimentalist Governance ein vielversprechender Ansatz. Die Idee konzentriert sich auf den regulatorischen Rahmen von Governance, der durch fünf aufeinander folgende Elemente beschrieben wird: Diskussion der gemeinsamen Wahrnehmung gemeinsamer Herausforderungen; Definition von Zielen und Messsystemen; Umsetzung; Bericht und Überprüfung; Überprüfung des Messsystems, des Entscheidungsprozesses und der Interessenvertreter (Sabel and Zeitlin 2012b; Sabel and Zeitlin 2010a).

Durch die Anwendung des Ansatzes von Sabel und Zeitlin auf geographisch-funktionale Regionen unterhalb der EU-Ebene stellt die vorgestellte Forschung regionale Aspekte in den Mittelpunkt des Interesses. Mit dem Fokus auf den regulatorischen Rahmen von Governance, die Entscheidungsprozesse und die beteiligten Institutionen in der geographisch-funktionalen Region von FRM wird das Konzept von Experimentalist Governance erstmals in der Politikwissenschaft analysiert. Das Fokus erweitert die ursprüngliche Idee und führt zu einer Fokussierung auf die Region. Damit wird Experimentalist Regional Governance (ERG) als theoretischer Analyserahmen abgeleitet, der sich auf die Untersuchung des regulatorischen Rahmens von Governance, der Entscheidungsprozesses und der beteiligen Institutionen in geographisch-funktionalen Regionen konzentriert. In der vorliegenden Untersuchung werden die Annahmen und Thesen von Sabel und Zeitlin analysiert, indem die Entscheidungsprozesses des Flughafenausbaus in der geographisch-funktionalen FRM analysiert werden.

Auf der Grundlage eines fundierten wissenschaftlichen Ansatzes der Inhaltsanalyse werden mehr als 400 Dokumente mit über 31.000 Seiten analysiert – einschließlich Entscheidungen, Memoranden, Protokolle, Berichte und Transkripte. Dabei werden der regulatorische Rahmen von Governance, ddie Entscheidungsprozesse und die beteiligen Institutionen beim Flughafenausbau des Frankfurter Flughafens unter Anwendung des theoretisch abgeleiteten Konzepts von ERG nachgezeichnet.

Die Ergebnisse der vorgestellten Forschung haben drei wesentliche Aspekte. Zunächst werden die Annahmen und Thesen, die in den fünf Aspekten des theoretisch abgeleiteten Konzepts von ERG liegen, an der sozialen Realität in der Metropolregion FRM getestet, nämlich an den Regulierungsstrukturen und Entscheidungsprozessen der offiziellen Verfahren zum Ausbau des Frankfurter Flughafens seit den 1960er Jahren – dem Raumordnungsverfahren und dem Planfeststellungsverfahren – sowie dem Mediationsverfahren seit 1997. Zweitens werden die am Ausbauprozessen beteiligten beteiligen Institutionen dargelegt. Abschließend wird die Forschung zum Thema Governance Learning Ergebnisse liefern. Insgesamt trägt die vorgestellte Forschung zu einem besseren Verständnis von Governance in geographisch-funktionalen Regionen wie FRM bei.

# List of acronyms and symbols

| | |
|---|---|
| ACCF | Air Cargo Community Frankfurt e.V. |
| AG | Aktiengesellschaft (stock corporation) |
| AoP | Approval of the Plans |
| ASEAN | Association of South East Asian |
| BallrG | Gesetz zur Stärkung der kommunalen Zusammenarbeit im Ballungsraum Frankfurt/Rhein-Main (Law on promoting communal cooperation in metropolitan Frankfurt/Rhine-Main) |
| BARIG | Board of Airline Representatives in Germany |
| BMVBW | Bundesministerium für Verkehr, Bau- und Wohnungswesen (German Federal Ministry of Traffic, Construction, and Housing), 1999-2009 |
| BMVI | Bundesministerium für Verkehr und digitale Infrastruktur (German Federal Ministry of Traffic and digital Infrastructure), since 2013 |
| BNatSchG | Bundesnaturschutzgesetz (Federal Nature Protection Act) |
| BVerwG | Bundesverwaltungsgericht (Federal Administrative Court) |
| BVerwGE | Entscheidungen des Bundesverwaltungsgerichts (Decision of the Federal Administrative Court) |
| CCFTF | CargoCity Frankfurt Task Force |
| CDU | Christlich Demokratische Union Deutschlands (Christian Democratic Union of Germany) |
| DABB | Dialogforum Airport Berlin Brandenburg |
| DAG | Deutsche Angestellten-Gewerkschaft |
| DFS | Deutsche Flugsicherung GmbH |
| DGB | Deutscher Gewerkschaftsbund |
| DLH | Deutsche Lufthansa AG |
| EC | European Commission |

| | |
|---|---|
| ECJ | European Court of Justice |
| ERG | Experimentalist Regional Governance |
| et al. | et alii (and others) |
| etc. | et cetera (and the rest) |
| EU | European Union |
| e.g. | exempli gratia (for example) |
| FAG | Flughafen Frankfurt/Main AG |
| FDP | Freie Demokratische Partei (Free Democratic Party) |
| FFR | Forum Flughafen & Region |
| Fraport AG | Fraport AG Frankfurt Airport Services Worldwide |
| FRM | FrankfurtRhineMain |
| GG | Grundgesetz (Germans Basic Law) |
| GMO | Genetically Modified Organism |
| GVBl. I S. 427 | Gesetz über den Umlandverband Frankfurt (Law of the Umlandverband Frankfurt) |
| HENatG | Hessisches Naturschutzgesetz (Hessian Environmental Law) |
| HessVGH | Hessischer Verwaltungsgerichtshof (Hessian Higher Administrative Court) |
| HLPG | Hessisches Landesplanungsgesetzes (Hessian Land Planning Act) |
| HMULV | Hessisches Ministerium für Umwelt, ländlicher Raum und Verbraucherschutz (Hessian Ministry of Environment, Rural Areas and Consumer Protection), 2003-2009 |
| HMWEVL | Hessisches Ministerium für Wirtschaft, Energie, Verkehr und Landesentwicklung (Hessian Ministry of Economics, Energy, Transport, and Regional Development), since 2014 |
| HMWVL | Hessisches Ministerium für Wirtschaft, Verkehr und Landesentwicklung (Hessian Ministry of Economics, Transport, and Regional Development), 1995-2014 |

| | |
|---|---|
| HMWT | Hessisches Ministerium für Wirtschaft und Technik (Hessian Ministry of Economics and Technology), 1969-1991) |
| HOLM | House of Logistics & Mobility (HOLM) GmbH |
| HOLM e.V. | HOLM (House of Logistics and Mobility) e.V. |
| HVwVfG | Hessisches Verwaltungsverfahrensgesetz (Hessian Administrative Procedure Act) |
| ICT | Information and Communication Technology |
| IHK | Industrie und Handelskammer (Chamber of Commerce and Industry) |
| IMF | International Monetary Fund |
| LEPH2000 | Landesentwicklungsplan Hessen 2000 |
| LRM | Logistik RheinMain. Die Wissensinitiative |
| LuftVG | Luftverkehrsgesetz (Air Traffic Law) |
| LuftV-ZustVO | Verordnung zur Bestimmung von luftverkehrsrechtlichen Zuständigkeiten (Regulation on the Determination of Air-Traffic-Rights Responsibilities) |
| Mediationsgruppe | Mediationsgruppe Frankfurt Flughafen |
| MetropolG | Gesetz über die Metropolregion Frankfurt/Rhein-Main (Law of the Metropolitan Region Frankfurt/Rhine-Main) |
| MRM | Mobilität RheinMain |
| NAFTA | North American Free Trade Agreement |
| NFM | Nachbarschaftsbeirat Flughafen München |
| NGO | Non-Governmental Organization |
| NUTS | Nomenclature of Territorial Units for Statistics |
| OECD | Organisation for Economic Co-operation and Development |
| PDIA | Problem Driven Iterative Adaptation |
| Planungsverband | Planungsverband Ballungsraum Frankfurt/Rhein-Main |

| | |
|---|---|
| PlanvG | Gesetz über den Planungsverband Ballungsraum Frankfurt/Rhein-Main (Law of the Planungsverband Ballungsraum Frankfurt/Rhein-Main) |
| RDF | Regionale Dialogforum Flughafen Frankfurt |
| Regionalverband | Regionalverband FrankfurtRheinMain |
| ROG | Raumordnungsgesetz (Regional Planning Law) |
| RoP | Rules of Procedure |
| ROV | Raumordnungsverordnung (Regional Planning Regulation) |
| RP | Regierungspräsidium (Regional Council |
| RPD | Regional Planning Decision |
| RPP | Regional Planning Procedure |
| RPS2000 | Regionalplan/Südhessen2000 |
| RPU | Regionale Planungsgesellschaft Untermain |
| RSE | Road Scheme Evaluation |
| SAARC | South Asian Association for Regional Cooperation |
| SPD | Sozialdemokratische Partei Deutschlands (Social Democratic Party of Germany) |
| SSCI | Social Science Citation Index |
| TTIP | Treaty on European Union Transatlantic Trade and Investment Partnership |
| UVF | Umlandverband Frankfurt |
| UVPG | Gesetz über die Umweltverträglichkeitsprüfung (Law on Environmental Impact Assessment) |

# 1 Introduction

*Over the next century, urban regions will furnish the context of social, economic, and political life for a growing majority of world's citizens.*
(Sellers and Hoffmann-Martinot 2005: 11).

Following their predicted death in the 1980s, cities and geographic-functional regions such as metropolitan regions are becoming increasingly popular again (Newton 2012) – economical and space-wise, but especially regarding their regulatory structure of governance and political decision-making processes, because political boundaries in which involved institutions operate do not necessarily follow geographic borders. The regulatory structure of governance, the decision-making processes, and the institutional arrangement in geographic-functional regions are the focus of the following research.

Today, half of the world's population already lives in cities. Moreover, it is assumed that during the 21st century an increasing number of people will live in geographic-functional regions (Kübler 2012: 402). In Germany already 70 percent of the population and worldwide 3.75 billion people live in geographic-functional regions (Walter-Rogg 2013: 109; WB 2014). The top 50 metropolitan areas are larger than 80 countries of the world (Newton 2012: 409). Within the Organisation for Economic Co-operation and Development (OECD), metropolitan regions are not only home to over half of the world's population, but also centers of economic activity.

Geographic-functional regions such as the metropolitan region of FrankfurtRhineMain (FRM) are recognized as places of economic development and innovation. Around the globe, these metropolitan regions are essential economic engines for the global economy. This makes them vital for the growth and the sustainability of national economies, whereby they define the wealth of their home countries (Chattopadhyay 2011; see footnote for further details).[1] Geographic-functional regions account for more than 50 percent of the total gross domestic product (Ahrend and Schumann 2014b; see footnote for further details).[2] According to new economic geography, this is because economic actors strive towards urban agglomeration to benefit from agglomeration effects (Scott 1998; Storper 1995; Storper 1997b). New institutional economics states that spatial proximities lead to lower transaction costs and support cooperation (Coase 1937; see footnote for further details).[3] For these reasons, metropolitan areas are growth- as well as competitive-oriented regions and they promote economic integration (Federwisch 2012: 44). Thus, metropolitan regions are more productive than less-populated and less-agglomerated areas. Consequently, these geographic-functional regions hold vital importance.

However, this only reflects one side of the coin: as melting pots for social and environmental issues, they are also disintegrated.[4] Fragmented structures to handle regulatory structures of governance, as the sum of all co-existing forms of collective regulations on social issues, hamper productivity and shatter the advantage of agglomeration at the same time (Ahrend, Gamper, and Schumann 2014). In Germany, the metropolitan region of FRM is seen as a region with room for improvement to enhance the fragmented governance structures (Blotevogel 2002: 348). These circumstances make geographic-functional regions and especially

1 Knieling (2009: 2); Kujath (2005: 9); Slack and Chattopadhyay (2013: 1). In cases where more than three sources are relevant, the further results are presented in the respective footnotes for better reading.

2 Ahrend, Gamper, and Schumann (2014); Combes, Duranton, and Gobillon (2011); Brezzi and Veneri (2014: 3); European regions with a population of over 500,000 inhabitants grow approximately 0.2 percentage points faster on a per capita basis than agglomerations with fewer inhabitants (Ahrend and Schumann 2014b: 11f). In absolute numbers Blotevogel (2005) counts 500,000 inhabitants for the center of a metropolitan region and at least 1 to 1.5 million inhabitants for a metropolitan region (Blotevogel 2005: 644).

3 Federwisch (2012: 43); Knieling (2009: 3); Krugman (1991); Williamson (1975).

4 Beside the economic productivity it is also worth to consider that metropolitan regions in particular in developing countries are also centers of poverty and environmental degradation (Slack and Chattopadhyay 2013: 1)

FRM an interesting field for political scientists in which to analyze the regulatory structure of governance, decision-processes, and institutional arrangements.

In the meaning of Mayntz (2003), the regulatory structure of governance is understood as the reconstructable decision-making process of private and non-private institutions regarding the sum of all co-existing forms of collective regulation of social issues. Following Geißel (2013), the institutional arrangement is the surrounding environment of institutions the decision-making processes are embedded in. The institutional arrangement embraces the lower levels and extends beyond the inclusion of trade unions and employers' representatives (Geißel 2013: 10f). Thus, decision-making is negotiated between individuals coming from administrative and nonadministrative institutions as well as citizens (Geißel 2012b: 165).

The research on metropolitan governance dates back to the beginning of the 1930s. (Kübler 2012: 404f; Lackowska 2011a). Since then, metropolization has become a key concept to describe the evolution of geographic-functional regions in economically-leading countries (Ascher 2003: 54). Even though in 1899 reference was made to Groß-Frankfurt and it was discussed as one region, almost no comprehensive and systematic information on the regulatory structure of governance, decision-making processes, and institutional arrangements in geographic-functional regions is available (Ahrend et al. 2014: 5; Piecha 2005: 41).[5] Regarding this, Zürn (2008) states that the analysis of the sum of all co-existing forms of collective regulations on social issues is only at the very beginning (Zürn 2008: 568). Additionally, in some cases contemporary governance research on geographic-functional regions simply speaks of governance failure, especially regarding the metropolitan region of FRM (Hoyler, Freytag, and Mager 2008a; Lackowska 2011a: 111).

This disregard is significant because metropolitan regions – as presented – are important and they will become increasingly relevant for society in the future. Researchers and the United Nations Commission on Human Settlement state that metropolitan regions "will furnish the context of

---

5 The OECD working paper of Ahrend et al. (2014) has to be understood as a first quantitative attempt.

social, economic, and political life for a growing majority of the world's citizens" (Sellers and Hoffmann-Martinot 2005: 11). Hence, politicians ascribe importance to the spatial planning of those regions. Accordingly, the promotion of metropolitan regions is considered as an adequate method to strengthen the growth and international competitiveness of regions (Blotevogel 2010: 7; see footnote for further details).[6] Moreover, metropolitan regions are partly accepted as an administrative unit. As a tool for political and spatial planning, metropolization is supported equally at the European and country-level (Federwisch 2012: 42; Knieling 2009). According to the *Ministerkonferenz für Raumordnung* of April 2005, Germany recognizes eleven European metropolitan regions, including FRM (Blotevogel 2005: 642; Federwisch 2012: 16).[7] Furthermore, FRM is also legally recognized in Germany.[8] There is evidence that especially the German economy relies more on metropolitan regions and the agglomeration of industries in local productive systems compared with other countries of the European Union (EU) (Le Galès and Trigilia 2004: 331).

The aforementioned points and the challenges regarding the regulatory framework of governance of metropolitan regions and the decision-making process were stated and repeated at the *Tag der Metropolregionen* of May 5, 2015, in Frankfurt am Main.[9] "We need more pragmatism!," Benjamin R. Barber stated during the discussions regarding the governance of metropolitan regions or global cities. In this respect, he expressed the importance of a pragmatic governance approach towards geographic-functional regions in the future. This pragmatism seems to be important

---

6 Danielzyk and Oßenbrügge (2003: 34f); Kujath (2005: 9); Nelles and Durand (2014: 106–108).

7 Berlin-Brandenburg (1995, year of recognition), Bremen-Oldenburg (2005), FrankfurtRhineMain (1995), Halle-Leipzig (1997), Hamburg (1995), Hannover Braunschweig-Göttingen-Wolfsburg (2005), Munich (1995), Nürnberg (2005), Rhine-Neckar (2005), Rhine-Ruhr (1995), and Stuttgart (1995).

8 *Gesetz über die Metropolregion Frankfurt/Rhein-Main* (MetropolG), March 8, 2011.

9 *Wenn Bügermeister die Welt regierten, Tag der Metropolregionen*, Tuesday, May 5, 2015, IHK Frankfurt am Main.

for the regulatory structure of governance and decision-making processes, and it helps to overcome the challenges of fragmentation.

The presented research follows the appeal to improve policy analysis and contributes to a better understanding of governance in geographic-functional regions, by historically reconstructing the regulatory structure of governance, its decision-processes, and institutional arrangements during infrastructure projects with supra-regional characteristics. It intends to help close the previously identified gaps in current governance research. The extension of Frankfurt Airport represents the most promising example, because its documentation is most thorough and transparent, unlike in other infrastructural projects (Gohl and Meister 2012: 55).

By analyzing the official proceedings of the extension process of Airport Frankfurt at the outset of the 1960s and the mediation process starting from 1997, the presented research lays out a comprehensive and detailed overview of the regulatory structure of governance, its decision-processes, and institutional arrangements. It also offers indications of governance learning.

## 1.1 Presentation of the problem

Today, the classical definition of Max Weber (2006) no longer seems to fit regarding cities and the agglomeration of economic action and prosperity (Weber 2006: 833-855). This challenge – as introduced before – is clarified in the following paragraph by presenting the more detailed research problem.

Contemporary cities cannot be characterized as relatively closed settlements with one centered "*Lokal*markt" (Weber 2006: 835),[10] representing a political and economic focal point of the agglomeration. European cities are currently characterized by diversity in terms of their spatial structure, networks, social and functional characteristics as well as modes of governance. Western cities have certainly urbanized extensively in the last century (Ascher 2003: 53). As agglomerations, they have outgrown the long-lasting administrative borders of municipalities (Ahrend et al.

10 Text highlighted in original.

2014: 5).[11] Cities have turned into geographic-functional regions with one (monocentric) or several (polycentric) major centralities. They embrace different "territorial systems of economic, social and cultural exchange" (Kübler 2012: 402). According to Ascher (2003), these regions are characterized by the extension of the old conurbations and the fading of the overall urban density, while increasing their suburban areas at the same time. Furthermore, residential areas and urban fabric are more dispersed. Thus, the distinction between urban and suburban areas has partly disappeared and new peripheral polarizations are coming to the fore (Ascher 2003: 55). The following example of FRM graphically illustrates this phenomenon (Fig. 1.1).

From a political perspective and following a geographic-functional definition, the challenge is that metropolitan regions are not necessarily bound to administrative or institutional borders. They comprise different local, regional and – in some cases – national administrative units (Blotevogel 2002: 345; Kübler 2012: 403). The spatial agglomeration of those geographic-functional regions is defined regarding political, economic, social and cultural characteristics (Bartels 1970: 21f; see footnote for further detail).[12] As a result and in contemporary political practice, boundaries of proximity are not comparable with initial political borders, the common limits of a regulatory governance structure (Bathelt and Glückler 2011: 6; see footnote for further detail).[13] Due to this functional character, it is possible that the jurisdictions and the spheres of influence partially overlap. For this reason, no local administration can address all relevant issues, especially if the geographic-functional region is polycentric. Additionally, it is not clearly evident, which the

---

11 However, the discussions about the importance of cities is finding its way back into academia (B. R. Barber 2014; Ketels 2015).

12 Blotevogel (1999: 44); Koch (2013: 399); Sinz (2005: 919f); For a detailed presentation of the classical geographic-functional understanding see Lauschmann (1976). Contemporary efforts by the EU try to match the geographical-functional region with the political levels (Eurostats 2015a).

13 M. E. Porter (1998: 79); (Saxenian (1996); Scott et al. (2008: 27); Here region is recognized and highlighted as an operational level, which is relevant for social action. This phenomenon is discussed in the field of new regionalism in-depth (Fürst 2007: 355).

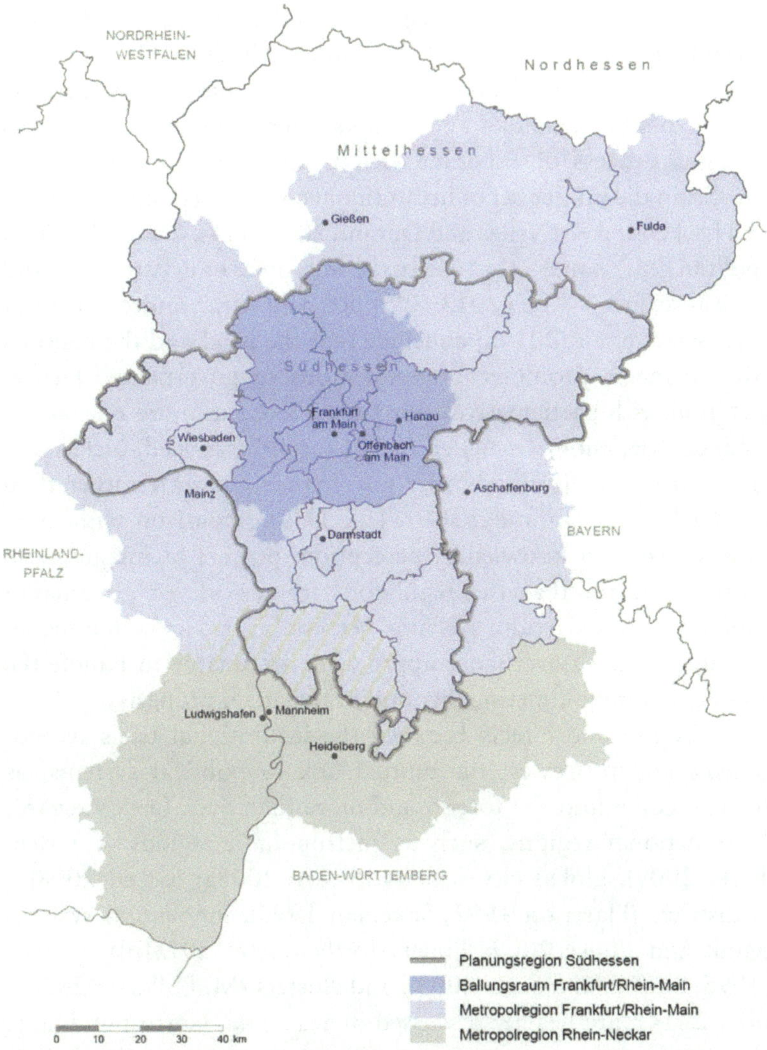


*Figure 1.1: Geographic-functional region of FRM, RegFNP – cartographical illustration (RPDA and Regionalverband 2011: 3).*

responsible authority is *ex ante* (Ahrend and Schumann 2014a: 4; Ahrend et al. 2014: 4). For this specific reason and due the to ascertainable lower productivity with an increasing number of involved administrative institutions, metropolitan regions require sound political coordination (Ahrend, Gamper, and Schumann 2014; Chattopadhyay 2011: 23; Nelles and Durand 2014: 106). This is even more valid for polycentric regions with cross-border links and actions. Here, "governance [...] is built through a process of experimentation, innovation, error, and re-evaluation within the unique set of institutional constraints imposed by the fragmented [...] context" (Nelles and Durand 2014: 114). Under the cloak of metropolitan governance, it is "one of the hot topics of urban policy and urban political science" (Koch 2013: 399). Regarding the improvement of democracy, Geißel et al. (2014b) conclude that the local and the regional level is the strongest promoter of innovations in governance. Hence, experimentation with participative approaches becomes more relevant in the governance discussion (Geißel et al. 2014b: 497). Indeed, Nelles and Durand maintain that "[c]ross-border governance is an art rather than a science" (Nelles and Durand 2014: 119). This comparison with tasks and processes based on knowledge, perception, powers of imagination, and intuition indicates that the regulatory framework of governance of geographic-functional regions is not yet established in a fact-based manner. Contemporary governance approaches are unable to handle the challenges of geographic-functionality and pragmatism properly.

Reciprocal action and effects between the institutional basis of economic systems and processes, the mutual link to political systems, as well as society, come into the foreground more than ever. In this regard, geographic-functional regions, such as metropolitan regions (Gehrlein 2009; Phares 2009), global city regions (Sassen 2008a; Sassen 2008b), industrial districts (Harrison 1992; Saxenian 1996), innovation systems (Blättel-Mink and Ebner 2009; Ebner (forthcoming) 2015[b]), milieus (Storper 1995; Storper and Scott 2003), and clusters (M. E. Porter 2008a; M. E. Porter 2014), are heavily discussed in academia. Correspondingly, besides the EU, further examples such as the Association of South East Asian (ASEAN), the North American Free Trade Agreement (NAFTA), the South Asian Association of Regional Cooperation (SAARC) and others show the importance of geographic-functional regions and regionalism (Pierre 2000b: 4f).

Besides the historical aspect of urban sprawl, the functional aspect of the specialization of regions and spatial mobility, geographic-functional regions are characterized by global and regional – or local – phenomena at the same time (Kübler 2012: 403). This connection was already made by Robertson in 1995 with his portmanteau word "glo*c*alization" (Robertson 1995).[14] Glo*c*alization describes changes occurring due to the interconnection of global and local phenomena (Robertson 1995; Robertson 2012). This facet is apparent in metropolitan regions. Consequently, globalization leads to the re-evaluation of regions (J. Esser and Schamp 2001: 9). Both phenomena – global versus local – have to be reflected in the regulatory structure of regional and metropolitan governance (Lackowska 2011b: 21). They are highly relevant in the context of the "local" extension of a "global" airport (Geis 2005: 130). This point is also apparent from the perspective of spatial embeddedness.[15] The cooperation of actors is not isolated and footloose. At least structurally, cooperation is embedded in the particular regional environment (Salet, Thornley, and Kreukels 2003: 11). Furthermore, spatial embeddedness links globalization with localization – or regionalization. Geographic-functional regions are aligned globally but also embedded in their national, regional, and local environment (J. Esser and Schamp 2001: 11). Challenges of governance occur where traditional jurisdictions are questioned.[16] The subnational administrative level becomes more insistent and metropolitan regions – cities, industrial districts, and clusters – put themselves into the global arena, bypassing other institutions (B. R.

14 Text highlighted in original. For the concept of glo*c*alization see also Swyngedouw (1992) and Swyngedouw (1997). For the debate of globalization and local governance in Germany see Rüdiger and Konegen (2006).

15 Spatial embeddedness see, e.g. Grabher (1994c), Storper (1997b), Storper (1997a), and Storper (2009).

16 Deregulation of the financial markets and the growing volatility of global capital withdraw administration's traditional capacities to control the local economy is one extensively discussed example. Among others, this phenomenon can be observed by the example of value-guaranteed funds. Although the overall market has grown since 2009 already funds focusing on Europe decreased, whereas funds with a global investment universe increased significantly. Funds focusing on emerging economies doubled from 2009 to 2013 (Sorg et al. 2014).

Barber 2014).[17] Under those circumstances, the administration of previous jurisdictions loses its position as the main level of market governance (Crouch and Trigilia 2003: 227; J. Esser 1999; J. Esser and Schamp 2001: 12f). At the same time, societal networks also challenge administrative decisions (Pierre 2000b: 1; Salet, Thornley, and Kreukels 2003: 5–11).

Also, today's "Globalia" (H. Simon 2012) is regarded faster than in the past, whereas the associated changes are seen as being stronger and deeper. Different topics are linked to globalization: politics, public policy, the environment, the economy, labor division, stock markets, culture, religion, ideology, consumption, media, education, sports, health, and technology (Ritzer 2007: 177–538). Glassmann and Voelzkow (2003) highlight that globalization undermines the well-established chains of collaboration in local production systems. They lead to the change of the higher governance system (Crouch and Trigilia 2003: 219). Moreover, Crouch et al. (2004) identified changes in governance modes in different local economies. In regions of machinery industry, a shift from traditional structural organized systems towards more networked systems is identifiable,[18] whereas high-tech regions tend to come up with new district-like forms of local production systems (Le Galès and Trigilia 2004). Indeed, Koch (2013) identified three mechanisms of change for governance: the functional perspective highlights exogenous factors that transform actors' interests, the institutional perspective identifies path-dependence and "moments in which existing institutions are challenged" (Koch 2013: 402) as the promoter for endogenous incremental changes, and the constructive perspective sees political dilemmas and new beliefs as triggers for the change of governance (Koch 2013: 401–405). Accordingly, governance is confronted with the challenge to react to the changing framework conditions, both exogenous and endogenous (Ahrend, Gamper, and

17 A good example here are the comments of New York's former mayor Michael R. Bloomberg at Massachusetts Institute of Technology on November 29, 2011, where he said: "I have my own army in the NYPD, which is the seventh largest army in the world." He was quoted further: "I don't listen to Washington very much, which is something they're not thrilled about." (DM 2011).

18 Contemporary this phenomena is heavily discussed by the keyword 'Industry 4.0' (Hermann, Pentek, and Otto 2015).

Schumann 2014: 5; Crouch and Trigilia 2003; Salet, Thornley, and Kreukels 2003).

At this point, the change of institutions has to be taken into account. According to Ebner (2008), this is one of the major challenges in recent institutional theory (Ebner 2008: 5–8). Challenges in the "Changing Governance of Local Economies" (Crouch et al. 2004) are a part of the presented research. Thus, the issues are raised and questions regarding the regulatory structure of governance and the decision-making processes in geographic-functional regions are accentuated. Today, it becomes necessary to discuss fruitful approaches of governance and policy analysis for geographic-functional regions, because it is to assume that both, the regulatory structure of governance and the decision-making processes, will become increasingly complex in the future.

Lackowska (2011a) as well as Zimmermann and Heinelt (2002) do not confirm a total absence of a regulatory structure of governance within the metropolitan region of FRM. However, Lackowska (2011a) – based on Jassop (2002) – certifies governance failure for FRM. From the perspective of Lackowska (2011a) and Zimmermann and Heinelt (2002), local politics has to be aligned and harmonized with the functional character of a metropolitan region. In this case, Blotevogel (2002) sees backlog demand, especially for FRM.[19] According to him, the metropolitan region of FRM only exists on paper (Blotevogel 2002: 348). Numerous loose and uncoordinated networks of private and non-private institutions as well as unavailable capacity to coordinate joint actions raise "doubts about characterizing them as engaging in governance" (Lackowska 2011a: 111). Nevertheless, the challenges relating to the complexity and fragmentation of political, economic, and social actors, as well as the resulting diversification of power coalitions is not FRM specific; rather it is a major challenge in metropolitan governance and apparent in other regions worldwide (Merk 2014: 6; Salet, Thornley, and Kreukels 2003: 3). Due to the question of legitimacy, the institutional arrangement within metropolitan regions in the "shadow of hierarchy" (Héritier and M. Rhodes 2011) is also problematic, because it is often not

19 Beside the metropolitan region of Rhine-Ruhr.

democratically legitimized. Furthermore, if no efforts are made to link the civil society to the regulatory structure of governance and the decision-making processes, it obstructs participation due to its non-transparency (Zimmermann and Heinelt 2002: 87).

Moreover, geographic-functional regions include more than one jurisdiction at once. FRM includes at least three federal states and five counties. Each of them acts as a representative and its stronghold representing local (self-)interests (Lackowska 2011a: 79). Furthermore, FRM includes several cities claiming supremacy. The geographic-functional region is represented by two state capitals: the city of Wiesbaden, for the Federal State of Hesse and the city of Mainz, for Rhineland-Palatinate. Darmstadt is the seat of the Regional Council, and the city of Frankfurt am Main represents the economic powerhouse of the region. Based on a long-lasting fragmentation in former times and the absence of territorial history, several authors have diagnosed a missing common identity for FRM (Böderlin 2000: 539).[20] This separation is additionally fostered by the characteristic of the German administrative system of townships and municipalities. *Inter alia*, local administrations are in competition to attract firms, because through settlement business tax money is raised (Lackowska 2011a: 80f). These conclusions are considered by Stephans and Wikstrom (2000) and their more general metropolitan reform approach. Conflicts between local authorities and pertinent shortcomings in service provision lead to the lack of a metropolitan-wide political perspective (Lackowska 2011b: 23), which is also identifiable for FRM.[21]

Regarding an appropriate governance system, this draws a picture of incoordination and governance failure (Hoyler, Freytag, and Mager 2008a; Lackowska 2011a: 111). It is assumed that this is related to the inability of previous governance concepts to capture the picture of

---

20 E.g. Freund (2003), Hoyler, Freytag, and Mager (2008a), and Schaffer and Scheck (2006).

21 The point of completion was already made for the 1970s and 1980s more generally: "Beyond the 'Limits of Growth' the intensification of inter-municipal competition for migration gains and purchasing power flows leads to municipal profiling strategies that mainly focus on the correction and embossing of specific 'images' of urban spatial settings [...]" (Durth 1987: 161).

decision-making processes in the geographic-functional regions. In the works of Lackowska (2011a) and Zimmermann and Heinelt (2002), it is apparent that there is always the question of a necessary central institution,[22] although the metropolitan governance discourse in general is affected by the mismatch between the institutional and functional territory (Lackowska 2011b: 21). It appears that common concepts of governance – especially those with a narrow understanding (in delineation of government)[23] – are unable to handle the challenges and the changes of geographic-functional metropolitan regions.

However, on the one hand, contemporary governance approaches are in need to include and handle those challenges of change – relating to proceeding globalization. On the other hand, they have to handle a gaining regionalization and the supremacy claim of "Metropolis". Notably, a geographic-functional region requires specific undertakings for a coherent governance system (Lackowska 2011b: 23). As presented, this holds especially true for FRM as the most globalized metropolitan region of Germany (Blotevogel 2002: 349).

Based on the experimentalist governance approach (Sabel and Zeitlin 2012a; see footnote for further details),[24] experimentalist regional governance (ERG) is theoretically derived. It is assumed that it provides answers to the contemporary and upcoming challenges in geographic-functional regions. It seems to be a sound analytical approach to analyze the regulatory structures of governance, decision-making processes, and the institutional arrangements in geographic-functional regions. As Sabel and Zeitlin (2010a) have exemplified, regarding the EU and their initial approach, experimentalist governance aspects – "common problem solving by a certain mechanism of goal setting," "individual

22 As the *Regionale Planungsgesellschaft Untermain* (RPU) in the 1960s and 1970s, the *Umlandverband Frankfurt* (UVF) from the 1970s to the 1990s, or the *Planungsverband Ballungsraum Frankfurt/Rhein-Main* (*Planungsverband*) in the 2000s, and since 2011 the *Regionalverband FrankfurtRheinMain* (*Regionalverband*), which is identified by the OECD as the metropolitan governance body of FRM (Ahrend et al. 2014: 27).

23 E.g. Benz et al. (1999) and Stroker (1998).

24 Sabel and Zeitlin (2012b); Sabel and Zeitlin (2010a); Sabel and Zeitlin (2010b).

implementation on country level," "reporting, and renegotiation of goals and measurement systems" – are identifiable in regional regulatory structures of governance and decision-making processes. Furthermore, Sabel (2004a) and others[25] identify some forms of experimentalist governance within national parliaments, courts, and constitutions (Sabel and Zeitlin 2010b: 17). For Germany, the planning law for infrastructure projects with supra-regional characteristics has to be mentioned. Under these circumstances, the analysis of the regulatory structures of governance, decision-making processes, and the institutional arrangements by the ideas of Sabel and Zeitlin is perceived as rewarding for political analysis.

Although geographic-functional regions – like metropolitan regions – appear as a blueprint for environments without a central, final decider, Sabel and Zeitlin (2012b) have not referred to those regional agglomerations at all.[26] However, if the conclusion of Sabel and Zeitlin (2012b) is correct, such geographic-functional regional agglomerations are the perfect environment to analyze the regulatory structures of governance, decision-making processes, and the institutional arrangements by the theoretical derived analytical approach of ERG.

25 Dorf and Sabel (1998), Sabel and W. H. Simon (2004), and Zeitlin (2005).

26 One might argue that the EU is a poor example to demonstrate experimentalist in environments without a central, final decider. According to the Treaty on the European Union, the Treaty on the Functioning of the European Union, and the Charta of the Fundamental Rights of the European Union, the EU is the Guardian of the Treaties and ensures that EU law is properly applied (Art. 17 (1), 9d (1) Treaty on European Union) by individuals, national authorities and other EU institutions (Seeger 2008: 88f). The European Commission (EC) can impose sanctions and take actors, administrative and nonadministrative, to the European Court of Justice (ECJ) (Hustedt et al. 2014; Wonka 2008). Especially the Treaty of Lisbon (2007/C 306/01) was established to gain the EU's capacity for action and institutional development to foster decision-making processes and the implementation of agreed actions (Bauer and Baumann 2008: 189; Seeger 2008). However, this normative question is neglected at this point and it is argued here that it should be of interest for further research.

## 1.2 Governance of geographic-functional regions and policy analysis – naming the white spots

In light of the general problem presented, this paragraph outlines the relevance, the stage of research, and the research gap in the policy analysis.

According to the "Oxford Handbook of Governance," "governance has become an important concept and [...] probably one of the most important manifestations of the rise of neo-institutionalism in social science" (Levi-Faur 2012: 3). Nevertheless, as presented above, current research on factors supporting an efficient regulatory structure of governance is at the very beginning (Ahrend et al. 2014: 5; Zürn 2008: 568). As already indicated, this holds for regional and metropolitan governance. This lack of definitions reaches its limits regarding the analysis of the decision-processes (Ahrend et al. 2014; Fürst 2010; Zimmermann and Heinelt 2002). In this context, only one research study is identifiable for Germany. Walter-Rogg (2013) analyzed the metropolitan governance of the geographic-functional regions of Hamburg and Central Germany (Walter-Rogg 2013).[27] Focusing on the differences between Eastern and Western Germany, her analysis identified similarities and differences regarding financing strategies in both regions (Walter-Rogg 2013). However, statements regarding the regulatory structure of governance and decision-processes in geographic-functional regions remain opaque and their analysis is still vague in her analysis. Therefore, in her conclusion Walter-Rogg (2013) requests a "future research on metropolitan governance in general – but of course with regard to the special case of co-operative arrangements over several state borders [...]" (Walter-Rogg 2013: 140).

These identified shortcomings are addressed in the following chapters. The presented research analyzes the regulatory structure of governance, decision-processes, and institutional arrangements in the geographic-functional region of FRM. The analytical approach of the theoretical

27 Walter-Rogg (2013) defines Central Germany corresponding to the metropolitan region of Halle-Leipzig which was acknowledged by the *Ministerkonferenz für Raumordnung* in April 2005 as one of the metropolitan regions of Germany.

derived concept of ERG is applied. Its five aspects – "discussion of shared perceptions of joint challenges," "definition of goals and measurement systems," "implementation," "report and review," and "review of the measurement system, the decision-making process, and the stakeholder" – are traced in the regulatory structure of governance in the geographic-functional region of FRM. The five aspects of ERG present a suitable analytical framework for governance analysis. Hence, the approach makes it possible to empirically analyze the regulatory structure of governance, decision-processes, and the institutional arrangements in geographic-functional regions by sound content analysis. Based on the experimentalist governance approach of Sabel and Zeitlin (2012b) this shows that decision-processes in geographic-functional regions are identifiable and analyzable by ERG and policy analysis. Here, governance is not an art with accidental characteristics simply based on knowledge, perception, powers of imagination, and intuition; rather the regulatory structure of governance in geographic-functional regions is traceable according to the reciprocal theoretical concept of ERG.

Regarding experimentalist governance, it has to be noted that this research path is very young. Aside from the initial research of Sabel[28] – the analysis regarding the EU, conducted by Sabel and Zeitlin (Sabel and Zeitlin 2010a; Sabel and Zeitlin 2012a), as well as the contemporary specification of the concept by Sabel and Zeitlin (2012b) and De Búrca, Keohane, and Sabel (2014) – only limited, non-comprehensive contributions to experimentalist governance are identifiable.[29] This oversight of experimentalist governance is also shown by a query of the Social Science Citation Index (SSCI) database.[30] The query of "experimentalist governance" in all research fields and years has revealed 37 results.[31]

---

28 Dorf and Sabel (1998), Sabel (2004a), Sabel (1996), and Sabel and W. H. Simon (2008).

29 E.g. Brassett, Richardson, and W. Smith (2010), Eckert and Börzel (2012), Nance and Cottrell (2014), Overdevest and Zeitlin (2014), and Pollak and Slominski (2009).

30 The SSCI provides access to the bibliographic and citation information from 3,000 of the world's leading social sciences journals across 50 disciplines.

31 The analysis was conducted on September 22, 2014. The database of the SSCI contains 170,394,093 references. The terms "governance" and

Those came from the field of jurisprudence mainly. Here, the sub-field of European law discusses experimentalist governance as a new form of governance (Dawson 2011). Some administrative lawyers[32] refer to this phenomena as anomalous administrative law (Sabel and Zeitlin 2012b: 11). Far and foremost, the question of legitimacy ('shadow of hierarchy') is raised and discussed from a law perspective. Stemming from this field, Sabel and Zeitlin have identified characteristics of experimentalist governance for several policy fields at the European level, such as financial markets, competition, air security, energy regulation, environmental protection, and others (Sabel and Zeitlin 2010a; Sabel and Zeitlin 2010b).

Turning to the emerging question of governance learning, Fabrizio Gilardi and Radaelli (2012) state that with the reviews of Dobbin, Simmons, and Garrett (2007), Freeman (2008) and Grin and Loeber (2007) there are "only a handful of empirical studies on [governance] learning. Some authors – such as James and Lodge (2003) and Volden, Ting, and Carpenter (2008) – even express a kind of frustration about the theoretical leverage (Fabrizio Gilardi and Radaelli 2012: 155). The presented analysis will offer insights into governance learning by examining the "processes of updating beliefs" (Fabrizio Gilardi and Radaelli 2012: 159) in the regulatory structure of governance during the mediation process and the official proceedings of the regional planning procedure (RPP) and the road scheme evaluations (RSE) at Frankfurt Airport. On the one hand, in a common approach, combining comparative public policy and different modes of governance, governance learning

---

"experimentalist governance" were searched by topic containing all years (1864–2014). The named research fields are predetermined by the SSCI. The basic search by the topic of "governance" taking all years and all research fields into account has shown 109,288 results, which were cited up to 2,950 times per article. The search for "experimentalist governance" has counted 37 results cited up to 97 times. Of those 37 articles 31 referred to the research field of government law, 18 to public administration, six to business economics, three each to environmental science and ecology as well as international relations, two each to philosophy, sociology, and social science and other topics, and one each to geography and urban studies. A single article might refer to different research fields (SSCI 2014a; SSCI 2014b).

32 E.g. Kingsbury, Krisch, and R. B. Stewart (2005) and Cohen and Sabel (2006).

refers to problem definition, the achievement of results, goals, and implicit as well as explicit knowledge. On the other hand governance learning includes hierarchy, the 'shadow of hierarchy' or "the shadow of law" (Fabrizio Gilardi and Radaelli 2012: 159), facilitating coordination, and voluntarism. Furthermore, governance learning is focused on compliance, emulation, and herding, as well as dense communication and reflexivity (Fabrizio Gilardi and Radaelli 2012: 159).

As stated, geographic-functional regions become increasingly important. Compared with the national and international sphere, the research at the subnational level to analyze the regulatory structure of governance and its decision-processes has been neglected in the past (Lütz 2010: 144). This could be the result of the missing definitions in regional or metropolitan governance, which leads to the obstacle of these governance approaches not being able to embrace geographic-functional regions, especially those perspectives that follow a narrow understanding of governance (Ahrend et al. 2014; Fürst 2010; Zimmermann and Heinelt 2002). Indeed, there is some analysis available. However, governance research has to catch up in this field, which holds especially true for FRM. Besides the metropolitan regions of Munich and Stuttgart, FRM – with its financial district and logistics hub – is one of the most important geographic-functional regions in Germany. According to the gross value added, it is also one of the most productive metropolitan regions in Europe (Ahrend, Gamper, and Schumann 2014: 7; Lindner et al. 2014; Blotevogel 2002: 348). Nevertheless, compared with other regions in Germany and Europe, the metropolitan region of FRM is neglected in the research field of governance (Lackowska 2011a: 79).

In their handbook on regional governance Kleinfeld et al. (2006a; 2006b) concentrate on London, Lugano, the region of Flandern within Europe and the German cities of Berlin, Hamburg, Hannover, Munich, Rhine-Neckar, Rhine-Ruhr and Stuttgart (Kleinfeld, Plamper, and Huber 2006a; Kleinfeld, Plamper, and Huber 2006b). Similarly, Adamascheck and Pöhl (2003) focus on the geographic-functional regions of Hannover and Rhine-Neckar. Although there is more analysis available in the field of human geography, the picture regarding FRM is nearly the same. H.-D. Schulze (2006) neglects the metropolitan region of FRM in his anthology, while Breuer (2012) only examines the region of Saxony. Pütz

(2004) analyzes the regional governance structures of Munich regarding the settlement policy. Further prominent examples of regional governance research in Europe cover Britain (Fürst and Knieling 2002; R. A. Rhodes 2001) and Northern-Italy (Putnam 1993).

The literature specific to Germany refers mostly to Baden-Württemberg and its car industry (Grabher 1994b; M. E. Porter 1998: 78; Storper 1997b: 24, 50) or the areas of Hannover and Stuttgart as illustrative examples of prosperous metropolitan reform (Lackowska 2011a: 79).[33] Furthermore, J. Esser and Schamp (2001) of the former Collaborative Research Center Networking as a Competitive Factor at Johann Wolfgang Goethe University focused on the standardization and processes of intercommunication of metropolitan regions (König 2001: 7).

For the metropolitan region of FRM, only limited and scarce analyses are available. Regarding the presented research, four analyses can be named as comparable, whereas two of them solely analyze the actors' landscape of governance. With reference to the example of Frankfurt Airport, Knippenberger (2012) focuses on the change of actors' capacities to handle governance conflicts (Knippenberger 2012: XXX, 127). From a spatial perspective, she situates her research in the field of airport cities and not in the context of metropolitan regions (Knippenberger 2012: 17f).[34] Becker-Ritterspach (2015) compares Frankfurt Airport and Berlin-Brandenburg Airport by concentrating on the delays of the implementation of large infrastructure projects. Taking a law perspective, Schäfer (2011) focuses on the extension process and its procedures, especially the mediation process. In the governance discourse Zimmermann and Heinelt (2002) analyze governance from a broader point of view. They merely describe the governance structure of metropolitan FRM and its institutional landscape. Comparing institutional arrangements, they remain at a general level, thus, not analyzing the decision-processes in depth. In the same manner, Lackowska (2011a) describes FRM and its governance structure. Although she adds a historical component,

33 In the field of corporate governance and questions regarding corporate management Germany was of interest, e.g. Janka (2011).

34 See, e.g. Ashford, Mumayiz, and P. H. Wright (2011) and Kasarada (2010) for the concept of airport city or aerotropolis.

she exclusively refers to the rises and falls of involved actors in the overall metropolitan system. Furthermore, the analysis of Langhagen-Rohrbach (2005) proceeds in the same direction and solely presents actor constellations.

Focusing on the regional competitiveness and applying the diamond model of M. E. Porter (2014) on different sectors in the metropolitan region of FRM, Ebner and F. Raschke (2012) and F. W. Raschke (2009) – the latter with a focus on logistics – have analyzed different clusters or business agglomerations in the metropolitan region of FRM. In their analysis, decision-making processes and the institutional arrangement have only played a minor role.

Even though Ludwig et al. (2008) have taken FRM into account in their short anthology, the paragraph of Krüger-Röth (2008) – which refers to FRM – stepped back behind the other analysis of metropolitan regions. It mainly represents a short description of FRM and its then-metropolitan governance body, the Planungsverband.

Furthermore, regarding very special topics, a few undergraduate theses deal with governance in the region of FRM superficially.[35]

Thus, it should be noted, that research on governance in the metropolitan region of FRM is identifiable. However, this research landscape is two folded: on the one hand, there are existing research studies, analyzing governance broadly by focusing on the metropolitan region. After all, these research studies neglect decision-making processes and governance learning. On the other hand, there are studies focusing on very specific phenomena, but disregarding the wider context of governance and the sum of all co-existing forms of collective regulations of social issues. This gap was already identified by Lackowska (2011b). She confirms that during "[…] the whole [of the] 20th century the practice of governing metropolitan affairs lagged behind the real challenges shaped particularly by urban sprawl" (Lackowska 2011b: 22).

35 E.g. Hahn (2012) and Zunic (2007).

Contrary to this disintegration, the following research reconstructs and analyzes the broader regulatory structure of governance, the decision-making processes, and the institutional arrangements as well as governance learning. This will be a major contribution to the closure of the identified research gap. Nevertheless, the focus on the extension process of the airports and the respective institutional arrangement represent a very specific phenomenon.

Due to the prevailing research goal of logistics and mobility, those research fields focus on technical-oriented solution strategies and economics. Accordingly, quantitative research designs significantly outweigh qualitative approaches (Wilde 2014b). Notably, focusing on sustainability and the aging society, political and social science slowly enter into mobility research.[36] It is celebrated as a new gateway to old topics (Holz-Rau and Scheiner 2015: 3), whereas the field of logistics is nearly untouched by political and social science. Considering that logistics is the backbone of the economy and society, this omission of social and political issues is short-sighted. Both research fields should be connected.

The presented research contributes to this, linking political science with logistics from a qualitative research perspective.

## 1.3 Contribution to contemporary political science

The presented research analyzes the regulatory structure of governance, its decision-making processes, and the institutional arrangements of metropolitan FRM by the theoretically derived concept of ERG. It represents the initial analysis to apply the theoretical concept of ERG to regulatory structures of governance and certain decision-making processes and accordingly. Here, the first overall contribution is the initial broader analysis of decision-making and the regulatory structure of governance by the theoretical concept of ERG from a political science perspective. As outlined before, only a few studies – mainly in the field of jurisprudence – analyze, explain, and define experimentalist governance. The aim is to advance the research on governance and institutional arrangements

36 E.g. Dangschat and Segert (2011), Hunecke (2015), Scheiner and Holz-Rau (2015), Schwedes (2014), and Wilde (2014a).

in geographic-functional regions. The presented research continues and advances the analytical discussion in political science. Initially, the analytical approach of ERG is structured to make it functional for policy analysis. Subsequently, the theoretical concept of ERG is applied as a theoretical derived analytical approach for the very first time to analyze regulatory structures of governance, its decision-making processes, and the institutional arrangement in geographic-functional regions. Consequently, the presented research contributes to advance the analysis in the field of political science. This is necessary to gain deeper insights into regulatory structures of governance.

Analyzing the regulatory structure of governance and decision-processes by the theoretical concept of ERG contributes also to the policy analysis showing that ERG is an appropriate life cycle model. The theoretically derived concept of ERG is perceived to be a functional analytical approach to analyze the regulatory structures of governance and decision-making, especially in the context of (polycentric) metropolitan regions with different, partly equal institutions and actors. Thus, the presented research contributes to advancing policy analysis, especially in this field. It contributes to providing a deeper insight into the collective-cooperative, non-dictatorial control of public and societal issues, as requested by the general governance discourse (Mayntz 2009a: 13). Hence, this research is an attempt to fill the previously identified white spots. Additionally, by integrating the embeddedness perspective, the presented research overcomes the methodological challenge of policy analysis. It combines the institutional and structural aspects of power with mechanisms of process-oriented problem-solving. At the same time, it recognizes the structural embeddedness into institutional arrangements in the regulatory structure of governance in a certain spatial context – e.g. geographic-functional regions.

Frankfurt Airport is the research object in the analysis, whereby it advances research on governance specifically for metropolitan FRM. First, it helps to overcome the simple identification of governance failure in FRM. Second, it contributes to close the gap in metropolitan research of FRM in political science.

Furthermore, the presented research provides first starting points to link political science with researches in logistics. The gathered data opens the field of logistics to contemporary political science and further qualitative analysis.

## 1.4 Structure

After this introduction, the second chapter presents the fundamental understanding of governance as a regulatory structure (Sects. 2.1 and 2.2). Here, the multilevel aspects of governance (Sect. 2.3) as well as the contemporary approaches of regional governance (Sect. 2.4.1) and metropolitan governance (Sect. 2.4.2) are laid out and their limitations are elaborated. This broader introduction of governance sharpens the understanding of governance as a process and illustrates the shortcomings of the previous governance discussion regarding the analytical framework of ERG.

Based on the theoretical foundations of governance the chapter further clarifies the underlying and initial concept of experimentalist governance. The manifest and latent content presented by Sabel and Zeitlin (2010a) – who reconstructed the experimentalist governance on the European level – systematizes the analytical approach. Hence, the theory-laden framework is structured by the manifest and the latent content. Thus, the five aspects of ERG as the analytical framework are defined and outlined (Sect. 2.6.2).

The chapter closes, with the presentation of the research-guiding question and further subordinated research questions. On the one hand, these questions address the traceability of the assumptions of ERG in FRM. On the other hand, they address the field of governance learning. Here, the hypothesis which occurs regarding the presented research will also be introduced.

The presented research will follow a sound scientific approach. The method, which follows the approach content analysis, will be presented in the third chapter.

The official documents of the hearing and planning authorities of the airport extension process constitute the golden thread of the analyzed data.

The fourth chapter analyzes the case of the extension of Frankfurt Airport. In the beginning, the general administrative setting is clarified (Sect. 4.1). Subsequently, the context of the geographic-functional region of metropolitan FRM is laid out (Sect. 4.2.1) and the institutions considered as relevant are identified (Sect. 4.2.2). Finally, the RSE of Runway 18-West (Sect. 4.3.3), the mediation process (Sect. 4.3.4), the RPP (Sect. 4.3.5), the following RSEs between 2002–2004 and 2004–2007 (Sects 4.3.6 and 4.3.7) as well as the additional supplementary adjustment plans (Sect. 4.3.8) are reviewed and the analytical framework of ERG is applied. The decision-making processes as well as the institutional arrangements are examined by reconstructing the official proceedings of the regional planning history. Each of the paragraphs presents the findings of the analysis by its interim findings.

Thereafter, the discussion in the final chapter summarizes the findings of the interim-conclusions presented (Sect. 5). Based on the findings, the areas of "metropolitan governance bodies" (Sect. 5.2) and "governance learning in FRM" (Sect. 5.3) are discussed in more detail.

After detailing the areas *ex ante* identified, the discussion is closing with further promising research areas (Sect. 5.5. The research fields of innovative democracy and the entrepreneurial state are discussed to present the starting points for future analysis in both subjects. At the same time, several further fields of interest for future analysis are named.

# 2 Governance

*Governance blurs the distinction between state and civil society.*

(R. A. Rhodes 2001: 57).

After the following initial overview, this chapter presents the theoretical background of governance as a regulatory structure for decision-making more broadly. First, this shows how the government versus governance discussion handled the challenges of the change of "Government to Governance" (Blumenthal and Bröchler 2006). Retracing this transformation from planning and steering to govern, sets basic understandings for the challenges of governance research, in regard to jurisdiction and the regulatory structure of governance in geographic-functional regions. As Blumenthal (2005) noted, governance, as an analytical research agenda, means a change in the perspective, not the change of the research subject (Blumenthal 2005: 1175). This shift of the perspective, Mayntz (2009a) speaks of a change of the research paradigm, is presented. Regarding the specification of governance, this comprehensive presentation is necessary. It leads to the clarification of causal connections, about the development of Sabel and Zeitlin's theoretical experimentalist governance approach, which is the foundation of deduced ERG (Gschwend and Schimmelfennig 2007: 16). Second, it sets the basic understanding of governance as a regulatory structure, which the presented research is following. Finally, this chapter presents the identified gaps in governance research, regarding geographic-functional regions more broadly.

The chapter paves the way to the research framework and the empirical part of the presented research.

The phenomenon of the regulatory structure of governance emerged by the awareness of the society and the economy, that planning and steering by politics were insufficient approaches to handle economic and social issues as well as challenges evoked by globalization and regionalization. Its theoretical evolution developed parallel between the research fields of administration, international relations, institutional economics, management studies, and the field of international cooperation and development (Grande 2009: 79; Zürn 2008: 553). It is assumed, that the term "governance" comes from steering or piloting. Linguistically, it comes from the Greek expression *kybernan* which means "to pilot." It is translated into the Latin word *gubernare*. Thereto, the meaning is not only metaphorical, it is linked to the basic idea to steer politics (Levi-Faur 2012: 5; Mayntz 2009b: 13; Peters 2012: 20).[1] This idea of interventionism goes back to the absolutist territorial states of Europe (Mayntz 2003: 73).

As a research field, governance was nearly meaningless within academia until the 1950s and 1960s (Levi-Faur 2012: 5; Mayntz 2009b: 15). However, coming from the area of economics, already Coase (1937) pointed out, that action is shaped by organizational structures. Finally, it was Williamson (1979) who introduced governance as a regulatory structure more broadly. Coordination of action was no longer interpreted as just governed by hierarchies. Non-market phenomena and institutions, such as family, clans, associations, and later networks, were taken into account (Mayntz 2009b: 14; Schuppert 2008: 16).

The growing interest in markets and hierarchies, as organizational forms steering administrative and economic actions, brought up an increased interest in governance research (Levi-Faur 2012: 5; Mayntz 1996: 148; Mayntz 2009b: 14).

The final breakthrough in political and social science can be determined for the 1970s (Mayntz 2006: 11f). Today, governance is a buzz word, not only within political and social science but also within applied politics and the administration (government) (Mayntz 2009a: 9). It "[...]has become probably one of the most important manifestations of the rise of Neo-Institutionalism in social science " (Levi-Faur 2012: 3).

Starting from Williamsons' basic idea, that action is no longer just routed by hierarchies and pure market mechanisms, two principal meanings of governance are identifiable. On the one hand, governance is used

---

1 Within the English literature, the common term is "governance", but to describe the evolution of the governance theory, the presented research uses the term of "steering" (Mayntz 2009b: 29). Against the background of the German expression *Steuerungspolitik* this is considered as useful.

to describe a nonhierarchical mode of cooperative control. Administration and private actors are coordinated by a structure of "mixed public-private networks" (Mayntz 2009b: 13). This research path focuses on problem-solving through structures of cooperation outside hierarchical frameworks. It describes the narrower understanding of governance. On the other hand, in a broader understanding, governance is used more generally to describe the coordination of actors, through social order or deliberation.

Due to the phenomena of globalization and the ongoing divestment or enlargement of clear common political borders and jurisdictions, the concept of governance even in its broadest understanding comes under pressure.

Now, the question has to be raised: how can challenges regarding a faster and more complex world, where decisions on the global level have deeper impacts on the regional or the local level (and *vice versa*) and were former clear jurisdictions became geographic-functional, be described and pictured by the regulatory structure of governance. In front of the background of Dewey (2012) and his idea, that there is no absolute certainty of the correctness of decisions (fallibilism), already formulated in 1927, Sabel and Zeitlin (2012b) developed a concept which seems to tackle the challenge – experimentalist governance. Dewey (2012) argued that ends and means are mutually transformative. Serving democracy, ends and means have to be reconsidered as conditions within a changing environment (Sabel 2012: 36). Research indicates, that aspects of experimentalism in handling these challenges are identifiable on the supra-national level, but also on the national level (Sabel and Zeitlin 2010a; Sabel and Zeitlin 2010b; Sabel and Zeitlin 2012b).

Nevertheless, before ERG as a theoretical derived concept to analyze governance in geographic-functional regions can be presented, it is necessary to discuss the general understanding of governance as a regulatory framework.

## 2.1 From government to governance

Before and after World War I and II, administrations planned economic and social development from a legislator's perspective, to reach certain goals. In particular the goals were the (re-)establishment of the national state and its economic development (Esteva 1993: 98f; Menzel 2010: 12; Stockmann 2010: 2). The state was a guarantor of public welfare and the central instance of control (Mayntz 2003: 73). Until then, fostered by generations of philosophers and politicians, the understanding, that the administration and the society are segregated, was common (Brikinshaw, Harden, and Lewis 1990: 1). Political action was characterized by certain steering mechanisms. Due to that, academia and empirical analysis paid attention to policy planning and implementation.[2] As it was political reality academia adopted also the top-down perspective of the legislator in its research designs. The administration, as the democratically elected "steering committee," aimed at the planning and steering of developments, which embraced issues and challenges of the whole society (Mayntz 1996: 148; Mayntz 2006: 12; Mayntz 2009b: 15).[3]

Discussions on law and regulations came up, revealing that certain forms of regulative policies came to nothing. The diagnosed reason was, that planning and control are somehow in the need of cooperation. As a result, the claim of the administration to be the core instance of control and the guarantor of public welfare, which goes back to the times of the Enlightenment, fell.[4] Sheer planning was seen as too simple to handle social reality.[5] It was recognized, that demands between policy fields differ remarkably. This perception led to a change of perspectives and a significant widening of the initial concept of planning (Mayntz 1996: 148–150; Mayntz 2009b: 15).

---

2 E.g. J. Esser, Naschold, and Väth (1972), Luhmann (1971), and Mayntz and Scharpf (1973).

3 Shonfield (1968) exemplifies this by comparing the examples of France and Great Britain with their efforts in controlling several parts of their national economy (e.g. mail services or rail). Further, he illustrates examples of Austria, Germany, Italy, The Netherlands, and the United States.

4 At this point, connections with the change of the position of the city, as defined by Weber (1999), to conurbations and metropolitan regions, are apparent. The city is not the focal point or guarantor of welfare anymore.

5 Counter positions can be found in capitalism, e.g. J. Esser (1975), Häußermann (1977), Hirsch (1974), C. Offe (2006), and Poulantzas (1977), as well as in new pluralism, e.g. Dunleavy and Brendan (1988) and Lindblom (1965) (J. Esser 1999: 120f).

Succeeding politics were now characterized by steering. The 1970s were characterized by discussions about the inability of the administration, to react on economic and social challenges by its hierarchical model of order, e.g. mail services, telecommunication, and public transportation (Mayntz 2003: 74; Pierre 2000b: 4). Beside missing its political goals in those times, politics of control (*Steuerungspolitik*) also wasted economic resources (Mayntz 1979: 59, 62f). The identified challenges were attributed to the modernizing society, which belittled the regulative scope of action (Mayntz 1979: 76). Within the field of social policy research, actors and their actions came into the fore. The research perspective changed from the presumptive "subject of steering" to the "object of steering" and its governability (Mayntz 2009b: 15f; see footnote for further detail).[6] The empirical policy research identified several demands of successful policy steering, e.g. the right of intervention, financial resources, functional executive mechanisms, as well as the capability to consensus (Mayntz 2009b: 53). Now, the focus was no longer the question of steering and control capacities. Due to the fact, that politics of control have to assume a "will of steering" of its target group, academia focused now on the controllability of the "object of steering" (Mayntz 2009b: 32, 53). Furthermore, the top-down perspective was enlarged by adding the bottom-up perspective (Mayntz 1996: 148–150; Mayntz 2006: 12f; Mayntz 2009b: 15f). Under those circumstances, the behavior and the structural characteristics of different fields of regulation were now included in the analysis (Mayntz 2009b: 30).

With that, the way was paved for the second step towards the integration of non-administrative institutions and their collaboration on policy-development and -making as well as its implementation. The result of the transformation of the paradigm of political steering was a model of a cooperative administration or the cooperative state,[7] where the distinctive differentiation of subject and object has vanished. Political

6 Mayntz (2006: 12); Mayntz (1987: 93f); Mayntz (1979: 55).

7 See Kay (2013) and Schuppert (2011) for a jurisprudential perspective and Benz (1997), J. Esser (1999), and Holtkamp, Bogumil, and Kißler (2006) for a political science perspective of the concept of the cooperative state. See Geißel and Joas (2013) for the contemporary discussion of cooperative governance.

steering and agenda setting was no longer restricted to administrative institutions. Rather, all actors who argue for and try to realize a public interest or good in a broader sense, have now to be at least linked to the political decision-making and steering process (Mayntz 1996: 150f; Mayntz 2006: 13; Mayntz 2009b: 18). This change was accompanied by deregulation, privatization, contractions in civil services, "and the introduction of 'managerialism' in the public sector" (Pierre 2000b: 1).

By the inclusion of nonadministrative institutions, not only the hitherto valid guiding principle of hierarchical control and its regulatory structure was questioned. The terminology changed from steering to governance (Mayntz 2006: 13; Mayntz 2009a: 11). In this context, corporatism highlights "the delegation of state power to quasi-governmental organizations, quasi-non-governmental organizations and private associations" (Brikinshaw, Harden, and Lewis 1990: 8).[8]

Mayntz (2006) illustrates that this terminological change was not only semantical. On the one hand, it is the expression of real changes of institutions and processes in policy development as well as social regulation within modern nation-states (Mayntz 2003: 72). On the other hand, the issue of the "subject of steering" and the conception of central policy control on the European level and moreover on the global level (global governance) is *reductio ad absurdum*. This is because broad discussions on the global level have always to be linked to the national context of the subject, at the latest when it comes to legislative action (Mayntz 2006: 14, 16; Mayntz 2009a: 11).[9]

However, steering and governance do not exclude each other. Rather, the actor-centered institutionalism tries to combine them. According to Mayntz and Scharpf (1995), this approach highlights the integration of administrative and nonadministrative institutions into a common frame-giving environment. This point of view allows overcoming the perspective

8 For the discussion of non-governmental organizations (NGOs) see e.g. Brühl et al. (2001) or Frantz and Martens (2006).

9 The current crisis in Greek is a good example. National governmental action and law-making are discussed on the European level with the European creditors as well as on the international level with the World Bank. The demands of the European creditors and the World Bank have to be transferred to national legislation.

of the legislator too (Mayntz and Scharpf 1995: 43–47). Laws and service regulations, set-up by the administration, are just one form to coordinate society in a regulatory framework (Zürn 2008: 563).

Here Blumenthal (2005) describes four major changes in the research perspective. First, governance in dissociation to the government concentrates more on nonhierarchical systems of problem-solving. Societal self-control becomes important. Second, as already Mayntz (2009b) noted, the distinction between subject and object of steering blurs. Third, the view changes from an actor focus to an institutional focus. The institutional arrangement comes into the fore. Finally, governance does not always require a problem to solve. Governance also takes the divergent sets of rules between different institutions into account (Blumenthal 2005: 1170–1173).

This change over history is also described by the transformation from government to governance (Blumenthal and Bröchler 2006; Haus 2010). The hierarchical understanding of government was widened. Now, governance is always seen in connection with government, and "Governance without Government" (Rosenau and Czempiel 1992) is recognized as an uncommon exception. Indeed, there are changes in government research identifiable, which allow to include more than the analysis of autonomous action of an administration. The picture which is drawn on democracy, hierarchy, and the concentration of problem-solving by law, has changed by the incorporation of new institutionalism. New dimensions of the formation of political objectives, networks, and the process dimensions are included in the governance (Bröchler and Blumenthal 2006: 13–16).

This phenomenon, the transformation from hierarchical and closed to federate and open, is also identifiable for organizations. Until the 1980s, goals were set by headquarters and subordinates followed these goals by labor division. Like identified for the administrative level, organizations from the 1980s on became more network-oriented (network organization). Goal-setting of the headquarters was shaped by the lower organizational levels (Sabel 2007: 107–110).[10]

---

10 In that time, the research of Womack, D. T. Jones, and Roos (1990) and their prescription of Lean Production was pointing the way. Without essential

However, to avoid nit-picking and the objection that governance just replaces common concepts,[11] the presented research understands government as an administrative action, which is still hierarchical or describes the action between equal administrative units. With that, governance is not a counter concept to government. Rather, the government, as an administrative action, can be subsumed under the broader understanding of a regulatory structure of governance. Both are not detached (Bröchler and Blumenthal 2006: 16). Government and planning are still relevant within the continuum of collective control and coordination of social action ('by government'), at least when it comes to legislative action (Bröchler and Blumenthal 2006: 16; Mayntz 2003: 72f; Pierre 2000a: 242). The necessary analytical and empirical link between government and governance is still given.

This indicates a broader understanding of a regulatory governance structure, which includes administrative and nonadministrative actors in heterarchical relations. The broader understanding of governance is now presented in more detail from a political scientist's perspective.

## 2.2 Governance – a political scientist's perspective

The contemporary policy analysis works on questions beyond hierarchy and the intervention state of control as well as beyond models of self-regulating markets and Smiths "invisible hand". With that, research moves between the Marxist theory of a communist society without administrative control and the utopia of a peaceful self-regulating economic society. The attentiveness changed to neo-corporatist decision-making structures, which can be characterized as the integration of organized negotiation processes between the administration and other interest groups (Sebald

---

changes of the former core concept, it was promoted as Lean Management by Pfeiffer and Weiß (1992) as a management concept.

11 According to Blumenthal (2005) this discussion of old wine in new bottles holds especially true for political scientists in Germany. Jessop (1998) and Pierre and Peters (2000) critically discuss the benefit of the term "governance" (Blumenthal 2005: 1150).

and Straßner 2004: 41).[12] However, tendencies of denationalization, which were discussed especially concerning the United States (US), were not certifiable (Mayntz 1996: 150f).[13] Rather, there was a real change of paradigm or at least from the perspective (Blumenthal 2005: 1170–1173; Mayntz 1996: 163; Mayntz 2009a: 11).

This change was heralded by the modernization theory within social science. In the course of the social modernization of democratic societies, which is in close relation to the argument of T. Parsons (1976), the administration was neither robbed of its functional necessity nor its influential capacity. Discussions within political academia reveal that there is no doubt about this. The change of the social structure did not destroy the administrative influence or power. The administration, within a multilevel system, no longer just carries out the hierarchical steering actions of the government. It still has possibilities to influence traditional administrative fields and to establish new ones. Moreover, further responsibilities are ascribed to the administration (Dingeldey 2008: 318; see footnote for further detail).[14] "Hollowing out the state" (R. A.

12 Based on its long-lasting state theory, this phenomenon was apparent and researched in particular in Germany. At least since the discussion of the failing regulations for the health system and the negotiations between the German administration and involved associations the phenomena of governance were established in a broader sense (Lehmbruch 1988; Mayntz 1996: 152; Mayntz 2009b: 18). Regarding corporatism Weßels (2000) described a German model of interest representation, even Germany itself is not a typical corporatist country (Weßels 2000: 18). For further readings on neo-corporatism see, e.g. Alemann (1981), Alemann and Heinze (1981), Crouch and Streeck (2006), Mayntz (1992), and Streeck and Schmitter (1985).

13 Even if Evans, Rueschemeyer, and Skocpol (1997) requested to "Bring the State Back In," it must be pointed out, that especially in the US and their understanding of law and state administration, distrust against the administration has always played a role (Haller 2002: 37–52; Sorg 2007: 7–10). "The founding fathers placed the utmost importance to prevent the state from interfering with social development" (Kalberg 2000: 948, translated by the author). The contrary can be identified in France. Here skepticism prevails over intermediate organizations and more emphasis is put on the control function of the administration (Mayntz 1996: 152).

14 Mayntz (2009b: 55); (Mayntz 2003: 78); Mayntz (1996: 151, 163–165); Sassen 2002; Menzel (1998) critically discusses the field of contradiction of

Rhodes 2001: 17) is not ascertainable. This is supported by the fact, identified by analyzing the privatization processes in Britain, that public employment fell only slightly by the privatization of public services (R. A. Rhodes 2001: 47). This phenomenon is identifiable for Germany. Many public officers are still employed in former state-owned enterprises, e.g. *Deutsche Bahn AG* or *Deutsche Telekom AG*. Hence, not the burden has changed, the modalities did. The state takes now the task of a guarantor (*Gewährleistungsstaat*), which has no longer the responsibility of execution.[15] The administration is now in charge to guarantee and enable the provision of common goods by private actors (Dingeldey 2008: 318; Kooiman 2010: 118–123; R. A. Rhodes 2001: 51). Mayntz (2009b) even speaks of an overtaxing of national administration and not of an abolition (Mayntz 2009b: 58–61).

> In this context Mayntz (2003) defines governance as "the sum of all co-existing forms of collective regulation of societal issues: from the institutionalized civic self-regulation through various forms of cooperation between non-private and private actors to sovereign actions of state actors" (Mayntz 2003: 72).[16]

With that, governance characterizes a regulatory structure of different forms of purposive regulations of collective circumstances, both formal and informal. Further, the regulatory structure of governance is distinct, but not diametrical to hierarchical rules (Mayntz 2009a: 9; Mayntz 2009b: 13). On the one side, governance requires reluctant interests *per definitionem* (Mayntz 2009a: 13). On the other side, governance presupposes the intention of the actors, to cooperate and demands systems of negotiation (Mayntz 2003: 79; Mayntz 2009b: 11). Institutions or

---

"Globalization *versus* Fragmentation." In a globalized "*Weltinnenpolitik*", he argues for the empowerment of the United Nations (UN) to handle global challenges, according to the subsidiarity principle, which, in the end, will lead to an international-democratically legitimized monopoly on the use of force (Menzel 1998: 242–262).

15 A good example here is the privatization of the execution of a prison sentence (Cho 2014; Nibbeling 2001; Wadle 2013). Administration entrusts private actors with sovereign tasks. In return, it guarantees certainty of justice and that the entrusted tasks are carried out according to a given law (Nibbeling 2001: 244–259).

16 Translated by the author.

forms of regulation vary from purely administrative to purely private (Mayntz 2003: 74).

In addition to this broader political science definition, which represents the basic understanding of governance as a regulatory structure in this research, there are several further attempts to define governance. In the following they are outlined, to sharpen the former basic definition.

Transaction cost theory elaborated an even broader understanding of a regulatory structure of governance. Above all, Williamson (1975) analyzed markets and hierarchies as alternative systems of economic organization and the "invisible hand." Here, as Mayntz (2009b) argues, governance is understood as a structure of various modes of cooperation between private actors and social order. This understanding was further enlarged.[17] Beside additional social phenomena, such as family and associations, especially networks were included in the systematization of governance interpretation (Kenis and Raab 2008: 135; Mayntz 2009b: 14).

This goes along with the definition of Le Galès and Voelzkow (2003) in the field of local production systems and the production of local collective competition goods. To them, governance refers to a regulatory structure of all institutions, coordinating and regulating actions among actors within an economic system (Le Galès and Voelzkow 2003: 5f.).

Coming from a network perspective, R. A. Rhodes (2001) more emphasizes the self-organization of inter-organizational networks, which are denoted by interdependence, exchange, rules, and the autonomy from the administration. Thereto, for him, the regulatory structure of governance is non-hierarchic in principle (R. A. Rhodes 2001: 15, 46–60). This non-hierarchic principle and the involvement of private actors in decision-making processes is also pointed out by Héritier (2002). However, she also stresses the "ultimate possibility for hierarchical intervention by public actors" (Héritier 2002: 3).[18]

---

17 E.g. Hollingsworth and Lindberg (1985), Powell (1990), and Kenis and V. Schneider (1991).

18 Also the global discussions within the 1980s and the link to normative characteristics refer to the environment and processes to coordinate administrative and social actors. The World Bank introduced the concept of good governance as an answer to the mismanagement of the governments of its credit taking

There are also more narrow definitions of governance. K. Offe (2008) and Risse (2008) criticize the broad understanding of governance. For K. Offe (2008) it is too broad, that a regulatory structure of governance includes each form of social order. He pleads to sharpen the regulatory structure regarding the private sphere, where reactions based on social and administrative order are spontaneous as well as in the direction of the administration's basic competence of government (K. Offe 2008: 64). Not quite that fare goes Risse (2008). He also argues to sharpen the concept of governance, but he contends to understand the regulatory structure of governance in the same manner as Mayntz (2003), as cited above. For Risse (2008) this represents a governance concept of middle range, which is characterized by actions and relations. It is more concentrated on structures and the environment and less on the actors. Thereto, he focuses on institutions and the institutional governance arrangement (Risse 2008: 150).

Stroker (1998) argues in dissociation of government. "The essence of governance is its focus on governing mechanisms which do not rest on recourse to the authority and sanctions of government" (Stroker 1998: 17).

In the field of international relations, also several definitions of governance are traceable. Zürn (2008) follows the definition of Mayntz (2003).[19] For him, regulatory the structure of governance is "the sum of collective regulations which aim at a specific problem situation or a particular social situation and are justified concerning the collective interest of the affected group" (Zürn 2008: 554).[20] Based on Kooiman (1993), Le Galès (1997) defines governance by processes of coordination between administrative and nonadministrative actors, which discuss

countries. In the 1980s, this mismanagement was identified as the *causa malus* for the country's incompetence to pay back their loans from the 1950s and 1960s. Good governance became one of the main criteria for further credit applications (Benz 2003a: 20f) and is one of World Banks' main exports (Schuppert 2008: 13).

19 According to Zürn (2008) also Benz (2006), Schuppert (2007) and other authors use a similar definition of governance (Zürn 2008: 554).

20 Translated by the author.

common tasks and goals in an uncertain environment (Le Galès 1995: 59; Le Galès 1997: 241).

The EU itself, based on its white papers (EC 1993; EC 2001), defines governance as "[...] the state's ability to serve the citizens. [...] Governance refers to the rules, processes, and behaviour by which interests are articulated, resources are managed, and power is exercised in society" (EC 2003: 3). Also, the UN defines governance as "the exercise of economic, political and administrative authority to manage a country's affairs at all levels. It comprises the mechanisms, processes and institutions through which citizens and groups articulate their interests, exercise their legal rights, meet their obligations and mediate their differences" (UN 2006: 3).[21]

In front of the above-presented definition, the regulatory structure of governance serves several functions. If C. Offe (1987) speaks of the indispensability of a competent and responsible authority, and Theisen (1995) describes the state as a future-oriented trustee, who is necessary when vested interests came into the fore,[22] the administration becomes the specialist and the representative of the general public (C. Offe 1987: 317; Theisen 1995: 122). Legitimized by a democratic constitution, the control in the sense of intentional influence of social processes remains the particular function of the political-administrative system. To that end, not the task of the administration has changed (Mayntz 1996: 156f). What has changed is how the task is managed, namely through delegation and cooperative negotiation (governance). This interpretation

---

21 Here it has to be mentioned, that the UN has a normative character. In development cooperation, both in practice and theory, governance is characterized normatively as good governance, defined by the World Bank. Thereto, good governance is "epitomized by predictable, open and enlightened policy-making; a bureaucracy imbued with a professional ethos; an executive arm of government accountable for its actions; and a strong civil society participating in public affairs; and all behaving under the rule of law." (WB 1994: viii). For further readings see, e.g. Audretsch (2009), BMZ (2009), Dolzer, Herdegen, and Vogel (2009), Nuscheler (2010), and Rothstein (2012).

22 According to Theisen (1995) this is especially the case when the precept of survival is touched, because those aspects have to be discussed in regard to human beings not yet able to negotiate (Theisen 1995: 122).

underlies a conception of a state, in which the administration does not use its power for particular or vested interests. Politics represents action in the meaning of the public interest (Mayntz 2006: 12; Mayntz 2009b: 35). Furthermore, governance emphasizes more horizontal than vertical approaches (Héritier 2002: 2; Mayntz 1996: 163–165). Empirical policy research not only uncovered control and steering defects. Studies in this field identified new forms of organized coordination of conditions, which were in need of regulation (Mayntz 1996: 151, 155–157). They confirmed, that further institutions do matter for the political steering capacity (Putnam 1993; Weaver and Rockam 1993). The perspective changed from an actor-centered to an institutionalist point of view (Mayntz 2006: 16).

To handle a continuously modernizing society, also the administrative system and its decision-making process have to modernize and differentiate itself (Héritier 2002: 1; Mayntz 1996: 156f). The administrative system and its actors have to be included or at least be linked to sectoral or regional policy networks, which might evolve from outside the system.[23] This holds especially true, when those sectoral and regional networks promote vested interests with a quality similar to a public good, for example, the development or extension of huge private infrastructure projects, which provide a public service or are of special interest (e.g. airports, public transport systems, or hospitals). Based on its legitimacy, the administration is the central instance to negotiate and "govern" the different interests in the regulatory structure of governance. Without those negotiations, the political decision-making process would disintegrate and fall back into its sectoral and regional networks (Mayntz 1996: 156f).

Thus, the regulatory structure of governance and, therefore, the process of cooperative negotiation can be characterized by social self-regulation and direct influence of administrative actors. The administration is not unitaristic. It is a comprehensive wickerwork of sometimes hierarchically linked actors as well as formal (forced by law) and informal (forced by reciprocity) control. The degree of institutionalization depends on regularity and intensity. Where conflicts, relationships, or other negative

23 These policy networks comprise actors, which are autonomous and interdependent at the same time. The actors and parties within the network can have mutually contingent and divergent interests (Marin and Mayntz 1991: 18).

externalities have a constant character, the level of institutionalization is probably high, e.g. collective wage negotiations, whereas new and spontaneous challenges are often informal and not well institutionalized. Thereto, a regulatory structure of cooperative social-political governance should especially evolve, where organizational forms of mediation are not yet installed or traditional regulatory structures of authority and well-known methods were unsuccessful to solve the challenges. With this, the administration can also have an impact on self-regulated cooperation, namely through law and service regulations (Mayntz 1996: 159–161 Mayntz 2006: 15; Mayntz 2009b: 18). In addition, the regulatory structure of governance has an implicit duplex character. On the one hand, governance refers to an action-guiding environment, the institutional governance arrangement. On the other hand, it concerns the regulatory (governance) process itself (Mayntz 2006: 15).

With the broader definition of Mayntz (2003), at least three categories of governance-settings can be identified. First, there are sovereign actions. In this case, the administration is in charge at each level. The governance processes take place between different administrative actors. Horizontal and vertical cooperation only evolve between different public units, e.g. ministries or departments. Second, there is the cooperative state. It is characterized by network relations between administrative and nonadministrative actors. As *primus inter pares*, the administration can delegate the realization and the policy development process to private actors. Within political networks, administrative and nonadministrative institutions interact cooperatively.[24] Those political networks can be formal and informal (Mayntz 2003: 77). In those cases, where the administration participates, Mayntz (2009b) speaks of mixed public-private policy networks (Mayntz 2009b: 17).[25] The third category is characterized by institutions of private or civil self-regulation. Those self-regulations can occur voluntarily by chance (e.g. optional quality standards for members

24 For Germany this is especially the case, when the administration, umbrella organizations, and other private actors negotiate, e.g. collective wage negotiations (Mayntz 2003: 76).

25 In this context other authors speak of public-private-partnerships (Zangel and Zürn 2004: 15; Zürn 2008: 558f).

of an association), or they are based on administrative order (e.g. as in the case of medical associations in Germany). In all three cases, decisions are based on negotiation and are not simply enforced by political power, even in the first case (Mayntz 2003: 77).

Following Mayntz (2003) further, a second general system to analyze governance structures is applicable. This systematization focuses on challenges. Regarding negotiation processes four challenges are identifiable. The first challenge is the possibility of a blockade. This challenge is particularly high, where decisions have to be forced and an agreement has to be unanimous. If the observed pressure is high and a delayed or suboptimal decision will have (high) negative externalities for the negotiating parties, this challenge is easy to overcome. The second challenge occurs about suboptimal compromises. Negotiations based on to-give-and-take have the inherent possibility of suboptimal results. They can be real costs or normative. The third challenge is conclusions based on costs for third parties, which are not part of the negotiations. In this case results for the negotiating parties might be easier to achieve, but the agreement has negative externalities for others. The last challenge is represented by decisions with little or insufficient cohesiveness. Here, agreements cannot be enforced effectively and might fail *ex post* (Mayntz 2003: 79f).

This approach of governance, as a surrogate to hierarchical control, is also used by Kooiman (1994) and R. A. Rhodes (2001) for "political steering" on the (sub-)national level, by Bulmer (1993) for the European level of policy-making, and Rosenau and Czempiel (1992) and Smout (1998) follow this understanding within international relations (Mayntz 2009b: 13). The common denominator of governance is, that governance points on processes of action, coordinating different actors in regard to collective action and decision-making (Blumenthal 2005: 1153; Bröchler and Blumenthal 2006: 8; Fürst 2010: 50).[26]

As the broader definitions of governance imply, the regulatory structure of governance appears on different levels. It is, as already R. A. Rhodes

26 For a further comprehensive listing of (further) definitions of governance see Panebianco (2013) pp. 48f.

(2001) described, identifiable on the micro-, the meso-, and the macro-level. Based on the example of Britain he argues, that administration is no longer monocentric. "[T]here is not one but many centres linking many levels of government – local, regional, national and supra-national" (R. A. Rhodes 2001: 3).

In the following paragraph, the multilevel approach of the regulatory structure of governance is exemplified, to then explore it at the meso-level – regional and metropolitan governance – as the focus of the presented research in particular.

## 2.3 Multilevel governance

When the processes and the responsibilities of a regulatory structure pass beyond the levels, it is spoken of multilevel governance. Regulatory structures of multi-level governance systems emerge where jurisdiction is formally separated by the different levels, but tasks, issues, or challenges are interdependent through the levels. The decision-making process and, therefore, tasks, rights, and obligations as well as resources have to be coordinated between the different levels within the regulatory structure. This coordination is organized by or between different institutional "arenas defined by rule-based contexts" (Benz 2010: 112). The categories of federalism (state and federal state) do not mark the background. The combination of miscellaneous institutions characterizes the specific context of the institutional governance arrangement of the regulatory structure. Hence, multilevel governance refers to the processes and the mechanisms of a coordinating regulatory structure within or between the various levels, not necessarily based on hierarchy (Benz 2010: 111–116).

### Micro-level

The micro-level is described by corporate governance, a regulatory structure coordinating the institutional environment of an organization, e.g. a firm (Lütz 2006: 35; M. Wright et al. 2013: 17).[27] Initially, the

27 Even some researchers argue that corporate governance can be traced back to the East India Company (founded on December 31, 1600), e.g. Motlanthe (2010), it was in the mid-1970s and the discussion of an agency-theoretical perspective by Jensen and Meckling (1976) and the elaboration the transaction

concept of corporate governance mainly referred to legal questions about the involvement of the board of directors. Questions of the relationship between shareholders and the firm's managers followed (Eberle 2010: 160). The 1980s and 1990s were marked by a focus on the regulatory structure coordinating the relationship between the management and shareholders of a firm. Ongoing liberalization of the market and the growing internationalization of institutional investors such as insurance companies or pension funds further promoted the interest in corporate governance (Cheffins 2013; Eberle 2010: 165).

Today, issues of target orientation of firms, their societal responsibility (Corporate Social Responsibility), the rules of transparency and financial reporting, provision and incentive systems for the management, as well as the scope and organization of supervisory boards are discussed in the context of corporate governance (Eberle 2010: 156; Schuppert 2008: 16). This holds especially true for Germany (Eberle 2010: 167). Coming from a more critical perspective of political economy Nölke, Overbeek, and Apeldoorn (2007) identify "an increasing openness of corporate governance regulation to market forces" (Nölke, Overbeek, and Apeldoorn 2007: 217).

Nevertheless, most researchers have a common denominator. The regulatory structure of corporate governance determines formal and informal rules of how authority and power are shared among the stake- and shareholders of a firm. The former perspective represents the insider, the latter the outsider point of view. On the one hand, according to Deeg (2006) and Eberle (2010), inside oriented stakeholder systems tend to have a strong commitment to the stakeholders, especially to the firms' employees (Deeg 2006: 84f; Eberle 2010: 162). On the other hand, the internal view highlights the control function of the management and supervisory board which monitors the management (Eberle 2010: 157).

Summarized and in a narrow understanding of political science, the regulatory structure of corporate governance comprises hierarchical frameworks, financial models, and the organization of labor within a

---

cost approach by Williamson (1979) when the broader, contemporary discussion of corporate governance was initiated (Cheffins 2013: 46–49; Eberle 2010; Padgett 2012: 203–205).

firm (Cheffins 2013: 59; Eberle 2010: 155; Mayntz 2003: 73). Those characteristics are embedded in the surrounding institutional environment of the meso- and the macro-level (Eberle 2010: 157). Gourevitch and Shinn (2005) exemplify the link between economic preferences and political institutions by various cases from Asia, Australia, and Europe, as well as the Americas. They conclude that "corporate governance arises from incentives created by rules and regulations" based on policy formulation processes (Gourevitch and Shinn 2005: 277).

**Meso-level**

The region represents the meso-level of governance. Coming from Marshall (1997), Weber (2011), Polanyi (2009), Chandler (2002), and Piore and Sabel (1984), academia's contemporary research agenda focuses on the regulatory structrue of regional production systems and relations. These regional production systems and relations are described by the concepts of metropolitan regions (Gehrlein 2009; Phares 2009), global city regions (Sassen 2005; Sassen 2008a; Sassen 2008b), industrial districts (Harrison 1992; Piore and Sabel 1984; Saxenian 1996), innovation systems (Blättel-Mink and Ebner 2009; Ebner (forthcoming) 2015[b]), milieus (Storper 1995; Storper and Scott 2003), or clusters (M. E. Porter 1998; M. E. Porter 2008b).[28]

All concepts have a common denominator. They describe an agglomeration of institutions within a certain space, whereas space or the spatial region is not defined by political borders. Rather the region is defined as geographic-functional. This emphasizes, as also formulated by polarization theory, the interdependencies and the interconnections of actors. Moreover, the system of a region is characterized by a specific structure of relationships between institutions, the institutional governance arrangement (Grabher 1994a: 61–66, 108). For Perroux (1968), this system is more characterized by sectors, rather than places (Harrison 1992: 473). Enriched by socio-cultural and political-administrative institutions, economic action is (re-)connected to the region (Grabher 1994a: 67). Thereto, regions and their institutional arrangements are neither

28 See Le Galès and Voelzkow (2003) for further concepts and contributors (Le Galès and Voelzkow 2003: 2).

restricted, nor comparable to political borders (Grabher 1994a: 46–49; see footnote for further detail).[29]

Within the German research tradition, the transformation of state-affiliated sectors and industries were analyzed (Lütz 2010: 144).[30] The American tradition focused on institutions, which coordinate actions within the branches and beyond its borders (Lütz 2006: 18; Lütz 2010: 137, 147).[31] This tradition goes along with the understanding of Geißel (2013), which is applied in the presented research. Following Geißel (2013), the institutional arrangement of a regulatory structure of governance is the surrounding environment of the decision-making process. It goes beyond the inclusion of trade unions and employers' representatives (Geißel 2013: 10f). Decision-making is negotiated between private and non-private institutions as well as citizens (Geißel 2012b: 165).[32] This indicates, that on the meso-level the regulatory structure of governance can be described as "competing institutions, overlapping jurisdictions, multiple identities, [and] territorial flux" (Cerny 2006: 694) or as "undeveloped, unstable tangled hierarchies" (Mayntz 2007: 7).

With that, the meso-level of governance is a regulatory framework characterized by decisions and decision-making processes. They take place between different actors, which have according to federalism, unambiguous jurisdictions. Jurisdictions are subject to negotiation processes and the unambiguity is dissolved.

---

29 Kiese (2008: 10); Loer (2007); M. E. Porter: (1998:79); Saxenian (1996); Scott (2008); etc.

30 E.g. Mayntz (2009b), Mayntz (2007), Mayntz and V. Schneider (1995), and Schmidt (2010).

31 E.g. Crouch, Le Galés, et al. (2003), Hollingsworth, Schmitter, and Streeck (1994), and Campbell, Lindberg, and Hollingsworth (1991).

32 Based on Crouch (2005) and Crouch and Farrell (2002), who argue that actors have alternatives and are able to combine different paths and institutions, Lütz (2010) identifies five central institutions. She argues, that even in minimal hybrid models, where governance is based on essential regulation, the market, firm hierarchies, the state, associations, and networks represent the basic forms and institutions of regional governance (Lütz 2010: 148f).

**Macro-level**

In the course of ongoing internationalization, governance research on the macro-level moved from the question of deficits of steering capacities to questions of globalization. According to Mayntz (2009b), the focal point of a critical analysis is the question of state autonomy and the capacity to act, both in regard to domestic and foreign policy (Mayntz 2009b: 54). This raises the question of if "Governance without Government" works. From an international relations perspective Rosenau and Czempiel (1992) acknowledged that critical remark. Rosenau (1992) concludes, that developments of new orders on the global level cannot successfully evolve without the counterparts on the national level (Rosenau 1992: 272).[33] Biersteker (1992) demonstrates in the same volume, that these phenomena of an international relations perspective are applicable for international economic order. After the negotiations for a new international economic order, aiming at the change of the Terms of Trade,[34] nearly all developing countries began to restructure their domestic economy in regard to neoclassical economic policy, based on the World Banks' demands formulated in their structural adjustment programs (Biersteker 1992; Milner and Keohane 1996: 24). This illustrates, that as a consequence of globalization researchers from different fields saw the sovereignty of the national state in danger. Especially influences of the global economy, rising global trade, international investment, and the influence on the national economy, the tax system, and the attributed shrinking capacity of the administration to act, were identified as the major challenges (Mayntz 2009b: 54). This phenomenon was discussed already in the 1970s by Copper (1972), who argued that "joining national markets, erodes the effectiveness of [...] policies" (Copper 1972: 164). Within political economy research in the 1990s, this was discussed by Strange (1996) and

33 Again, the crisis in Greek and the negotiations with creditors from Europe and the World Bank is a good example.

34 The Terms of Trade describe in a broader sense the trade relationships between countries. Based on Françoise Perroux and Gunnar Myrdal especially the Singer-Prebisch thesis (Prebisch 1950; Singer 1949) discussed the secular deterioration of ex- and import relations between the products of the industrialized nations and the countries of the so-called Third World. For further readings, see e.g. Nuscheler (2010) pp. 201–218.

Ohmae (1995), and Potter (1992) asked: "*are states becoming more or less autonomous within the global economy?*" (Potter 1992: 216).[35] In the research field of law, Kerwer (2004) discusses the impacts of the financial market regulations (Basel I), where banks became a part of the global banking regime by links through national regulatory authorities (Kerwer 2004: 219). This example shows, that the "economic denationalization," as Mayntz (2009b) refers to this phenomenon, is not always an exogenous fate. Furthermore, it is sometimes a conscious policy decision (Mayntz 2009b: 54f; Sassen 2002: 107). Reflecting further on phenomena, such as the global financial crises, there is no doubt that single states cannot handle such challenges.[36] In times of raising internationalization, and, therefore, international cooperation, a regulatory structure of global governance is a basic principle to handle international threats (Zürn 2006: 121).

Zürn (2008) characterizes global governance as an ideal type of international politics to govern. For him, global governance represents comprehensive political order, in which national administrations are embedded (Zürn 2008: 128f), even if it is marked with several structural challenges regarding juridification (*Verrechtlichung*) and socialization (*Vergesellschaftung*) as well as a normative discussion hidden behind (Zürn 2006: 138–141).

Summarizing the macro-level, aspects of "glocalization" and state autonomy come into the fore. The establishment and implementation of new orders on the global level cannot successfully evolve without the counterparts on the national, regional, and local levels. Within the raising internationalization, the regulatory structure of global governance is a basic principle of comprehensive political order, in which national administrations are embedded and global issues with regional challenges have to be handled.

The depiction of multilevel governance shows, that processes are located on different levels within the regulatory structure of governance.

---

35 Text highlighted in original.

36 Beside economic challenges Zürn (2006) refers to ecological threats, like the climate change, the hole in the ozone layer, and the declining biodiversity as well as security threats such as transnational terrorism and civil wars (Zürn 2006: 121).

Additionally, concerning the geographic-functional understanding of the region, it is apparent, that processes of the regulatory structure of governance are not enclosed in a certain level. Rather, the regulatory structure of governance has to be analyzed through the different levels, because jurisdictions overlap. This is most obvious on the meso-level where jurisdictions overlap the regulatory structure bottom-up and top-down. Thus, in the next paragraphs the theoretical concepts of regional and metropolitan governance, applying to the meso-level are discussed in detail. Gaps regarding the analysis of the regulatory structure of governance in geographic-functional regions are exposed.

## 2.4 Governance in geographic-functional regions

With increasing cooperation between administrative and nonadministrative institutions beyond the levels and the dissolution of Webers' "*Lokal*markt," as the economic and political focal point of a city, the phenomenon of governance in geographic-functional regions becomes more prominent. Nevertheless, the subnational region is still important for economic, social, and political (inter-)action. Two aspects have to be named. On the one side, several topics are too small to be discussed globally. On the other side, there are also regional discussions and decisions, which have far bigger externalities, than actual regional administrative borders embrace (Benz and Fürst 2003: 11; Bröchler and Blumenthal 2006: 9; Dose 2006: 25).

For those reasons, it is apparent, as it is also presented in the general governance discussion, that the regulatory structure of governance in local economies is subject to change Crouch, Le Galés, et al. 2003; (Crouch et al. 2004). The common understanding of the region, as a clearly determined spatial district with unequivocal jurisdictions, is jeopardized. This phenomenon becomes even more questionable if the region is not defined by law, but by "arbitrary" administrative and nonadministrative cooperation, embedded in an institutional arrangement (Benz et al. 1999: 47).

### 2.4.1 Regional governance

As in the general discussion and touched in the previous paragraph of multilevel governance, regulatory structures of regional governance also questions, how development processes can be realized in a more fragmented and more globalized world.[37] The trigger is an identified deficit within the administration and failure of mechanisms of the market economy to handle regional challenges. On the one hand, the regulatory framework of regional governance becomes apparent, when the deficits are too big and other mechanisms of decision-making processes fail (Fürst 2007: 353; Fürst 2010: 49).

On the other hand, regional governance comes into the fore, when regional cooperating institutions identify a comparative advantage regarding other forms of collective action. For administrative institutions, the advantages come from the possibility of more flexible and direct cooperation. Private institutions have an advantage regarding a direct link to administrative decision-makers (Fürst 2010: 55).[38]

To define the regulatory structure of regional governance, there has to be at least a common regional interest or a regional connection between the cooperating institutions. This means regional governance arrangements have to comprise territorial- and functional-oriented actors with a common public goal. It concerns regional administrative institutions, such as administrative districts and municipalities, as well as nonadministrative actors, such as firms, associations, and other civil society organizations. The former are territorial oriented, the latter functional (Fürst 2007: 354). Thereto, the regulatory structure of regional governance refers to space and functions (Fürst 2010: 51). As socio-economic systems, regional or local production systems "are complex and rarely embody one over-riding logic or ideal type" (Crouch and Trigilia 2003: 237).

There is no exact definition or an analytical framework of the regulatory structure of regional governance. However, Fürst (2010) identifies several aspects in regard to space and functions. According to him, regional governance is characterized by the interaction of

37 For the delineation of comparable concepts like Corporatism, research on urban regimes, Political Network Theory, Milieu Studies, and participation see Fürst (2007), pp. 356f.

38 This direct link is also characterized by linking social capital. Linking social capital describes the direct hierarchical link between administrative and nonadministrative actors from the micro-, the meso- and the macro-level (Sorg 2009: 67; Woolcock 2001).

different administrative and nonadministrative institutions, coming from the administration, the economy, and civil society. These institutions are interdependent. This interdependency leads to a fuzzy differentiation between the three spheres of politics, economy, and society. The borders between the different systems within the regulatory structure are often not identifiable. The involved institutions are connected by mutual informal negotiation processes and formal proceedings. Furthermore, the regulatory structure of regional governance is embedded in an institutional environment of rules of the game, which shape and channel interactions. Here, processes of learning are very important.[39]

This vague definition and a missing analytical framework make it hard to analyze the regulatory structure of regional governance empirically (Fürst 2010: 50–54). It describes a complex framework of control within spatial regions (Benz and Fürst 2003: 12), which have to handle certain challenges but also have several inherent opportunities.

Due to challenges to define regional governance, all attempts are in close relation to the general definition of governance. No one of them has become prevalent.

Following Fürst (2010), regional governance can be defined as a close and enforcing system of regional cooperation between administrative and nonadministrative institutions, where formal structures are limited or not available. The concept aims at regional structures and processes of coordination, in which the distinction between the "subject of steering" and the "object of steering" gets fuzzier (Fürst 2007: 353–355). From the spatial science point of view and further following Fürst (2007), the regulatory framework of regional governance refers to a network-like inclusion of administrative and nonadministrative institutions to advance regional development (Fürst 2007: 356).

In regard to processes of regionalization, Benz et al. (1999) define governance in delineation of government (regime of state-owned institutions).[40]

39 For the importance of the aspect of learning in a certain regional environment, see also the concept of relational economy and embeddedness (Bathelt and Glückler 2002; (Bathelt and Glückler 2003; Bathelt and Glückler 2011; Glückler 2000).

40 This confrontation is typically within political science (Blumenthal 2005: 1151).

They define regional governance as the formal and informal structure of control and rules, coordinating administrative and nonadministrative institutions. This coordination is not necessarily hierarchical (Benz et al. 1999: 50f).

Benz (2003b) describes regional governance as the coordination and control of regional processes within complex structures (Benz 2003b: 505). It is characterized by regional self-controlled coordination of the region, its actors (administrative and nonadministrative), and actions (Benz 2003b: 515–517).

Coming from the field of geography, Danielzyk and Reitzel (2003) define regional governance as a control system, which coordinates administrative and nonadministrative institutions. It embraces formal and informal elements and is not exclusively organized hierarchically. For them, the core element of regional governance is, the trustful interaction of administrative and nonadministrative institutions, within the regional control system (Danielzyk and Reitzel 2003: 513).

Also coming from the field of geography, Kieling (2003) emphasizes practical relevance. For him, regional governance represents a regulatory structure linking administrative and nonadministrative institutions, within regional systems of self-organization. Regional governance describes a system of cooperative and network-based control systems, coexists with traditional hierarchically oriented control systems (Kieling 2003: 471).[41]

The regulatory framework of regional governance is not aligned with a specific outcome. The advantages for actors participating in the governance process are not clearly identifiable upfront. Each regional governance system has "political opportunity structures" (Fürst 2010: 55). According to Benz et al. (1999), this is the transformation to the supporting state. Fürst (2010) speaks here of the "activating state" (*aktivierender Staat*) and Schuppert (2005) of the "warranty state" (*Gewährleistungsstaat*). All have in common, that the state requests more personal responsibility. As a countermove, the state provides a regulatory structure and institutional framework for coordination (Fürst 2010: 55). The claim is to integrate private institutions and their growing

41 For a comprehensive listing of further definitions of regional governance see Panebianco (2013) pp. 58f.

position as co-producers or even veto players.[42] A further opportunity is the change of developmental policies to decentralized development concepts. The paradigm shift from the "competition of locations" to the "competition of regions,"[43] and the need of interconnectedness within a society characterized by labor division, was forced by the need of change and innovation (Fürst 2010: 55f.).

Nonetheless, this relation to space is also not without challenges for the emergence of regulatory structures of regional governance. Whereas location is in close relation to a traditional municipal policy or local affairs within clear political borders, the region encompasses a geographic-functional definition of spatial agglomeration in regard to political, economic, social, and cultural characteristics. This is important, because the boundaries of proximity, as they are identifiable in industrial districts or clusters, are not necessarily comparable to political borders (Bathelt and Glückler 2011: 6; see footnote for further details).[44] The challenge is, that the local administration is legitimized within political borders, however, private institutions are not inevitably bounded to this regulatory context (Fürst 2007: 354, 362f). V. Schulze (2013) takes up the claim regarding safety and security questions in local settings. She concludes, that safety and security topics are sovereign tasks, but corresponding policy-making processes should take administrative and private institutions into account, and, therefore different levels and different spaces (V. Schulze 2013: 241).

This aspect is closely connected to the phenomenon, that the distinction between administration and society gets fuzzier. Private and non-private

---

42 Against Tsebelis (2002), who defines only several administrative institutions (members of the government) as veto players (Tsebelis 2002: 2), a veto player here is defined as an institution, which is not necessarily democratically legitimized, but which has the power to make use of a veto, e.g. because of resource-based supremacy.

43 This phenomenon was also recognized on the international level by the comment of Bill Clinton, that nations are "like a big corporation competing in the global marketplace" (Clinton 1993), because the competitiveness of nations (or regions) has become a valuable social and economic good.

44 M. E. Porter (1998: 79); Saxenian (1996); Scott et al.: (2008: 27); Here, region is recognized and highlighted as an operational level, which is relevant for social action. This phenomenon is discussed in the field of new regionalism in depth (Fürst 2007: 355).

institutions have to cooperate more, because the challenges cannot be handled by one sphere alone, e.g. in public-private-partnerships (Fürst 2007: 354f). However, concerning institutions and their constellation, it is difficult to identify the relevant stakeholders of a certain topic, meaning who are, by which characteristic, the relevant actors or mandate holders. On the one side, they are commonly chosen bottom-up by the involved institutions themselves. This is the case when actors agglomerate around a common project. On the other side, institutions agglomerate, when collective action is promoted or enforced by the administration (Fürst 2010: 60f).[45] These aspects make cooperation between the two spheres not only favorable. Sometimes this cooperation is a must, e.g. in situations where there is a need for democratic legitimacy but private veto players are involved. In those situations, traditional policy-making reaches its limits. Furthermore, based on this administrative and nonadministrative labor division, it is often not clear, if administrative law is applicable in the negotiation process (Fürst 2007: 354f). Another challenge, which has to be mentioned in regard to institutions, is in close relation to democratic legitimacy. Institutions established by private cooperation can lead to *de facto* pre-judicative decisions, which are interest based and selective. Decisions made in this environment call democratically authorized institutions into question and might have a defective effect on democracy (Fürst 2007: 361f; Fürst 2010: 61f). Börzel (2008) and Héritier and Lehmkuhl (2008) refer to this as governance in the "shadow of hierarchy." In addition, Scharpf (1997) defines common negotiations between state authority and institutions of the civil society as self-organization in the "shadow of hierarchy" (Scharpf 1997: 197–205).[46] Nevertheless, these preliminary decision systems are not always negative. They can have a positive effect because decisions are based on broader professionalism and are more well-balanced (Fürst 2007: 362).

---

45 Fürst (2010) refers here to promotion programs such as Horizon2020 or the Cohesion Fund of the EU to promote common research cooperation or common infrastructure projects.

46 Scharpf (1997) refers to wages and working negotiations of trade unions and the *Deutsches Institut für Normung* (DIN) as examples. In both cases, binding rules of the game are negotiated by private actors (Scharpf 1997: 200).

Basically, regional governance represents an open form of governance. Negotiation processes are embedded in regional regulatory structures and institutional arrangements, which reduce uncertainty. Finally, there is a challenge with regard to the topics. The regulatory structure of governance itself is restricted to certain topics because topics and, therefore, the results are framed by the knowledge and the capacity of the involved institutions and the available cosmopolitan knowledge and their tacit knowledge (Fürst 2010: 55–63).

All attempts of definition have in common, that regional governance describes complex regional regulatory structures of control. Following the analytical approach of Benz and Fürst (2003), several elements are important: organizational constellations, network-like control structures, regional administrative actors, the governance-mix of different control mechanisms, including regional management structures (Benz and Fürst 2003 12f). Within political science, this phenomenon is especially identifiable on the regional level. Here, politics are pushed onto a level, which is incompatible to common political borders. The administration has to organize tasks (and itself) above the local level and below the national level simultaneously within a regulatory structure of multilevel governance. This contradicts, that regional policy-making should take place within public institutions with unambiguous jurisdictions. New institutions are the result, e.g. planning groups or regional promotion committees (Benz et al. 1999: 47; Blumenthal 2005: 1155f.). These newly established institutions are oriented at geographic-functional topics and challenges, which actors have in common, not on political borders (Bröchler and Blumenthal 2006: 9).

In contemporary research, all this is discussed also in connection to regulatory structures of metropolitan regions and metropolitan governance. Although regional governance is closely related to metropolitan governance, in the following paragraph the concept of metropolitan governance is presented. At this point, the concept of regional governance represents a broader framework of metropolitan governance. The regulatory structure of metropolitan governance focuses more on polycentrism.

### 2.4.2 Metropolitan governance

Cities, as spatial agglomerations of economic, social, political, and administrative activities, represent melting pots and incubators for innovation and change (Kübler 2012: 402). As stated, contemporary cities cannot be characterized as relatively closed settlements with a centered "*Lokal*markt" anymore. They have turned into geographic-functional regions with one (monocentric) or more (polycentric) major centers. These geographic-functional regions embrace different "territorial systems of economic, social and cultural exchange" (Kübler 2012: 402) and coordinate globally linked value chains, power, and decisions (J. Esser and Schamp 2001: 12). Geographic-functional regions are characterized by global and regional phenomena simultaneously (Kübler 2012: 403).

Following the geographic-functional definition, metropolitan regions are not necessarily bounded to administrative borders or institutional settings. They comprise "of a multitude of different local, regional and sometimes even national politico-administrative units" (Kübler 2012: 403). Both, administrative and nonadministrative institutions are embedded in the institutional arrangement and regulatory structure of governance (Zimmermann and Heinelt 2002: 9f.). Regarding the regulatory structure of metropolitan regions or conurbations, local authorities within their administrative territory are hardly addressed separately. According to Zimmermann and Heinelt (2002), especially the debate in Germany assumed that conurbations are defined by their scope of duties and, therefore, dependent on different spatial layouts. This geographic-functional understanding is also comprehensible on the refusal of the *Regionalstadt*, which refers to clear administrative borders and jurisdictions (Zimmermann and Heinelt 2002: 10).[47]

---

47 The concept of the *Regionalstadt* was introduced after World War II, but the administrative reform failed in the 1970s due administrative resentments (Ganser 2005: 20). 25 years after the refusal of the concept in FRM in 1971 the Frankfurter Allgemeine still calls the *Regionalstadt* a bogey ("The ever-present bogey of Regionalstadt", FAZ 1996, translated by the author). For the concept and the suggestions for the implementation in Frankfurt see Frankfurt (1971).

This allows the conclusion, that the regulatory structure of metropolitan governance in Germany is characterized by geographic-functional arrangements, including administrative and nonadministrative institutions in heterarchic and hierarchic decision-making constellations. Referring to Dahl and Tufte (1973), Zimmermann and Heinelt (2002) conclude that "[d]ifferent problems require political units of different sizes" (Dahl and Tufte 1973: 135) and a flexible political geometry is preferable in metropolitan governance (Zimmermann and Heinelt 2002: 18).

Nevertheless, as already for regional governance, it is claimed for metropolitan governance, that an exact definition is a *desideratum*. According to Ahrend et al. (2014) and Zimmermann and Heinelt (2002) it also lacks of an analytical framework (Ahrend et al. 2014: 5; Zimmermann and Heinelt 2002: 28). The conclusion of Crouch and Trigilia (2003), that regional production systems "are complex and rarely embody one over-riding logic or ideal type" (Crouch and Trigilia 2003: 237), is already apparent for the distinction between mono- and polycentric metropolitan regions or different definitions of the same region.

Even the challenges of metropolitan governance were identified already at the beginning of the 20th century, it took until the 1950s and 1960s before the discussion of regulatory structures of metropolitan governance entered academia more far-reaching (Lackowska 2011a: 22). Three approaches have to be named: the metropolitan reform tradition, with its early beginning in the 1930s, the public choice approach, emerged in the expiring 1950s, and new regionalism, turned up in the 1990s (Kübler 2012: 404f; Lackowska 2011a).

**Metropolitan reform tradition**

The metropolitan reform tradition[48] identifies fragmentation and its adjustment as the main challenge for metropolitan governance. The approach argues for the creation of a coherent system and the formation of a single metropolitan government, to overcome conflicts between local actors, especially administrative actors, and to handle the shortcomings

48 As former prominent researchers Studenski (1930) and Wood (1958) have to be named (Zimmermann and Heinelt 2002: 21).

in service delivery. It is further claimed, that negative externalities of fragmentation can be limited by establishing a new metropolitan governance body, which plans strategically for the whole region. This reduces segregation and addresses spatial mismatches. The central governance body addresses the whole region, reduces free-riding, and fosters joint metropolitan interests and a common identity (Lackowska 2011a). Nevertheless, the question of losing power and independence in regard to the existing local jurisdiction in the regulatory structure is the major obstacle for the metropolitan reform approach. "Existing local governments have always looked unfavorably upon the appearance of new autonomous and powerful political structures [...] " (Lefèvre 1998: 15), because this would call their authority and jurisdiction in question (Kübler 2012: 404; Lackowska 2011a: 22–24).

More recent studies in this research path emphasize, that even in highly fragmented metropolitan regions, at least informal regulatory structures and coordination systems between local authorities exist. It is argued further, that these informal agglomerations ensure better service provision (Ahrend et al. 2014: 5).[49]

### Public choice

Public choice[50] argues *vice versa*. Fragmentation is not seen as a challenge, which has to be changed. It secures a better functioning of local democracy. Smaller units with local policies are more able to adjust policies to the needs of citizens. It is argued further, that due to tacit knowledge local authorities are closer to the citizens' needs and, therefore, participation on the local level of the regulatory structure is higher. In smaller units, the public choice approach claims that social control is higher, trust stronger, and anomie less common. To govern geographic-functional regions, the public choice approach relies on market allocation and competition (Ahrend et al. 2014: 5; see footnote for detail).[51] Public

49 E.g. Aulich, Sansom, and McKinlay (2014), Bish (2001), and Martin and Hock Schiff (2011).

50 See Bish (1971), Ostrom, Tiebout, and Warren (1961), and Tiebout (1956) as then previous debate pioneers (Zimmermann and Heinelt 2002: 21).

51 Kübler: (2012: 404); Lackowska (2011a: 22–24); Zimmermann and Heinelt (2002: 21).

choice argues further, that a central metropolitan governance body would "abstract and distant from the ordinary citizen" (Lackowska 2011a: 24) and a "[g]argantua, as a result, may become insensitive and clumsy in meeting the demands of local citizens [...] " (Ostrom, Tiebout, and Warren 1961: 837). Dahl and Tufte (1973) argued in the same manner. They claim that specific challenges are in the need of specialized political unites (Dahl and Tufte 1973: 135).

The public choice approach was mainly criticized concerning the promotion of single-purpose associations and the ineffectiveness of voluntary joint arrangements because they are difficult to initiate (Lackowska 2011a: 25).[52]

**New regionalism**

In the 1990s new regionalism, as an approach linked to the phenomena of "glocalization" and international competition, came into the fore (Kübler 2012: 405; Lackowska 2011a: 27; Zimmermann and Heinelt 2002: 23). New regionalism is based on factors behind local government and global governance as well as its transition. It can directly be linked to a geographic-functional definition of the region (Zimmermann and Heinelt 2002: 23). In this research path, it is argued, that especially the economic actors, previously dependent on administrative borders and national markets, disassociate themselves from the regulatory structure (John 2001: 9f). Metropolitan regions come under pressure of globalized markets and international competition. Former approaches were not able to handle the challenges in regard to changes coming with. On the one hand, metropolitan regions have to establish external oriented policies in regard to international competition. On the other hand, inner-functional coherence of the whole metropolitan region has to be developed, to address global challenges successfully (Lackowska 2011a: 27f).

The main argument of the new regionalism approach contests both former research paths. The regulatory structure of governance, in the wider understanding of Mayntz (2003), "cannot only result from hierarchic decision making (metropolitan reform tradition) or competition (public

52 E.g. V. Jones (1942).

choice tradition)" (Kübler 2012: 405). The focus on fixed and unchangeable spatial settings stepped behind a geographic-functional understanding (Kübler 2012: 405; Lackowska 2011a: 28). Metropolitan regions, as it is argued by new regionalism, are not isolated. Rather, new regionalism overcomes formal structures of administrative jurisdictions in favor of the socio-economic system (Blatter 2007: 337). Geographic-functional regions create opportunities for global and local action simultaneously. Here, "[m]etropolitan governance is a topic that increasingly concerns state agencies at the regional, national and even supra-national level [...], but also business and civil society actors [...] " in regional economies (Kübler 2012: 406). And even there are multiple, sometimes overwhelming actor constellations, "metropolis is a single, indivisible economic entity that should not be cut up into smaller pieces" (Newton 2012: 414). For this reason, the regulatory governance structure of "Metropolis" has to handle redistributive conflicts (global versus local and center versus suburbs) and has to find solutions regarding challenges of action coordination (Zimmermann and Heinelt 2002: 13).

Nevertheless, also new regionalism did not remain without critique. On the one side, the rising complexity (compared to the two former approaches) of decision-making and the democratic mechanisms were criticized. On the other side, the reaching of joint agreements in such diverse setting was questioned (Lackowska 2011a: 29).

Zimmermann and Heinelt (2002) distinguish further between an optimistic and a pessimistic point of view in new regionalism. The optimistic view assumes that even society is influenced by the constraints of globalization, it still develops strategies of action orientation between social entities, even if these entities overlap spatially. The pessimistic point of view focuses on economic action and the adjustment of decision-making jurisdictions to higher- or lower-level units (Zimmermann and Heinelt 2002: 23–25). However, this distinction has to be neglected, because it represents an academic feud.

## 2.5 Interim conclusion – governance

As indicated in the theoretical discourse, the general governance research brought up two interpretations – a wider understanding and a more narrow understanding of governance. The former includes the sum of all co-existing forms of collective regulation of social issues with a regulatory structure of governance. The latter, as a counter concept of government, excludes hierarchical administrative decision-making processes.

Regional governance research is according to Blumenthal (2005), in the need of a conscientious use of the more narrow understanding in the demarcation to government. She argues, that within a region no jurisdiction and political-administrative or hierarchical structure is available. Further, she exposes that regional governance is especially in the need for bottom-up initiatives (Blumenthal 2005: 1158). This holds true also for metropolitan governance because there is no common definition. By focusing on the geographic-functional aspect of metropolitan governance, it appears that both concepts do not differ from each other and can be interpreted as equal for the presented research.

Anyhow, Blumenthal (2005) also refers to the fact, that the more narrow understanding of governance becomes problematic in empirical reality. This happens when decisions of the negotiation process between administrative and nonadministrative institutions within the regulatory governance structure have to be implemented legally (Blumenthal 2005: 1158). At least now, the administration has to be taken into account. Pierre and Peters (2000) and even representatives of the narrow understanding of governance, e.g. Stroker (1998), emphasis also the capacity of a political science perspective of governance to analyze complex interaction systems and regulatory structures beyond constitutional institutions (Blumenthal 2005: 1167).

Due to the transition problem and the geographic-functional definition of the region, the presented research follows a wider understanding of governance. This approach recognizes the changing realities in steering and control systems. In contrast to the more narrow understanding of governance, it does not factor out governmental aspects upfront. Thereto, it draws the more realistic picture of the political environment and avoids the hair-splitting discussion on governance *versus* government. Furthermore, as Blumenthal (2005) points out, the wider understanding has to be prioritized in the analytical perspective. While the more narrow understanding focuses on institutional and actors relations, the wider understanding tries also to emphasize proceedings of action within a regulatory framework. It does not ignore pure administrative decision-making, and, therefore, it has not to include it afterward or handles it with another concept. Thereto, structural and process dimensions are taken into account. The wider understanding of governance allows considering the combination of different and sometimes conflicting regulatory structures of governance (Blumenthal 2005: 1153–1159). By not dividing the research into political science sub-disciplines upfront, if this is possible at all, the wider research perspective takes a problem-oriented position into account. The question is, how a regulatory structure is organized today. This contributes to the change from planning and steering to governance, by the contemplation of network structures, which do not stop from political borders and are characterized by administrative and nonadministrative institutions (Jachtenfuchs and Kohler-Koch 2004: 98). Rather, this approach tries to keep implicit assumptions down. This allows establishing an analytical framework, which is usable for a bunch of applications. It does not just highlight the decision-making in the "shadow of hierarchy" (Blumenthal 2005: 1166). It represents a framework to analyze and understand (changing) decision-making processes (Stroker 1998: 18). Additionally, the problem-oriented framework acknowledges the consideration of interdisciplinary starting points, if necessary and useful (Jachtenfuchs and Kohler-Koch 2004: 98).

With that, different forms of purposive regulations of collective circumstances, both formal and informal, and administrative and nonadministrative institutions are taken into account. Further, the regulatory structure of governance is distinct but not diametrical to hierarchical rules and it does not represent a simple delineation to government. "Governance without Government" (Rosenau and Czempiel 1992) is an uncommon exception, especially for regulatory structures of metropolitan governance, when legislative actions become necessary and governance steps out of the "shadow of hierarchy."

In front of new institutionalism, governance and governance research includes more than autonomous actions of an administration. The picture which is drawn on democracy, hierarchy and the concentration of problem-solving embraces new dimensions of the formation of political objectives, networks, and the process dimensions.

However, there are still further challenges left. First, neither regional governance nor metropolitan governance provides a suitable analytical framework to analyze regulatory structures of governance and processes in geographic-functional regions. There is no answer on how the decision-making processes in regulatory frameworks of regional or metropolitan governance are defined and organized. Here, both approaches fall back on the general definition of governance. This goes along with the second gap of regional and metropolitan governance, the identification and integration of the relevant institutions into the regulatory structure of governance and its processes. Both research paths do not indicate what the relevant institutions of the geographic-functional region are. Regional governance only identifies some kind of spatial and functional focus where relevant actors are connected by mutual negotiation processes within the regulatory structure. Falling back on the general governance approach, it is stated for regional governance, that all actors arguing for a public interest have to be integrated and linked to the decision-making process. The integration process here is very vague and based on mere argumentation. The researches of regional governance themselves state, that this is a too uncertain definition, which hampers empirical analysis of the institutional arrangement.

This lack is also identifiable for metropolitan governance. Generally, the metropolitan governance approach provides three research paths, trying to argue for or against the relevance of a central, metropolitan governance body within the regulatory structure. The metropolitan reform tradition suggests having a metropolitan governance body as the central, final decider, whereas there is no answer who is represented in the metropolitan governance body and, therefore, in the decision-making process. Public choice argues for market allocation and completion, which also gives no concrete indication on the relevant institutions within the regulatory structure of governance. Additionally, this path might exclude relevant institutions, because they are not able to participate, because of their weak competitive position, e.g. local citizens' groups. New regionalism argues also for "Metropolis" as a single entity. However, it understands this single entity not as closed and isolated, but gives also no indication of the relevant and involved institutions.

Both, regional governance as well as metropolitan governance, lack of a suitable analytical framework to analyze the regulatory structure of governance, including decision-making processes and institutional arrangements in geographic-functional regions.

With that, the regional governance approach and the metropolitan governance approach, are insufficient to analyze the regulatory structure of governance in geographic-functional regions. Rather, within metropolitan governance new regionalism is "call[ing] for a differentiate repertoire of political organizations including a variety of loosely institutionalized arrangements such as […] interlocal coordination, public-private partnerships, and policy-specific arrangements" (Koch 2013: 400). Based on the experimentalist governance approach of Sabel and Zeitlin (2012b) the theoretical derived approach of ERG addresses these issues. On the one side, it presents frame-giving conditions of the regulatory structure of governance and decision-making processes. On the other side, it dissolves the challenges in regard to the identification and involvement of relevant institutions within the regulatory structure.

To present the theoretical deduced analytical framework of ERG, in the following paragraph Sabel and Zeitlin's concept of experimentalist governance is introduced.

## 2.6 Experimentalist regional governance

The phenomenon of governance emerged due to the awareness of the society and the economy that regulatory structures of planning and steering by politics were insufficient approaches to handle economic and social issues as well as challenges evoked by globalization and regionalization. Its theoretical evolution developed parallel between the research fields of administration, international relations, institutional economics, management studies, and the field of international cooperation and development (Grande 2009: 79; Zürn 2008: 553). It is assumed that the term "governance" comes from steering or piloting. Linguistically, it derives from the Greek expression *kybernan*, which means "to pilot." It is translated into the Latin word *gubernare*. Therefore, the meaning is not only metaphorical; rather, it is linked to the basic idea of steering politics

(Levi-Faur 2012: 5; Mayntz 2009b: 13; Peters 2012: 20).[53] This idea of interventionism dates back to the absolutist territorial states of Europe (Mayntz 2003: 73).

As a research field, governance was nearly meaningless within academia until the 1950s and 1960s (Levi-Faur 2012: 5; Mayntz 2009b: 15). However, coming from the fields of economics, Coase (1937) already highlighted that (economic) action is shaped by organizational structures. Finally, it was Williamson (1979) who more broadly introduced governance. The coordination of action was no longer interpreted as only being governed by hierarchies. Non-market phenomena and institutions, such as family, clans, associations and later networks, were taken into account (Mayntz 2009b: 14; Schuppert 2008: 16).

The growing interest in markets and hierarchies, as organizational forms steering administrative and economic actions prompted increased interest in governance research (Levi-Faur 2012: 5; Mayntz 1996: 148; Mayntz 2009b: 14).

The final breakthrough in political and social science can be determined for the 1970s (Mayntz 2006: 11f). Today, governance is a buzz word, not only within political and social science but also within applied politics and the administration (government) (Mayntz 2009a: 9). It "[...] has become probably one of the most important manifestations of the rise of Neo-Institutionalism in social science" (Levi-Faur 2012: 3).

Starting from Williams' basic idea that action is no longer simply routed by regulatory structures of hierarchy and pure market mechanisms, two principal meanings of governance are identifiable. On the one hand, governance is used to describe the regulatory structures of a nonhierarchical mode of cooperative control. Administrative and private actors are coordinated by "mixed public-private networks" (Mayntz 2009b: 13). This research path focuses on collective problem-solving through cooperation outside hierarchical frameworks. It describes the narrower understanding of governance. On the other hand, in a broader

53 Within the English literature, the common term is "governance," but in order to describe the evolution of the governance theory, the presented research uses the term of "steering" (Mayntz 2009b: 29). Against the background of the German expression *Steuerungspolitik* this is considered as useful.

understanding, governance is used more generally to describe the coordination of actors, through social order or deliberation within a regulatory structure.

In this context, Mayntz (2003) defines governance as:

> "the sum of all co-existing forms of collective regulation of societal issues: from the institutionalized civic self-regulation through various forms of cooperation between non-private and private actors to sovereign actions of state actors" (Mayntz 2003: 72).[54]

Accordingly, the regulatory structure of governance characterizes different forms of purposive regulations of collective circumstances, both formal and informal. Furthermore, governance is distinct but not diametrical to hierarchical rules (Mayntz 2009a: 9; Mayntz 2009b: 13). On the one hand, the regulatory structure of governance and the decision-making process require reluctant interests *per definitionem* (Mayntz 2009a: 13). On the other hand, they presuppose the intention of the actors, to cooperate and demands systems of negotiation (Mayntz 2009b: 11; Mayntz 2003: 79). Institutions or forms of regulation vary from purely administrative to purely private (Mayntz 2003: 74).

Due to the phenomena of globalization and the ongoing divestment or enlargement of clear common (political) borders and jurisdictions, the concept of governance even in its broadest understanding comes under pressure.

Now, the question has to be raised: how can governance handle the challenges regarding a faster and more complex world, where decisions on the global level have deeper impacts on the regional or the local level (and *vice versa*) and were former clear jurisdictions became geographic-functional. In front of the background of Dewey (2012) and his idea, that there is no absolute certainty of the correctness of decisions (fallibilism), already formulated in 1927, Sabel and Zeitlin (2012b) developed a concept which seems to tackle the challenge – experimentalist governance. Dewey (2012) argued that ends and means are mutually transformative. Serving democracy, ends and means have to be reconsidered as conditions within a changing environment (Sabel 2012: 36). Research indicates that

54 Translated by the author.

experimentalism aspects in handling these challenges are identifiable on the supra-national level, but also on the national level (Sabel and Zeitlin 2010a; Sabel and Zeitlin 2010b; Sabel and Zeitlin 2012b).

The theoretical fortune of experimentalism can be found in Dewey's' ideas formulated in "The Public and its Problems," released in 1927 (Dewey 2012). Based on the epistemological approach of fallibilism – which assumes that there is no absolute certainty[55] Dewey concludes that individuals and institutions learn from mistakes. In this understanding, democracy is an evolutionary process of continuous self-correction. For him, ends and means are mutually changeable (experimental) and they have to be adjusted based on pragmatism (Sabel 2012).[56]

Based on this, Sabel and Zeitlin formulated their assertions and assumptions of the experimentalist governance approach. Following Sabel and Zeitlin's theoretical deduced concept, experimentalist governance transforms diversity as an obstruction to integration as a chance (Sabel and Zeitlin 2010b: 6; Sabel and Zeitlin 2012b: 170). It is not restricted to certain policy fields with weak procedures, recommendation, action plans, and other non-binding guidelines.

The concept of experimentalist governance receives its contemporary greatest attention in decision-making processes and policy-making in the EU. Sabel and Zeitlin assert that within many policy fields, a regulatory framework of experimentalist governance aspects is partly well established (Sabel and Zeitlin 2010a; Sabel and Zeitlin 2010b). According to their research, the theoretic approach has already arrived at the European level in the arena of politics, appearing in fields where the EU has major legislative power (Sabel and Zeitlin 2010b: 10; Sabel and Zeitlin 2012b: 176). Among other areas, Sabel and Zeitlin identify a regulatory structure of experimentalist governance in the field of data privacy (Newman 2010), the financial (Posner 2010) and energy sector (Eberlein 2010), networked

55 Going back on Socrates: I know that I know nothing (Popper 1979: XXI). However, Socrates does not mean that he knows nothing, "he knows nothing *fine and good*" (Stokes 1997: 18, highlighted in original).

56 This pragmatism was also requested by Benjamin R. Barber at the *Tag der Metropolregionen* on May 5, 2015 in Frankfurt am Main – "We need more pragmatism!" he stated during the discussions regarding the governance of metropolitan regions or global cities.

competition law (Svetiev 2010), or in the field of food safety (Vos 2010) and GMOs (Dąbrowska 2010). Moreover, other supra-national institutions such as the ASEAN, the NAFTA, or the SAARC, come into consideration. In such agglomerations, actors face similar challenges in a common environment. They cooperate and can learn from each other to handle challenges.

Nonetheless, other researchers identify regulatory structures of experimentalist governance also at the country level. In the US they assume to identify regulatory structures of experimentalist governance in several fields, e.g. education, public health, and safety (O'Donnell and Sabel 2001; Sabel and W. H. Simon 2008). At the international level, the Poverty Reduction Strategy Papers of the International Monetary Fund (IMF), the cooperation between the EU and the IMF (Campbell-Verduyn and T. Porter 2014; De Búrca, Keohane, and Sabel 2014; Nance and Cottrell 2014) or the Millennium Development Goals and the post-2015 development process of the UN also have characteristics of a regulatory experimentalist governance structure. Furthermore, Sabel (2004a) and others[57] assume regulatory structures of experimentalist governance within national parliaments, courts and constitutions (Sabel and Zeitlin 2010b: 17).

The initial experimentalist governance approach assumes several consecutive key elements in the decision-making process. First, goals are defined and measurement systems are established by a provisional determination of a broad framework. Provisional means that the goals and the framework are not set up in an irrefutable manner *ad infinitum*. Second, in the following step, stakeholders are given the maximum freedom possible to achieve the goals. Third, in return to that wide discretion, stakeholders have to report and peer review[58] their activities in achieving these goals.[59] Finally, the review of the defined goals, the measurement system, and the decision-making processes is the last key element. If necessary, the group of stakeholders is enlarged and

57 Dorf and Sabel (1998); Sabel and W. H. Simon (2004); Zeitlin (2005).

58 Peer review or peer learning are mutually beneficial and involve the sharing of knowledge, ideas, and experience between participants (Boud 2001).

59 Regarding economic development (Sabel 1994) and experimental regionalism (Sabel 1996), Sabel referred to this aspect as learning by monitoring.

further relevant institutions are officially recognized within the regulatory framework and the decision-making process.

Thereto, the regulatory structure of experimentalist governance is characterized by reciprocal redefinition, repeated provisional goal-setting, and revision. It frames and structures the apparently fluid practices of network governance (De Búrca, Keohane, and Sabel 2014: 477; Sabel and Zeitlin 2010b: 3; Sabel and Zeitlin 2012b: 175).[60] These key elements have to be regarded as a combination of required functions, which can be executed by any possible variation of institutions (Overdevest and Zeitlin 2014: 25, 27). Under these circumstances – lasting for an indefinite period – relevant institutions are integrated into the regulatory structure and decision-making process. It forms the policy-making process in a participatory way (De Búrca, Keohane, and Sabel 2014: 477).

Sabel and Zeitlin presume, that as a mechanism for learning from a diversity of possibilities given in a geographic-functional region, the experimentalist governance approach suits heterogeneous but highly-interdependent regional settings. In this manner, Sabel and Zeitlin (2012b) also speak of a "directly deliberative polyarchy" form of governance. It is deliberative because relying on the better argument exposes settled practices that are taken for granted, long-lasting institutions and encrusted systems for reassessment, which are quite common in the discussion about the establishment of metropolitan regions.[61] It is directly deliberative because the decision-making process utilizes the tacit knowledge, experience, and reactions of involved actors.[62] Polyarchy results from the reciprocal process of learning, disciplining, and goal-setting between the institutions in an environment, not having a central, final decider, like a metropolitan region (Sabel and Zeitlin 2010b: 6; Sabel and Zeitlin 2012b: 170).

60 For network governance see, e.g. Boger and Zølner (2007), Eising and Kohler-Koch (2002); Marcussen and Torfing (2007), or Sørensen and Torfing (2008).

61 Habermas (1981) discusses the issue of the better argument in his *magnum opus* "The Theory of Communicative Action." According to Habermas (1981) the process of consensus-making is ultimately based on the authority of the better argument, because intercommunication follows rationally motivated communication (consensus) (Habermas 1981: 218).

62 Sabel (1996) referred to the direct involvement of actors already in his initial paper on experimental regionalism to advance economic development (Sabel 1996).

Furthermore, experimentalist governance can be interpreted as a pragmatist alternative to classical governance approaches (Sabel 2007: 140).

While most authors refer to the four elements as presented above[63], contemporary researchers around Sabel and Zeitlin follow a five-step approach. De Búrca, Keohane, and Sabel (2014) add a discussion of shared perceptions of joint challenges as the first element to the initial experimentalist governance design. Hence, they place more emphasis on the common problem orientation within the regulatory framework of experimentalist governance (De Búrca, Keohane, and Sabel 2014: 478).

From a broader perspective, Fabrizio Gilardi and Radaelli (2012) and Verdun (2012) understand the regulatory framework of experimentalist governance as an open method to coordinate "public policy based on the diffusion of information among social and policy networks at different levels of governance" (Fabrizio Gilardi and Radaelli 2012: 156).

In the EU and the US, as the most advanced research units, Sabel assumes that the regulatory structure of experimentalist governance typically works through public supervision. This also holds for the experimentalist regulation of private economic activity. In such systems, central administrations superintend the process without determining its results (Sabel 1996).

Nonetheless, in the fast-moving world of today's "Globalia," where (global) regulations become more pluralist, administrations fail in areas where reliable and sustained public engagement is needed, although clear jurisdiction is unavailable (De Búrca, Keohane, and Sabel 2014: 477; Sabel 2004a: 173).[64] Experimentalist governance responds to those failures of hierarchical command-and-control regulations (Overdevest and Zeitlin 2014: 27; Sabel and Zeitlin 2012b: 172).[65] In situations where actors identify essential challenges and are aware of the situation that their way of handling them will probably fail or will have tremendous transaction

63 E.g. Eckert and Börzel (2012), Overdevest and Zeitlin (2014) or Verdun (2012).

64 Forestry and the failures of the international community to establish common binding rules by the UN and national administrations versus the help out of private initiatives is an illustrative example, see Overdevest and Zeitlin (2014).

65 Some administrative lawyers, e.g. Kingsbury, Krisch, and R. B. Stewart (2005) and Cohen and Sabel (2006), refer here to that phenomena as anomalous administrative law (Sabel and Zeitlin 2012b: 11).

costs, they will engage in joint reciprocal actions of a regulatory framework of experimentalist governance (Sabel and Zeitlin 2012b: 172).

On the one side, a regulatory structure of experimentalist governance appears particularly well suited in domains where there is no overarching sovereign with the authority to set up common goals. On the other side, it also seems suitable where the diversity of the local and regional conditions as well as the institutions makes the adoption and enforcement of *ex ante* fixed rules less or unfeasible (Overdevest and Zeitlin 2014: 26; Sabel and Zeitlin 2012b: 177). Otherwise, as already identified within the broader governance approach of Mayntz (2003), there is the possibility of a blockade or the failure of the agreement *ex post*. Following Overdevest and Zeitlin (2014), both aspects can be summarized and subsumed under the phenomena of regime complexity (Overdevest and Zeitlin 2014: 23). Nonetheless, even the open approach of experimentalist governance is not free from obstructions. On the one hand, there might be too many actors with too many interests, which makes an initial agreement and goal-setting difficult. On the other hand, a single veto player might be able to block the process if he is unable to enforce his interests (Sabel and Zeitlin 2012b: 177). Furthermore, the approach of experimentalist governance is sometimes seen as problematic, because it moves away from traditional, well-established procedures of decision-making (Campbell-Verduyn and T. Porter 2014: 410).

However, it seems that even in areas where those obstacles were (or still are) structural – like in the education system, where the educational success of the students strongly correlates with their social and economic background – a regulatory structure of experimentalist governance has a positive impact. No single actor can enforce the preferred solution. Power is distributed more or less equally. A regulatory structure of experimentalist governance leads to a kind of "strategic uncertainty" (Sabel and Zeitlin 2010a: 8; and Sabel and Zeitlin 2012b: 180). Strategic uncertainty describes a situation in which stakeholders have not finally defined their goals or how to achieve them, although regarding problem-solving actors must discover both in a common arrangement (Overdevest and Zeitlin 2014: 26). Here, "parties will prefer joint exploration, even if this leads to a change in preferences" (Sabel and Zeitlin 2010a: 8; Sabel and Zeitlin 2012b: 180).

Sabel and Zeitlin found evidence from the educational systems of New York and Finland, that a regulatory structure of experimentalist governance leads to problem-solving and the improvement of the education systems, even in areas where obstacles are deeply rooted within the institutional structure (Sabel and Zeitlin 2012b: 180f).

Criticism of the concept of Sabel and Zeitlin (2012b) concerns its conceptual design, its normative implications and its legitimacy (Eckert and Börzel 2012). Börzel (2012) criticizes that unlike previous research such as deliberative supra-nationalism, network governance, or new modes of governance, a regulatory structure of experimentalist governance does not necessarily require the simultaneous participation of administrative or non-administrative actors (Börzel 2012: 379). Fossum (2012) argues *vice versa*. He criticizes that experimentalist governance as a methodological approach in decision-making does not even distinguish between administrative and nonadministrative institutions (Fossum 2012: 394).

Here, in the first place, it is important, that the not-mandatory involvement of private institutions is strongly associated with the general governance definition. The general definition of a regulatory governance structure includes the sum of all co-existing forms of collective regulation of social issues (Mayntz 2003: 72; Zürn 2008: 554). This embraces purely "intergovernmental relations" (R. A. Rhodes 2001) or policy-making processes "by government" (Zürn 2008: 558). With governance "by government," Zürn (2008) characterizes structures of hierarchically-exercised order and power by the administration within a specific territory (Zangel and Zürn 2004: 15; Zürn 2008: 558f). Sabel and Zeitlin (2012b) refer to this by recognizing private institutions in the regulatory structure experimentalist governance and the participation in the decision-making process upfront the final (purely) administrative goal-setting process. At the latest, this is the case when the regulatory structure of experimentalist governance steps out of the "shadow of hierarchy" and legislative proceedings are initiated. On the one hand, in this manner, Sabel also argued together with Simon that the center of the regulatory structure of experimentalist governance could also be the national state and its sub-levels – a government agency and its departments. On the other hand, further public and private institutions with their local units can be the focus (Sabel and W. H. Simon 2008: 79). For this purpose, by focusing on the institutional arrangement, the theoretical framework of experimentalist governance clearly distinguishes between administrative and nonadministrative institutions, although it provides the possibility to include administrative and nonadministrative actors in the decision-making process as equal partners with clearly identifiable jurisdictions.

Only when it comes to the question of legitimacy, which is always a challenge in regard to the regulatory structure of governance, the issue of administrative or nonadministrative has to be examined in depth. The question is, how decisions, conclude on "private" knowledge, are legitimate. Here it can be argued, that "experimentalist or networked decision making is legitimated [...] by forms of dynamic accountability" (Sabel and Zeitlin 2010b: 17). Dynamic accountability enables (or forces) political actors to discuss and benchmark their proposals and decisions, based on more information about alternatives than without it (Sabel and Zeitlin 2012b: 175). They can rely on tacit knowledge,[66] which in the sense of Habermas (1981) leads to better context-specific results based on the superior argument.

Börzel (2012) also calls attention to the "shadow of hierarchy." She argues, that a regulatory structure of experimentalist governance contributes to democratic deficits instead of solving them. On the one hand, it "destabiliz[es] entrenched forms of authority in some areas while[, on the other hand it,] inceas[es] problems of democratic representation in others" (Börzel 2012: 381). Nevertheless, she further argues, that "[t]he threat [of deliberation] to hierarchically impose collectively binding decisions helps to unblock deadlock and prevents voluntaristic defection" (Börzel 2012: 380f). Therefore, she exposes the two sides of the coin of experimentalist governance. Moreover, she argues that a regulatory structure of experimentalist governance plays a prominent role besides other forms of governance (Börzel 2012: 380–382). In this context, the suggestions raised by Fossum (2012) and Verdun (2012), to analyze the nature of representation in democracy through the lenses of the theoretical approach experimentalist governance and not to integrate experimentalism into existing, forms of representative democracy also, have to be

66 Or non-cosmopolitan knowledge (Storper 1997a; Storper 1997b).

mentioned. They argue – as Sabel and Zeitlin (2012a) interpreted their argument – that this might contribute to the identification of new sources of legitimacy, which replying to Fossum's and Verdun's considerations – also gives importance to the integration of experimentalist governance into other forms of governance (Sabel and Zeitlin 2012a: 419). This integration is also promoted by Börzel (2012).

Fossum (2012) further criticizes, that experimentalist governance does not provide sufficient and clear benchmarks to classify the improvement or learning of the governance (Fossum 2012: 395). Here, it has to be mentioned, that these benchmarks are part of Sabel and Zeitlin's experimentalist governance approach itself. They have to be set during the negotiation process. Additionally, regarding the participatory process, only persons who take part can define the exact benchmarks and thus the learning efforts are. Fossum (2012) himself argues, that "[i]solated instances of poor indicator and benchmark quality need not to be an argument against [experimental governance]" (Fossum 2012: 396), if that is the case, the experimentalist governance approach shows that they are part of the revision (Fossum 2012: 396).

### 2.6.1 Experimentalist governance and governance learning

Deduced from the political philosophy of John Dewey (2012), experimental means that policies have to be subject to constant revision (Sabel and W. H. Simon 2008: 78). At the bottom of this approach lies the "recursive review of implementation experience" (Sabel and Zeitlin 2012b: 169) of policy frameworks in different regional contexts (Overdevest and Zeitlin 2014: 25; Sabel and Zeitlin 2012b: 169). It reflects routines of "learning by monitoring" and "continuous improvement" (Sabel and W. H. Simon 2008: 80).[67] Towards the traditional hierarchical logic of government, Sabel and Zeitlin's theoretical concept of experimentalist governance follows the learning logics of horizontal designs in multilevel settings.

67 Coming from management theory the concept is also well known from lean production and lean management. Womack, D. T. Jones, and Roos (1990), who analyzed the car industries of North America and Europe, which were not competitive to Japan, new ideas in car manufacturing, were one of the first researchers analyzing lean-production.

Based on learning from the benchmark of different approaches to reach certain goals and advancing them in various contexts, the concept of experimentalist governance is in close relation to reflexive social learning (Sabel and Zeitlin 2012b: 169). Coming from a social science perspective Giddens (2012) argues that "[t]he reflexivity of modern social life consists in the fact that social practices are constantly examined and reformed in the light of incoming information about those very practices, thus, constitutively altering their character" (Giddens 2012: 38). Beck, Giddens, and Lash (1994) comprehensively presented the concept of reflexivity, which is based on society-wide paradigmatic changes, influencing policy, social interaction and institutional behavior (Campbell-Verduyn and T. Porter 2014: 411; Fabrizio Gilardi and Radaelli 2012: 156). Fabrizio Gilardi and Radaelli (2012) identified three further mechanisms of learning. Beside "reflexive social learning," there is "instrumental learning" (Fabizio Gilardi 2010; Radaelli 2009) characterizing rational and evidence-based policy-making, "political learning" (Boswell 2008; P. J. May 1992) describing ways of proceeding for policy control, and "symbolic learning" (Powell and DiMaggio 1983), pointing to the "symbolic usage of learning" to justify policy (performance) (Fabrizio Gilardi and Radaelli 2012: 156–159).

Against this background, combining comparative public policy and different modes of governance is a promising starting point to examine governance learning in geographic-functional regions. On the one hand, governance learning refers to problem definition, the achievement of results, goals, and implicit as well as explicit knowledge. On the other hand, governance learning includes hierarchy, the "shadow of hierarchy" or "the shadow of law" (Fabrizio Gilardi and Radaelli 2012: 159), facilitating coordination, and voluntarism. Furthermore, governance learning is focused on compliance, emulation and herding, as well as dense communication and reflexivity (Fabrizio Gilardi and Radaelli 2012: 159). All of these items also characterize experimentalist governance.

For this purpose, single parts of the regulatory structure of experimentalist governance are not new in policy analysis (Fabrizio Gilardi and Radaelli 2012: 156). They can be found in the approaches of new public management (O'Donnell and Sabel 2001; Sabel 2004a: 177–181) or new modes of governance (Héritier and M. Rhodes 2011) and they are identifiable within cluster theory (Sabel 2004b). However, as Börzel (2012) notes, previous research has missed all experimentalist aspects (Börzel 2012: 378–380).[68]

Still, these aspects of a regulatory structure of experimentalist governance have been presented by the theoretical concept of Sabel and Zeitlin (2010a). Sabel and Zeitlin assert that they can be retraced by the analysis of the regulatory structure at the EU level in the fields of data privacy (Newman 2010), the financial sector (Posner 2010), the energy sector (Eberlein 2010), networked competition law (Svetiev 2010), ecology and environment (Homeyer 2010), food safety (Vos 2010, genetically-modified organism (Pollack and Shaffer 2010), anti-discrimination law (De Búrca 2010), justice and home affairs (Monar 2010), freedom, security and justice (Schutter 2010), and European neighborhood policy (Tulmets 2010).

Having set the theoretical foundation and presented the assumptions of the manifest content – the five aspects of experimentalist governance – the theoretical derived framework of ERG is presented in the next paragraph.

### 2.6.2 Experimentalist regional governance: a theoretical deduced concept to analyze regulatory structures of governance

As presented, new regionalism is "call[ing] for a differentiate repertoire of political organizations including a variety of loosely institutionalized arrangements such as [...] interlocal coordination, public-private partnerships, and policy-specific arrangements" (Koch 2013: 400). Taking Sabel and Zeitlin's ideas as the basis and integrating the geographic-functional aspect, the theoretical concept of ERG is deduced from the concept of experimentalist governance. It provides a research framework to analyze regulatory structures of governance in geographic-functional regions. For the geographic-functional region of Europe, Sabel and Zeitlin (2010a) show at the EU level, that their concept is most suitable to analyze regulatory structures of governance, decision-making processes and institutional arrangements within a certain region, with no identifiable

68 Börzel (2012) counts to this the deliberative supra-nationalism (Joerges and Neyer 1988), informal governance (Christiansen and Piattoni 2003), soft modes of governance (Borrás and Conzelmann 2007), and innovative governance (Tömmel and Verdun 2009) (Börzel 2012: 378f).

final decider. Here, jurisdictions and the relevant stakeholders, such as institutions and individuals are not explicitly identifiable up front.

This also holds for smaller geographic-functional regions, such as metropolitan regions, which are researched in the presented analyses. Metropolitan regions appear as a blueprint for environments without a central, final decider. Geographic-functional regions like metropolitan regions can be described as an agglomeration of institutions within a certain space, whereas space or the spatial region is not defined by political borders. The region is defined as geographic-functional. The system of a region is characterized by a specific structure of relationships between institutions, the institutional arrangement (Grabher 1994a: 61–66, 108).

Regarding jurisdictions and the identification of relevant institutions that hampered the empirical analysis of the regulatory structure of governance by the regional and metropolitan approach as presented, the theoretical deduced approach of ERG overcomes the problem. The regulatory structure of governance and the decision-making processes are not restricted to the administrative institutions of a classical administrative system – e.g. the state. Based on fallibilist and pragmatic elements, it is also an answer to the challenges of change. Additionally, with the integration of the embeddedness perspective, it closes the gap between process-oriented policy analysis and classical political research, focusing on institutions and the structure of institutional arrangements. By placing a stronger focus on geographic-functionality, the theoretical deduced concept of ERG matches the demands formulated by new regionalism.

Therefore, coming from the EU level, the theoretical concept of ERG assumes to provide a suitable analytical framework to analyze smaller systems, such as metropolitan governance systems, including their processes and structures.

In the following, the manifest content and the latent content of the theoretical deduced concept of ERG will be presented. It is structured according to the asserted five aspects of experimentalist governance. This theoretical derived systematization allows the analysis of the regulatory structure of governance of the extension processes of Frankfurt Airport.

### 2.6.2.1 *Discussion of shared perceptions of joint challenges*

Following Sabel's and Zeilin's insights, the first aspect is mainly characterized by joint discussions and identifying common challenges. Those discussions are marked by openness, transparency, and a multi-actor approach across hierarchical or traditional political or economic borders, whereby specialized experts and expert groups, administrative officials, economic actors, lobby organizations, and civil society actors communicate and discuss on equal terms. These discussions can be both, formal and informal, and results are based on expertise and evidence (the better argument).

In an optimal case, the joint dialog through public and non-public consultations, further open meetings, or discussions on ICT platforms leads to heterarchical relations, the integration of all relevant stakeholders, and the common understanding of the challenge. Institutions already in place at this stage support recursive policy-making and learning by experience from the lower-level.

In addition to the identification and joint advancement (harmonization and mainstreaming) of common topics, this aspect is further characterized by the creation of a common regulatory framework of governance, the establishment of new institutions, or the promotion and fostering of institutions already in place. This includes the delegation of topics and the proactive, inclusive advancing of mandate holders to advance the whole decision-making process as well as increasing ownership and stimulating mutual learning, partly through the inclusion of the lower levels.

Information is pooled and circulated to all stakeholders (Sabel and Zeitlin 2010a).

### 2.6.2.2 *Definition of goals and the measurement system*

The second aspect is represented by the definition of high-level goals, common objectives – e.g. protection of the environment, support of civil society – and the programming of measurement systems in the decision-making process and the regulatory structure. One of the main tasks is to elaborate a practical and workable implementation plan, in a deliberative fashion. This plan contains the identification of common and available technical feasible regulations and the elaboration and publishing of legal (non-)binding rules. Reference documents are formulated and specific issues or regulatory proposals are discussed with the addressees, clarifying the approach regarding common goals and the measurement system. The final guidance documents are adopted by all stakeholders (Sabel and Zeitlin 2010a).

### 2.6.2.3 *Implementation*

The third aspect is the implementation. Stakeholders of the regulatory framework of governance are provided with the most possible freedom and autonomy for the implementation of the goals and measurement systems. The implementation process is mostly characterized by the operationalization of implementation. To achieve the common goals, the objectives and the measurement system are adopted or existing rules and approaches are adjusted. Actors and addressees have the possibility for interpretation, based on their tacit knowledge and best available practices in their environment.

Through the delegation and establishment of implementation capacities, lower-level authorities or experts are entrusted with more responsibilities and power. Nevertheless, soft guidance within the regulatory governance structure is needed to complete a binding overall framework and further harmonize directives and regulations to reach the common goals. Furthermore, the cooperation of implementation between relevant actors is further fostered by promoting implementation capacities and the design or redesign of administrative and/or nonadministrative objective, imperial organizations (Sabel and Zeitlin 2010a).

### 2.6.2.4 *Report and peer review*

The fourth aspect is characterized by reporting and peer reviews. Due to the most possible freedom, implementation actors have to report particularly to higher-level authorities within the regulatory governance structure. Furthermore, joint monitoring and evaluation as well as reporting by independent specialized administrative and/or nonadministrative organizations from all involved levels are executed, e.g., regulatory agencies, national focal points, working groups, companies, think tanks, universities, or local experts. This reporting is marked by monitoring achievements, supervision of the proceedings, and benchmarking. These assessments, audits, and peer reviews are carried out by inceptions, monitoring and evaluation missions, as well as questionnaires to stress the implementation processes, the goals, and the measurement systems of the regulatory framework. In an optimal case, these actions are institutionalized and periodically organized – e.g. annual, semi-annual, or monthly.

The results of the reviews include opinions and comments on the implementation process, the goals, and the measurement systems. Taken implementation activities have to be justified and are discussed with the affected stakeholders, to create an environment of continuous adjustment and learning.

Created information is circulated up- and downwards the governance levels and is fully available to all stakeholders, to include additional opinions and relevant scientific issues (Sabel and Zeitlin 2010a).

#### *2.6.2.5 Review of the measurement system, the decision-making processes, and the stakeholder*

The fifth aspect embraces the review of the measurement systems and the decision-making processes as well as the inclusion of further relevant stakeholders of the regulatory governance structure. It is characterized by the identification and mapping out of areas that need adjustments. Regarding the results of the peer reviews and reports, the intended goals and the measurement systems are reassessed, questioned, and cross-checked. In this context, policy objectives and procedures of political programming, legislation, and best available practices are recursively reviewed. Suitable proposals for the amendments are presented to (higher) authorities, and – if imperative – legal action is initiated. If necessary, the goals, the implementation guidance, or the measurement systems are reformulated, to integrate new information, newly developed best available technologies, innovations, and proceeding.

After the revision, further stakeholders are brought into the process and, if necessary, structures are further formalized (Sabel and Zeitlin 2010a).

| | |
|---|---|
| Discussion of Shared Perceptions of Joint Challenges | • Open, transparent multi-actor discussions across hierarchical and traditional borders (formal and informal)<br>• Identification of topics, best available techniques, joint advancement, and common understanding of challenges<br>• Delegation of mandate in certain topics; proactive advancing of mandate holder to increase ownership<br>• Establishment of heterarichical relations and integration of stakeholders<br>• Institutions already in place support recursive policy-making and learning by experience from lower level<br>• Provision, exchange and pooling of information |
| Definition of Goals and the Measurement System | • Setting high-level goals and common objectives<br>• Programming of measurement system<br>• Establishment of practical and workable implementation plan and feasible regulations<br>• Discussion proposals with addressees<br>• Formulation of reference documents, recommendations, and elaboration of legally (non-)binding rules<br>• Provision and discussion of guidance documents to clarify approaches |
| Implementation | • Operationalization of implementation to adopt or model the goals, the measurement system, and rules<br>• Entrust lower level units and delegate responsibilities and power to lower level authorities or experts<br>• Actors and addressees have most possible freedom and autonomy for implementation of goals<br>• Adjustment of existing rules based on tacit knowledge and subsequent experience<br>• Establishment of implementation capacities; design objective and imperial organizations, private/non-private<br>• Soft guidance to fulfill a binding overall framework and further harmonization of directives and regulations<br>• Fostering cooperation while implementation |
| Report and Peer Review | • Joint monitoring and evaluation, peer reviews as well as reporting<br>• Constitutional and collective regulatory dialogs; environment of continuous adjustment and learning<br>• Justification of taken activities and performance<br>• Institutionalization of periodical assessments, audits, and peer reviews<br>• Release and discussion of opinions and comments on implementation process, goals, and measurement system<br>• Process information up- and downwards; consultation of additional opinions and any relevant scientific issue |
| Review of the Measurement System, the Decision-making Processes, and the Stakeholder | • Reassessment, questioning and cross-check of intended goals and measurement system<br>• Recursive revision of policy objectives, procedures of political programming, legislation, and best practices<br>• Identification of areas that need adjustments, if necessary initiate legal action<br>• Present suitable proposals for amendments to (higher) authorities<br>• Take new developments and innovations into account<br>• If necessary bring in further stakeholders formally into the overall process |

*Figure 2.1: Five Aspects of Experimentalist Regional Governance.*

### 2.6.2.6 *The 5 aspects of ERG*

Having defined the manifest content based on Sabel and Zeitlin (2010a), the following scheme provides the systemized analytical framework based on a theoretically-sound research perspective. The latent content is sharp but comprehensively defined. Therefore, the framework represents a theoretical deduced and empirical verified analytical concept to analyze regulatory structures of governance and evolutionary decision-making processes in geographical-functional regions. It further copes with the demands of new regionalism. Against the background of regional and metropolitan governance, the theoretical deduced concept of ERG has a clear analytical framework and does not lack vague definitions. It makes it possible to analyze regulatory structures of governance, its decision-making processes, and institutional arrangements in geographic-functional regions.

## 2.7 Research questions and hypothesis

Against the presented background and assumptions of the theoretically derived concept of ERG as a consecutive approach, the major research question is:

> How is the regulatory structure of governance of the airport extension in the geographic-functional region of FRM organized?

In this context, further subordinated research questions are:

> How is the institutional arrangement of the airport extension in the geographic-functional region of FRM organized?
>
> Who are the key actors of the regulatory structure of governance?
>
> Are the key actors characterized by individuals or institutions?
>
> How are the key actors integrated into the decision-making processes?

In addition to the major research question, the question of governance learning emerges.

> Is governance learning – as a process of updating beliefs – identifiable in the decision-making process of the airport extension?

The research-guiding hypothesis reads as follows:

> The theoretical derived concept of ERG is a suitable analytical framework for analyzing the regulatory structure of governance, decision-making processes, and institutional arrangements in geographic-functional regions.

At the same time, the hypothesis of Sabel and Zeitlin (2012b) that experimentalist (regional) governance is suitable for environments without a central, final decider is tested against the geographic-functional metropolitan region of FRM.

After the theoretical foundations of governance as a regulatory structure and ERG are presented in the next chapters, the research question and hypothesis will be analyzed. Thereafter, the results of the analysis will be discussed.

# 3 Approach and method

*Google is not a synonym for research.*
(Brown 2009: 14)

Initially it was intended to combine the content analysis of text-based material (Früh 2011; Merten 2009; Rössler 2010) with additional structures interviews (Beatty 2003; Froschauer and Lueger 2009; Gläser and Laudel 2009), analyzing the cases and embracing the utmost comprehensive social reality. However, the latter approach proved unfeasible. Structured interviews were not applied mainly due to two aspects. First, the explorative interviews with affected individuals as well as project proponents are mostly marked by the discussions of the pros and cons of the extension. Thus, the focus on the regulatory governance structure of the extension and the decision-making processes is hardly possible, because the discussion was always taken back to the pros and cons by the interviewees.[1] Likewise, the responsible authorities have proved unfeasible interview partners. For example, already the request for text-based material was refused, and the support of a lawyer was necessary to gain access to the in-principle public data. Second, the numerous text-based materials in a nearly untouched field tend to supersede the interviews. Thus, the thorough analysis of the innumerous databases made structured interviews at this point superfluous. Moreover, considering these challenges as well as the research question, the analysis of the text-based material proves to be more productive. This data is

1 In the exploration phase of the presented research, I had extensive contact with project proponents during my work at House of Logistics and Mobility (HOLM). Throughout the data research phase of text-based material, I also had far-reaching contact with local citizens' groups and affected individuals. In both cases, discussions focusing not on the pros and cons were nearly impossible. In some cases, the discussions, both, with supporters and objectors became that personal, that they have become irrelevant for the intended research immediately. In some cases, it was even spoken of a criminal case (DFLD 2015).

better represented by decisions, memoranda, minutes, reports, as well as transcripts and less by individual statements. The standard of objectivity in the text-based material wins – in comparison – to the emotional discussion about the pros and cons of the extension of Frankfurt Airport. These documents offer a deep insight into the regulatory structure of governance, the decision-making processes, and the institutional arrangements, not yet made. Analyzing the comparable and objectified documents offers valid results. Furthermore, the content analysis allows to examining areas in which resource persons are unavailable (Früh 2011: 41; Schnell, Hill, and E. Esser 2013: 398). In addition, the presented research focuses on the institutional environment and its institutions as "'units of analysis' being considered" (Bryman 2012: 295), rather than individuals. Cognitive attitudes of individuals are less relevant for this very first deeper analysis of the regulatory structure of governance, the decision-making processes, and the institutional arrangements of the extension of Frankfurt Airport in the geographic-functional region of FRM. Hence, documents with more than 31,000 pages are the primary resource of information. These documents were researched in detail.

One the one hand, the official documents of the hearing and planning authorities of the airport extension process – the RPD, the AoPs, and the transcripts of the respective hearings – were analyzed in detail. One the other hand, the official documentation of the mediation – final reports and the deliverables of the task forces – were comprehensively examined. Both constitute the golden thread of the analyzed data. This sequence clearly shows the evolutionary character of the extension process. Thus, the respective regulatory structures and the decision-making processes in FRM are evolutionary. Even though the RPP and the RSE are administrative proceedings anchored in the field of administrative law, public participation is intended. Similarly, for projects with supra-regional character, as the extension of an airport, the engagement of citizens is obligatory.[2] However, public participation in such decision-making

2 According to §1 *Raumordnungsverordnung* (ROV) and §8 *Luftverkehrsgesetz* (LuftVG) the construction and extension of an airport represents such an infrastructure project with supra-regional characteristics and according to §§10, 11 III, and 15.3 *Raumordnungsgesetz* (ROG) the general public has to

processes is widely understood in German law, whereas it embraces everyone, except for public bodies (Spannowsky, Runkel, and Goppel 2010: §10 recital 42). Therefore, the analysis is not restricted to the responsible hearing and planning authorities. Available statements made during the official proceedings, letters – partly personal, and articles of daily newspapers and weekly magazines were also analyzed to broaden the few of the presented research and to get further insights.

The first part of the actual content analysis focuses on systematizing the content of the text-based material. Initially, content analysis in political science used to be utilized to analyze political propaganda (Lamnek 2010: 438; Schnell, Hill, and E. Esser 2013: 397). However, according to the "Documents of Life" (Plummer 1983) content analysis can include a whole bandwidth of life stories, such as articles, diaries, letters, newspapers and many others (Plummer 1983; Plummer 2005). Today, even photos of Facebook profiles and YouTube videos are analyzed by content analysis (Schnell, Hill, and E. Esser 2013: 397). Thus, the utilized method is a modern research approach to analyze data, regardless of which source the data comes. Additionally, it allows analyzing the content of old official documents of the extension process and retracing further historical dimension by analyzing current online-articles (Flick 2011: 409). It proves to be the most suitable approach for the conducted analysis.

In the presented research, especially decisions, memoranda, minutes, reports, and transcripts regarding the extensions of Frankfurt Airport are analyzed. Following Bryman and Burgess (1999), these types of documents represent primary sources (Bryman and Burgess 1999: XX). In opposition to the hermeneutical interpretation, the focus lies on general testimonies rather than semantics (Früh 2011: 50–52; Rössler 2010: 18).

Here, the content analysis is used as an empirical-transparent, objective methodology to systematize and account for the original material, according to the presented research hypotheses (Bryman 2012: 304; Früh 2011: 52). In political science, it is understood as the dialog between theory – the theoretical derived concept of ERG – and data (Gschwend and

be involved. Whenever it is referred to the wording of a law, it refers to the then valid version.

Schimmelfennig 2007). It standardizes phenomena of the social reality – the regulatory structure of governance, its decision-making processes, and the institutional arrangements of the extension of Frankfurt Airport in FRM – whereas the social reality is portrayed in the documents concerning the extension processes (Rössler 2010: 18). The complex content of the analyzed documents is reduced to the presented research questions and the central patterns are extracted at the same time. Next, the gathered data extracted from social reality is processed to be further analyzed and the regulatory governance structure and the decision-making processes of the airport extension are reconstructed (Flick 2011: 416; see footnote for further details).[3] Consequently, the five aspects of the theoretical derived concept of ERG are traced in the regulatory governance structure and the decision-making processes of the airport extension.

The available data is systematized according to the analytical approach (Bryman 2012: 297f; Flick 2011: 409; Früh 2011: 27). The five aspects of the theoretical derived concept of ERG – "discussion of shared perceptions of joint challenges," "definition of goals and measurement systems," "implementation," "report and review," and "review of the measurement system, the decision-making process, and the stakeholder" – are the unambiguous indications as manifest content. They are characterized as such by paraphrases of the latent content, which describes and defines the five aspects in greater detail (Früh 2011: 42; Rössler 2010: 24). The latent content is paraphrased systemizing the examples of EU governance presented by Sabel and Zeitlin (2010a).[4] The unambiguous terminology is based on the definitions and theoretical explanations presented. Inevitably, this leads to the basic ideas of the theoretical derived concept of ERG. These paraphrases of ERG are extracted from the original texts of experimentalist governance. Therefore, this approach focuses on the exact definitions of the five aspects of the theoretical derived concept of ERG.

---

3 Früh (2011: 63–65); Pötschke (2010: 41); Rössler (2010: 18f)

4 The examples are: European data privacy (Newman 2010); financial sector (Posner 2010); energy sector (Eberlein 2010); competition law (Svetiev 2010); environment (Homeyer 2010); food safety (Vos 2010); Genetically Modified Organisms (GMO) (Dąbrowska 2010); anti-discrimination (De Búrca 2010); justice and home affairs (Monar 2010); freedom, security, and justice (Schutter 2010); and European neighborhood policy (Tulmets 2010).

As a result, it is concluded from the manifest content to the non-manifest content (Merten 2009: 15).

Under these circumstances, the approach of the presented research is based on a rich framework applied to analyze the regulatory structure of governance and the decision-making processes in the geographic-functional region of FRM.

In dissociation to the classical approach of content analysis of Berelson (1952) – who focused on quantitative aspects – the applied research approach follows the contemporary definition of Früh (2011), who defines content analysis as an "[…] *empirical method for the systematic, intersubjective comprehensible description of content and formal characteristics of messages, usually with the aim to interpretive inference based on external communication issues*" (cited by Rössler 2010: 22).[5] With this background, the challenges of political analysis using this approach are recognized. As requested by Ingram and A. L. Schneider (2006), the methodology applied extends beyond quantitative analysis and bare system modeling (Ingram and A. L. Schneider 2006: 184).

A similar method is used to analyze the institutional arrangements. The analysis of the presented research focuses on the separation of private and non-private institutions, which are identifiable in all three metropolitan regions. Based on Geißel and the understanding of the cooperative governance, it is assumed that the decision-making is negotiated between administrative and nonadministrative institutions as well as citizens. The institutional arrangement extends beyond the inclusion of trade unions and employers' representatives. The lower levels within the regulatory structure of governance are empowered and the power is shared between private and non-private institutions (Geißel 2012b: 165; Geißel 2013: 10f). The focus of the presented analysis is institutions, yet private individuals are not neglected.

Furthermore, the institutions are divided into sub-categories, resulting from the content analysis. Specifications of the relevant airport actors of Frankfurt Airport are based on the research of Knippenberger (2012), whereas F. W. Raschke (2009) provides additional particulars in the field

5 Text highlighted in original. Translated by the author.

of logistics. Lackowska (2011a) as well as Zimmermann and Heinelt (2002) offer further indications on general institutions of the metropolitan region of FRM. They also have to be considered in the analysis.[6] In this analytical process, the role of further institutions and actors is clarified.[7] Thus, the presented research contains characteristics of the Grounded Theory (Bryant and Charmaz 2013; Glaser and Strauss 2005). According to Glaser and Strauss (2005), this enables the data gathering, the encoding, and the categorization parallel to the research process (theoretical sampling) (Glaser and Strauss 2005: 53–83). Nevertheless, with this approach, the conclusions are always based on the original material (Flick 2011: 409; Rössler 2010: 24).

Only with such codification and operationalization of the research question the regulatory structure of governance, the decision-making processes, and the institutional arrangements can be analyzed according to the analytical approach of the theoretical deduced concept of ERG in geographic-functional regions. On the one hand, this approach reduces the complexity of social reality. On the other hand, it allows classifying the original material according to the research interest (Früh 2011: 42; Rössler 2010: 19). With the inclusion of latent characteristics, the presented approach prevents the research from confirming the upfront assumptions (Rössler 2010: 24). This also gives the analysis its explorative character.

The content analysis of the regulatory structure of governance, the decision-making processes, and the institutional arrangements of the extension of Frankfurt Airport – as the most thorough and transparent comprehensible infrastructure project in Germany (Gohl and Meister 2012: 55) – is particularly based on 243 official documents and 168

6 The representation is led by the analysis of the mentioned former researches in this field (Ebner and F. Raschke 2012; Lackowska 2011a; F. W. Raschke 2009; Zimmermann and Heinelt 2002).

7 E.g. the RPU, the UVF, and the *Planungsverband* as the predecessors of the *Regionalverband*, the *Wirtschaftsförderung Frankfurt GmbH*, the *FrankfurtRheinMain GmbH*, the *Industrie- und Handelskammer* (IHK), the *Handwerkskammer* (HWK) or other major cities of the metropolitan region Darmstadt, Offenbach am Main, Mainz, or Wiesbaden.

articles of the three major daily newspapers.[8] Furthermore, it includes statements and press releases of the project proponents, local citizens' groups, national environmental associations, political parties, and courts. The main documents, the regional planning decisions (RPD) and the approval of plans (AoP) have 2,800 pages on average.

Overall, 12,737 files with a total of around 150,000 pages were screened during the pre-analysis of all data gathered. Some of the screened documents have several thousand pages (up to 8,369 pages per document). Alone the planning documents of the RSE Runway North-West/Terminal 3 of Frankfurt Airport alone comprise 65 files for the final version. For the final analysis, more than 31,000 pages were examined by drawing on computer-based content analysis.

Accordingly, the analysis is mainly based on non-structured text-based material, which confirms the explorative character of the analysis (Kuckartz 2009: 123–127).

In the first step, the documents were processed so that it was possible to upload them into the QDA program.[9] Due to the different sources and texts, this pre-step was necessary to conduct a lexical full-text query (Kuckartz 2009: 121–132).

In the second step, the lexical full-text query was conducted. Regarding the amount of text-based material and given that the material was not analyzed to that extent before, the lexical full-text analysis was the most promising approach. This approach allows analyzing text-based material by keywords and a combination of keywords. It makes it possible to analyze and organize the text-based material efficiently and effectively. This is especially the case for analysis with a high amount of text, as is the case with the presented research (Kuckartz 2009: 127f).

In this context, it was possible to generate a so-called Keyword-in-Context list – meaning a list of keywords based on the surrounding context in a text (Kuckartz 2009: 121). The keywords were extracted from the theoretical foundations of Sabel and Zeitlin (2012b), Sabel and

8 *Frankfurt Neue Presse*, *Frankfurter Allgemeine Zeitung*, and *Frankfurter Rundschau*.

9 The content analysis was conducted by MAXQDA 11. MAXQDA allows the inquiry of the text-based material by context-based keyword analysis.

Zeitlin (2010a) and De Búrca, Keohane, and Sabel (2014) (Annex A.1). In addition, the lexical full-text analysis makes it possible to identify the most prominent instructions. For this purpose, this approach makes it possible to analyze the governance arrangement. Accordingly, the institutions identified as relevant upfront can be confirmed or falsified (Kuckartz 2009: 128).

Nonetheless, the lexical full-text approach also has its limits. On the one hand, this approach is always based on syntax. Misspelling in the original texts, identification problems of umlaut marks, case sensitivity, and partly the hyphenation rules hampers the identification of relevant paragraphs. On the other hand, the approach allows identifying paragraphs with manifested content only. However, these limitations were encountered in the presented analysis.

Regarding misspelling, the most common mistakes were also included as keywords in the lexical full-text approach, e.g. non-double letters instead of double letters. In the case of an umlaut, the query also included the word without umlaut marks. Furthermore, case sensitivity was enabled and hyphenation was eliminated to the greatest possible extent in the first step upfront, namely the processing of the documents.

To break up the challenge regarding the manifest content, the lexical full-text approach was combined with the intellectual coding approach, as suggested by Kuckartz (2009). This extra effort also became necessary due to the unstructured character of the text-based material. The non-structured text-based material as described above made it impossible to analyze the text-based material by the lexical full-text analysis only (Kuckartz 2009: 127).

In fact, in the third step, the full-text analysis was combined with an intellectual coding approach. After the keywords had been identified in the text-based material, the relevant paragraphs were analyzed manually. The paragraphs were examined and the latent content was interpreted against the theoretical deduced concept of ERG and its five aspects (Flick 2011: 416; see footnote for further details).[10]) This opened the approach to identify further aspects according to the Serendipity principal,

10 Früh (2011: 63-65); Pötschke (2010: 41); Rössler (2010: 18f

meaning that not only the initially desired outcome and results are achieved (Merton 2004; Merton and E. Barber 2004). Moreover, *ex ante* unintended results come to the surface, e.g. the change of the terminology during the discussions to a partly juridical debate (Kuckartz 2009: 121). Again, this highlights the generally explorative character of the presented research.

Accordingly, the presented research is characterized by common and scientifically-acknowledged methods in policy analysis. Retracing the assertions and assumptions of the theoretical deduced concept of ERG in the regulatory structure of governance, the decision-making processes, and the institutional arrangements of the extension process, Frankfurt Airport exposes policy-making as an attempt to handle the challenges of society. In the classical definition of Scharpf (1973) policy-making is defined by four steps: "discussing challenges with the need of a solution," "define high-level goals," "discuss and develop alternative opportunities of action," and "agree on certain actions" (Jann and Wegrich 2009: 75). Hence, policy-making is understood as the logical array of problem-solving. Its analysis focuses on politics in terms of content and not on power-creation or power-sharing (Naßmacher 2010: 135). Mayntz (1982) has previously identified that policy-making focuses on problem-solving in disassociation with the institutional and structural approaches (Mayntz 1982: 74), whereas, in 1956 Lasswell presented seven categories, following a timely linear model, to analyze "The Decision Process" (Lasswell 1956): "intelligence as pooling and processing of knowledge," "promotion as the identification of alternatives," "prescription as the agreement on decisions," "invocation of the decision," "application as the enforcement by legislation," "termination as bringing the policy to an end," and "appraisal as evaluation and assessment of the performance in reaching the intended goals" (Jann and Wegrich 2009: 78f; W. Parsons 1995: 447). Based on Lasswell's ideas, further life cycle models were developed (Jann and Wegrich 2009: 80).[11] The policy cycle is one of the

11 E.g. Brewer (1974), Anderson (1975), Mayntz (1977), Jenkins (1978), and J. P. May (1978).

most recognized models in this context (Blum and Schubert 2011: 104–144; see footnote for further details).[12]) It symbolizes an ideal description of a policy-making process and embraces six consecutive steps: "problem definition," "agenda-setting," "policy-formulation," "implementation," "evaluation" and "termination" (Jann and Wegrich 2009: 85f). First, this approach structures the research by time (Buhr and Schmid 2012: 243). Second, as sequential access, it is associated with general political research approaches. The aspect of history matters in political analysis, because social constraints like regulatory structures of governance are not understandable without socio-historical connections (Kappelhoff 1992: 221, 237; Tilly 2006: 417–437). These characteristics are similar to the concept of ERG.

Following a reciprocal approach, the theoretical deduced concept of ERG indicates a process characterized by five consecutive aspects, structuring governance and the decision-making process by time. For this reason, the criticism of life cycle models also applies to the analytical framework of the ERG concept. The first major critique is the sequential process and the exact division of the sequences. Here, it is questionable whether the theoretical deduced concept of ERG could be empirically utilized (Jann and Wegrich 2009: 102f). By contrast, Blum and Schubert (2011) argue that an exact distinction is not necessary to have a usable life cycle model. As a heuristic approach, it has its *causa existentia*.[13]) However, this heuristic character has to be recognized during the analysis. It is not the goal to press social reality into the mold of the theoretical derived concept of ERG. Instead, the identified differences should be mentioned (Blum and Schubert 2011: 104). The second major critique is the problem of defining the single steps and their indicators (Jann and Wegrich 2009: 102f). For the concept of ERG, all aspects representing the manifest content are defined by unambiguous latent content. Although the assertions and assumptions of the theoretical derived concept of ERG are testified against certain contexts (mediation, RPP, and RSE), which is typical for a process-oriented policy analysis, it does not follow an

12 Buhr and Schmid (2012: 276–278); Jann and Wegrich (2009: 85–102); Naßmacher (2010: 135–153.

13 For another opinion see Sabatier (2007: 6f.

immanent top-down approach from a legislative perspective. As Jann and Wegrich (2009) argue, there is certainly the possibility that problem-solving might not be the focus of each stakeholder in the decision-making process (Jann and Wegrich 2009: 103). However, the concept of ERG merely integrates the bottom-up perspective, up- and downwards-oriented knowledge transfer and learning from lower-level experience. Accordingly, the reconstruction of the decision-making processes and analysis of the regulatory structure of governance by the assertions and assumptions of the theoretical derived analytical approach of ERG, follow a well-accepted approach in policy analysis.

Based on the classical understanding of Mayntz (1983), a most differentiated overview of the dynamics in the policy-making is presented in this research. The goal of the presented research is to undertake an extensive analysis of particular, historical, and spatial phenomena. The research describes and analyzes the complex processes in decision-making and attempts to embrace the utmost comprehensiveness of social reality (Mayntz 1983: 14). The orientation on a well-established methodological approach in policy analysis further contributes to the understanding and systematization of policy processes. With a focus on governance – meaning to analyze heterarchical forms of political decision-making between administrative and non-administrative institutions as well as private individuals – offers further insights into whether, how, and why the policy changes during its implementation. Additionally, it exposes the involved and relevant institutions in the regulatory governance structure (Jann and Wegrich 2009: 104f).

From the democratic-administrative perspective, the latter aspects are especially relevant (Jann and Wegrich 2009: 104f). The presented research takes this aspect into account by analyzing the institutional arrangement in which the governance process is embedded. The integration of the embeddedness perspective of Granovetter means that actions are structurally embedded in the social environment (Granovetter 1985; Granovetter 1990c; Granovetter 1992).[14] The assertions and assumptions

14 However, Karl Polanyi is called "The Father of the Embeddedness Concept" (B. Barber 1995: 396) especially Granovetter, representing the link to today's debate on embeddedness (Beckert 2007: 8f; Dale 2012: 196; Hasse and

of the theoretical deduced analytical framework of ERG make it possible to integrate the structural aspects (institutional arrangement) of policy-making. This will overcome the identified challenge of a missing link between the procedural character of policy-making and the structural perspective of political science.

The starting point of the presented analysis is set at the first attempt to extend Frankfurt Airport after World War II. First, from a theoretical perspective, starting in the 1960s seems promising due to the transition from control and steering to governance in the 1990s and the following extensions of Frankfurt Airport after the 1960s to 1980s (Levi-Faur 2012: 5f). With that, the historical timeframe depicts that identified transition. Second, this transition held strong relevance for the Hessian State Government in real politics (Geis 2005: 130). This circumstance is better understood, including the conflictual history of the extension.

In the following, the presented approach will be applied to answer the research question and to confirm the research hypothesis.

---

Krücken 2005: 77–85, 105 and others), focuses on the structural embeddedness of actors (Granovetter 1985). For further readings on Granovetter's approach see Granovetter (2005), Granovetter (1992), Granovetter (2002), Granovetter (1995b), Granovetter (1995a), Granovetter (1990a), Granovetter (1990b), Granovetter (1990c), Granovetter (1988), Granovetter (1983), Granovetter (1973), as well as B. Barber (1995), Bathelt and Glückler (2012), Beckert (1996), Beckert (2007), Dacin, Ventresca, and Beal (1999), Dale (2012), Hasse and Krücken (2005), Heidenreich (2012), Krippner (2001), Krippner and Alvarez (2007), Krippner, Granovetter, et al. (2004), and J. Stewart F. (2011).

# 4 The case: analyzing the extension of Frankfurt Airport since the 1960s

*Beim Bau und Ausbau von Flughäfen sollen Fachleute [...] vom frühen Stadium der Planung an zusammenarbeiten, und zwar ohne sich allzu starr auf eine bestimmte Lösung festzulegen*

(DIL 1955: 38)

At the beginning of this chapter, the overall frame-giving conditions are outlined. This sets the necessary basic understanding of the presented research. Initially, the basic structure of the German political system is briefly epitomized, to connect the paramount administrative surrounding to the geographic-functional region of metropolitan FRM. Following Geißel (2013) and Lütz (2010), the institutional arrangement of metropolitan FRM and the institutions identified as relevant regarding the regulatory structure of governance and the decision-making process of the airport extension are presented. First, the respective regulatory structure of governance and the decision-making process of each period of the airport extension – Airport extension, the 1960s–1980s, Mediation process, since 1997, RPP and RPD, 2000–2002, RSE and AoP 2002–2004, RSE and AoP 2004–2007, and Adjustment of plans, 2007 and beyond – are reconstructed historically. Here, the structure of the presented research is led by the example of Sabel and Zeitlin (2010a) and how the research items are presented in their anthology. The theoretical derived concept of ERG is utilized to analyze the regulatory structure of governance and the decision-making processes of the extension of Frankfurt Airport. As a result, this will prove that ERG – as a theoretical derived concept – is a sound scientific approach to analyze existing regulatory structures of governance and the decision-making processes in the geographic-functional region of FRM. The research follows the presented understanding of Mayntz (2003), that governance is recognized as the reconstructable decision-making process of private and non-private institutions regarding the sum of all co-existing forms of collective regulation of social issues (Sec. 2.2).

Second, the institutional arrangements are presented according to the theoretical derived concept of ERG. Following Geißel, the institutional arrangement is the surrounding environment of the existing regulatory structures of governance and the decision-making processes. It includes the lower-level units going beyond trade unions and employers' representatives. Decision-making is negotiated between private and non-private institutions as well as citizens (Geißel 2012b: 165; Geißel 2013: 10f). With a focus on non-private metropolitan governance bodies and private metropolitan logistics actors, the institutional arrangements and relevant institutions are outlined. The result is a clear picture of the institutional arrangements of each period of the airport extension. This will show that the theoretical derived concept of ERG counterfeits the challenge of a missing link between the structural perspective of political science and the procedural character of policy-making, which Mayntz already proclaimed in the 1980s (Mayntz 1983).

For each period, interim findings present the major findings regarding the research question and the research-guiding hypothesis. On the one hand, these interim findings present the results, proving that the theoretical derived concept of ERG is suitable to analyze the regulatory structure of governance and the decision-making processes in the FRM region. On the other hand, the institutional arrangements are pictured and the involvement and relevance of metropolitan governance institutions are outlined.

Subsequently, these interim findings are summarized and discussed in the following chapter (5), to present the overall context. Finally, also in the following chapter, in a kind of historical overall assessment, it is discussed whether the decision-making processes improved and if governance is learning in this manner.

## 4.1 The German federal system

The political system of the Federal Republic of Germany is organized as a federal, parliamentary democracy. Federalism is a certain principle to organize a country. Within political science, federalism is characterized by "the advocacy of multi-tiered government combining elements of shared-rule and regional self-rule" (Watts 2008: 8). It has both, segregative and cooperative elements, connecting the central state and the member states (Rudzio 2015: 337). As a general principle, it is anchored in the German Basic Law.[1] The contemporary federalism of Germany is characterized by a rational, territorial structure (Rudzio 2015: 326).

Germany comprises of smaller autonomous units (*Länder*), which have their state functions. Governmental tasks are divided between the federal government and the *Länder* so that both levels are responsible for specific (constitutionally defined) tasks. The autonomy of the *Länder* in the federal system of Germany is shown by the fact, that the responsibility for the structure of local governments lies with themselves. They have their legitimacy, rights, and responsibilities. Thus, each state has its state constitution and thus independent political institutions for the executive, the judiciary, and the legislature (Ahrend and Schumann 2014a: 11; Rudzio 2015: 321–354). Nevertheless, this has to be in line with German Basic Law.[2]

The further sub-federal state levels are represented by states counties (*Kreise*) and municipalities (*Gemeinden*). Typically, large cities (*Kreisfreie Städte*) are independent of a county and they combine the functions of a county and a municipality administration. This sub-ordination is also justified by the advantages of the rational, territorial structure. These advantages are spatial proximity and thus more efficient and demand-driven decisions, relief of higher levels, innovation (based on competition between the subordinated units), distribution of power and broader political participation (Ahrend and Schumann 2014a: 11; Rudzio 2015: 355–383).

Therefore, it is unsurprising that Germany is characterized by an unusual diversity in the governance structure of metropolitan areas, embracing the *Länder*, the *Kreise*, the *Gemeinden*, as well as the *Kreisfreie Städte*.

However, in some metropolitan regions of Germany, the central municipality of the urban agglomeration was merged with the surrounding

1 According to Art. 20 *Grundgesetz* (GG).
2 According to Art. 28 GG.

counties to so-called regional counties (*Regionalkreise*). While municipalities in these regions continue to exist, some of their traditional functions were taken over by the regional counties. Regional counties were created with the explicit aim to coordinate policies between the central city and surrounding areas (Ahrend and Schumann 2014a: 11).

For FRM, this is not the case. As presented (Ch. 1), the concept of a regional county could not be enforced in FRM.

## 4.2 FrankfurtRhineMain and the city of Frankfurt

With about 690,000 residents, Frankfurt am Main is the largest city in the State of Hesse and the fifth-largest city in Germany (Destatis 2014). As a center "Franconofurd" was first documented in 794. In the Middle Ages, Frankfurt am Main became one of the major urban centers of Germany. As a coronation city, it was the center of political life from the High Middle Ages until 1806. This characteristic was continued with the first meeting of the Federal Convention of the German Confederation on November 5, 1816. In addition, the first elected German Parliament in 1848/49 had its seat in Frankfurt am Main. After the Austro-Prussian War, Frankfurt am Main lost its status as an independent city-state and was annexed by the Prussians.

At the beginning of the 19th century, the city suburbanized several urban units and after World War I it had more than 500,000 residents. During this period, Frankfurt am Main also became a stock exchange (1879) and university city (1914) (ISG 2014b).

After the city center was completely destroyed in World War II, the city of Frankfurt am Main and its reconstructed airport obtained major attention for the airlift to Western Berlin (Huschke 2008: 167–169).[3]

Today, the European Central Bank and the financial district make Frankfurt am Main the major financial market district of Germany. Due to its central location – which was always an advantage throughout history –, it is also a European transport and logistics hub. Frankfurt Airport is the largest airport in Germany and according to passenger numbers the

3 The Berlin Airlift was officially launched on June 26, 1948, and lasted over one year until August 27, 1949.

12th largest worldwide (ACI 2014: 3). The central station of Frankfurt am Main is the central railway hub in Germany and the Frankfurter Kreuz – where the highway A3 and A5 are connected – is the most commonly used route in Germany. Furthermore, Frankfurt am Main is also known for the digital infrastructure. The DE-CIX is the world largest internet exchange point (Ebner and F. Raschke 2012: 72).

### 4.2.1 The geographic-functional region of FrankfurtRhineMain

As a geographic-functional region, metropolitan FRM can be defined as a cross-border region. Coming from the field of international economic relations, Perkmann and Sum (2002) define a cross-border region as “a territorial unit that comprises contiguous subnational units from two or more nation-states” (Perkmann and Sum 2002: 3). This definition can be transferred directly to the regional level. FRM comprises contiguous subunits of the federal states of Hesse, Rhineland-Palatinate, and Bavaria with their own jurisdictions and own democratically elected authorities and institutions.

Additionally, the three federal states acknowledged the metropolitan character of the region. With the 1st Rhine-Main-Conference on March 4, 1994, they expressed their willingness to cooperate and jointly promote the development of the metropolitan region of FRM. They agreed to minimize municipal competition and initiate a common approach, including all private and non-private institutions (HMLWLFN 1994).

This first approximation to the geographic-functional definition of FRM has to be enlarged and deepened by a spatial planning approach. As outlined above, metropolitan regions are characterized by geographical-functionality. These regions are defined by a highly-developed infrastructure and a dense concentration of specialized service providers (Baumheier 2007: 4; Blotevogel 2005).

According to their internal structure, metropolitan regions can be distinguished by monocentrism and polycentrism. Whereas monocentric metropolitan regions are dominated by one center – e.g. Berlin or Paris – polycentric regions embrace a group of metropolitan centers – e.g. FRM (Baumheier 2007: 4).

From a political-administrative perspective, a metropolitan region represents a more or less institutionalized cooperative area of private and non-private, regional institutions. On the one side, those institutionalized regions can be defined by the number of their inhabitants. The European METREX-Network counts 50,000 inhabitants for the center and 500,000 for the whole metropolitan region (Baumheier 2007: 5). Other researchers count 500,000 inhabitants in the center and 1 to 1,5 million inhabitants in the region (Ahrend and Schumann 2014b; Blotevogel 2005: 644).

On the other side, a further distinction is given by the functional aspect, which is given precedence in this research. Even though the academic discourse is not enclosed today, there is consensus – based on Blotevogel (2005) – that metropolitan regions have three functions (Baumheier 2007: 5f). The first function refers to decision and control. It is characterized by related, cooperative private, and non-private institutions. As the second function, the innovation and competition function is represented by the generation and distribution of knowledge, attitudes, values, and products as well as the creation of economic-technical, social, and cultural innovations. The third function is infilled by access to people, knowledge, and the markets (Blotevogel 2005: 644–646). Baumheier (2007) also sees a symbolic function, which is reflected in the architecture, the media, and the (metropolitan) culture (Baumheier 2007: 6).

Having set the concept of metropolitan regions, it should be mentioned that metropolitan regions face major challenges. On the one hand, these regions are challenged by the phenomenon of "glo*c*alization," meaning that they are affected by global and local issues. On the other hand, they are simultaneously confronted by the existing resentments of local and regional administrations regarding power and jurisdiction governance. As a result, metropolitan regions represent productive centers of economic activity, fragmented although governance structures – which are especially common in polycentric metropolitan regions – hamper productivity and partly shatter the advantage of agglomeration (Ahrend, Gamper, and Schumann 2014). Here, FRM is seen as a region with room for improvement. Blotevogel states that FRM as a metropolitan region only exists on paper (Blotevogel 2002: 348). Challenges regarding the complexity and fragmentation of political, economic, and social actors and the resulting diversification of power coalitions are identifiable.

Numerous loose and uncoordinated networks of private and non-private institutions, as well as unavailable capacity to coordinate joint actions raise doubts about a common governance approach in metropolitan FRM (Lackowska 2011a: 111). The resulting non-transparency obstructs participation. According to Zimmermann and Heinelt, no efforts are made to link the civil society to the decision-making processes (Zimmermann and Heinelt 2002: 87).

Furthermore, the metropolitan region of FRM includes at least five counties, each of which acts as representative and own stronghold, representing local (self-)interests (Lackowska 2011a: 79). Due to the long-lasting fragmentation and the absence of territorial history (Böderlin 2000: 539), several authors identify a missing common identity for FRM.[4] This separation is also fostered by the characteristic of the German administrative system of *Länder*, *Kreise*, *Gemeinden*, and *Kreisfreie Städte*. *Inter alia* local administrations in the metropolitan region of FRM are in competition to attract firms because through their settlement business tax money is raised (Lackowska 2011a: 80f). These conclusions are also considered by Stephans and Wikstrom (2000) and their more general metropolitan reform approach. Conflicts between local authorities and pertinent shortcomings in service provision lead to a lack of a metropolitan-wide political perspective, as it is identified for FRM (Lackowska 2011b: 23). Regarding an appropriate governance system, this draws a picture of incoordination and governance failure (Hoyler, Freytag, and Mager 2008a; Lackowska 2011a: 111).

Nonetheless, the metropolitan region of FRM can be defined by geographic, historic, and jurisdictive characteristics. Geographically, the region is formed by the Central Mountains of the *Taunus* in the North and the *Odenwald* and *Spessart* in the East. In the South the Rhine-Main Basin – today's Rhenish Hesse – and in the West the Rhine River shape the contours of the metropolitan region of FRM.

Historically, the municipal action program after World War I of Frankfurt's Mayor Ludwig Landmann in 1929 – with its strategic incorporation planning (e.g. the city of Höchst with *IG Farben*) and

4 E.g. Hoyler, Freytag, and Mager (2008a), Schaffer and Scheck (2006), and Freund (2003).

| State | Nuts 2 | | Nuts 3 | |
|---|---|---|---|---|
| Hesse | Darmstadt | DE71 | Darmstadt, Kreisfreie Stadt | DE711 |
| | | | Frankfurt am Main, Kreisfreie Stadt | DE712 |
| | | | Offenbach am Main, Kreisfreie Stadt | DE713 |
| | | | Wiesbaden, Kreisfreie Stadt | DE714 |
| | | | Kreis Bergstraße | DE715 |
| | | | Darmstadt-Dieburg | DE716 |
| | | | Groß-Gerau | DE717 |
| | | | Hochtaunuskreis | DE718 |
| | | | Main-Kinzig-Kreis | DE719 |
| | | | Main-Taunus-Kreis | DE71A |
| | | | Odenwaldkreis | DE71B |
| | | | Offenbach, Landkreis | DE71C |
| | | | Rheingau-Taunus-Kreis | DE71D |
| | | | Wetteraukreis | DE71E |
| | Gießen | DE72 | Gießen, Landkreis | DE721 |
| | | | Limburg-Weilburg | DE723 |
| | | | Vogelsbergkreis | DE725 |
| | Kassel | DE73 | Kreis Fulda | DE732 |
| Rhineland-Palatinate | Rheinhessen-Pfalz | DEB3 | Mainz, Kreisfreie Stadt | DEB35 |
| | | | Worms, Kreisfreie Stadt | DEB39 |
| | | | Alzey-Worms | DEB3B |
| | | | Mainz-Bingen | DEB3J |
| Bavaria | Unterfranken | DE26 | Aschaffenburg, Kreisfreie Stadt | DE261 |
| | | | Aschaffenburg, Landkreis | DE264 |
| | | | Miltenberg | DE269 |

*Figure 4.1: Jurisdictions of the geographic-functional region of FRM, Regionalverband – list of jurisdictions (Regionalverband 2012).*

predictive traffic planning (the Rhine-Main-Danube Canal expansion, the highway planning, and the construction of the airport) – anchored the economic position of Frankfurt am Main in the middle of Europe (Piecha 2005).

Today, from a jurisdictive perspective, FRM includes at least three federal states, the State of Hesse, Bavaria and Rhineland-Palatinate, five counties (NUTS-level 2) and 25 counties (NUTS-level 3) (Figs. 4.1 and 4.2).[5] Therefore, FRM embraces almost 15,000 square kilometers and about 5.57 million residents (Regionalverband 2012: 8).

Based on the understanding of the *Regionalverband*, the official planning authorities for FRM utilize an almost similar definition in the *Regionalen Flächennutzungsplan* (RegFNP) (Fig. 4.3).

This jurisdictive layout is also argued by the inter-connectivity and links between the knowledge-intensive business services of the region (Hoyler,

5 The Nomenclature of Territorial Units for Statistics (NUTS) is a geocode standard for referencing the subdivisions of countries for statistical purposes. The standard is developed and regulated by the EU (Eurostats 2015b).

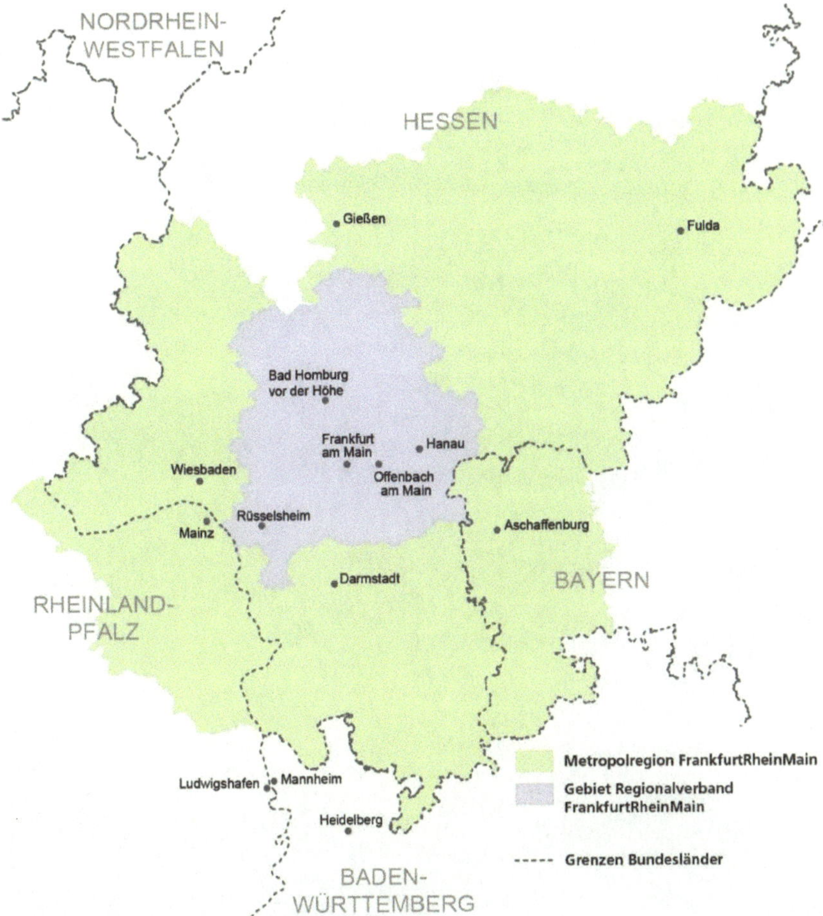


*Figure 4.2: Geographic-functional region of FRM, Regionalverband – cartographical illustration (Regionalverband 2012: 5).*

Freytag, and Mager 2008b). Moreover, the commuter movement expresses this geographic-functional definition (Fig. 4.4).

Besides this, some minor definitions also embrace these 25 counties. The Chamber of Commerce and Industry of Frankfurt am Main addresses an area within a radius of 75 kilometers, the Hessian *Christlich Demokratische Union Deutschlands* (CDU) promotes a *Stadtkreis* with 26

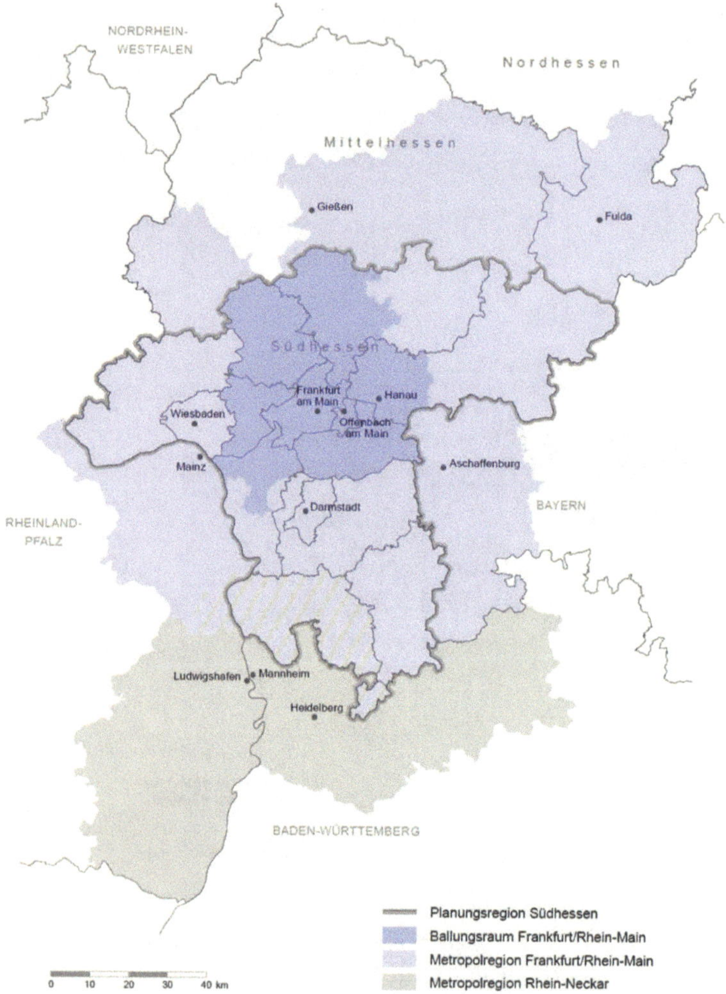


*Figure 4.3: Geographic-functional region of FRM, RegFNP – cartographical illustration (RPDA and Regionalverband 2011: 3).*

counties, and the Hessian *Sozialdemokratische Partei Deutschlands* (SPD) favors a *Regionalkreis* with around 5 million residents. Further solutions speak of an agglomeration of several divisions and local municipalities (Zimmermann and Heinelt 2002: 80).

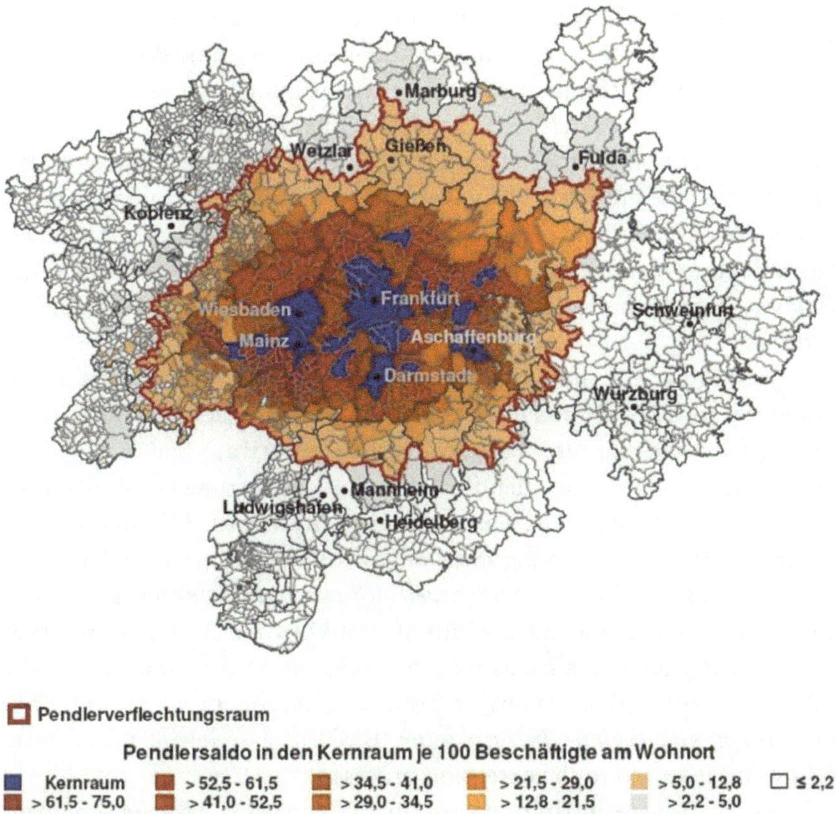


*Figure 4.4: Commuter interconnections in FRM (Prognos 2010: 10).*

This indicates that metropolitan FRM is not easy to define, whereby there are several, partly-contradicting definitions concerning what exactly the geographic-functional region of FRM is. Not having a final definition of the region – especially not defined by jurisdictions – takes previous governance approaches to their limits.

## 4.2.2 Institutional arrangement of FrankfurtRhineMain

Deduced from the geographic-functional definition and in line with the general governance approach – which includes the sum of all coexisting forms of collective regulation of societal matters – the institutional arrangement of FRM is characterized by private and non-private institutions. Non-private institutions are all administrative units. Private institutions are especially companies or other organizations, e.g. environmental associations and local citizens' groups.[6] This approach is also in line with the definition of the institutional arrangements identifiable at the meso-level, as the most relevant governance level for metropolitan regions.

However, as identified in the context of former governance approaches, it is difficult to identify the relevant institutions of a certain topic, namely who are – by which characteristic – the relevant stakeholders or mandate holders.

Concentrating on the extension process of Frankfurt Airport Knippenberger (2012) identified several relevant institutions. Institutions identified by her can also be distinguished by private and non-private characteristics. First, she identified the State Government of the Federal State of Hesse, represented by the *Hessisches Ministerium für Wirtschaft, Energie, Verkehr und Landesentwicklung* (HMWEVL)[7] and the *Hessisches Ministerium für Umwelt, Klimaschutz, Landwirtschaft und Verbraucherschutz*,[8] second the city of Frankfurt am Main, third, other state counties, municipality as well as *Kreisfreie Städte*, and finally, the Federal Republic of Germany, represented by the *Bundesministerium für Verkehr und digitale Infrastruktur* (BMVI),[9] as relevant non-private institutions involved in the extension process.

As major private institutions within the extension process Knippenberger (2012) identified the Fraport AG Frankfurt Airport Services Worldwide (Fraport AG),[10] the *Deutsche Lufthansa AG* (DLH), non-governmental organizations (NGO), and local citizen groups, lobby

6 One might argue that companies or organizations with a share greater than 50 percent of non-private shareholders might be called non-private. However, the presented research focuses on legal status, which defines the formal function. This is especially relevant for the case of the Munich Airport.

7 And its predecessors.

8 And its predecessors.

9 And its predecessors.

10 And its predecessor the *Flughafen Frankfurt/Main AG* (FAG); The company was officially named Fraport AG Frankfurt Airport Services Worldwide on June 11, 2001. Due to simplification, it is always spoken of Fraport AG in

organizations (chambers, trade associations, aviation associations, etc.) and unions, as well as churches as important institutions within the extension process (Knippenberger 2012).

Regarding the research question, this institutional setting has to be defined more specifically. On the one hand, further relevant institutions with a more regional focus are identifiable. On the other hand, there are several further institutions identifiable with a focus on Frankfurt Airport or the transport and logistics sector in general. The former was already highlighted by Lackowska (2011a) and Zimmermann and Heinelt (2002), the latter – as it can be argued – are all the companies within the spatial borders of the airport and are represented by Fraport AG as the economic site developer of Frankfurt Airport. Further logistics-related institutions, with a metropolitan character, are identifiable for FRM, which were partly outlined by F. W. Raschke (2009).

Based on this and the criticism that the metropolitan governance structure of FRM is diffused and frayed, it has to be admitted that the institutional landscape is virtually incalculable. Nonetheless, it is necessary to analyze the FRMs' institutional landscape. The following chapter presents a sharp picture of the *ex ante* relevant institutions. With a focus on metropolitan governance bodies, the result is a detailed institutional landscape of the relevant private and non-private institutions regarding the extension process of Frankfurt Airport.

#### *4.2.2.1 Non-private institutions*

Non-private institutions are marked by their jurisdiction for the topic of Frankfurt Airport. As presented, Knippenberger (2012) also names the BMVI as one of the relevant non-private actors in the extension process of Frankfurt Airport. According to German law, this inclusiveness is limited. Due to the federal system of Germany, the inclusion holds to the hierarchical relation between the national ministry-level and the state ministry-level. Nonetheless, the federal state has only the legislative authority on air traffic issues.[11] Although the *Luftfahrt-Bundesamt* is the

---

the presented research. Only in the presentation of the airport's history in Sec. 4.3 it is referred to the historically correct names.

11 According to Art. 73 I No 6 GG.

paramount authority at the national level, it does not have legislative authority on the construction and operation of airports.[12] The assessment and approval of the planning of airports is transferred to the federal state level (Hösch 2009: 75).[13] Accordingly, the HMWEVL is the central planning authority for Frankfurt Airport.[14] Due to these reasons, the BMVI and the national level are not of main interest in the following analysis. Additionally, the research focus is the metropolitan governance system, its institutions, and processes.

Regarding their legal status and the members, the RPU, the UVF, the *Planungsverband*, and the *Regionalverband* also have to be counted as non-private actors, all of which are represented only by administrative institutions (districts, municipalities, and cities).

**Non-private administrative actors**

*Hessisches Ministerium für Wirtschaft, Energie, Verkehr und Landesentwicklung (HMWEVL)*

As a traffic carrier, Frankfurt Airport is under the responsibility of the HMWEVL, which is the main ministry concerning airport issues (HMWEVL 2015b). The extension of Frankfurt Airport and the challenges with its enlargement is one of its main topics. It is anchored with its own sub-department (Aviation – Approval and Planning Permission) (HMWEVL 2015c).

The then-*Hessisches Ministerium für Wirtschaft, Verkehr und Landesentwicklung* (HMWVL)[15] was the responsible ministry for the RSEs and the AoPs of the A380 Shipyard in 2004 (HMWVL 2004) and of the Runway North-West in 2007 (HMWVL 2007).[16] Both were agreed in the RPD in 2002. It is also the central authority in the current extension phase of Terminal 3, which was also already agreed in 2002 (RPDA 2002b).

*City of Frankfurt am Main*

12 According to §2 *Gesetz über das Luftfahrt-Bundesamt.*

13 According to Art. 87 II GG, together with §§10 I and 31 II LuftVG; *Entscheidungen des Bundesverwaltungsgerichts* (BVerwGE) 4 C 51.89.

14 According to §8 LuftVG.

15 The predecessor of the HMWEVL.

16 According to §8 LuftVG.

Since the local government reform of 1972/77, the Frankfurt Airport is located in the city zone of Frankfurt am Main. Nonetheless, as presented, the jurisdiction is located at the federal state level. The city of Frankfurt am Main is engaged in the topic with the administrative body 79.01 (Higher-level Projects, Reports, and Regional Airport Affairs) of the Department of Environmental Protection (Frankfurt 2015b). Nonetheless, regarding the extension process, the city of Frankfurt am Main is limited to statements and reports (Frankfurt 2015c).

**Non-private metropolitan actors**

Especially four institutions appear as relevant regional planning intuitions in the historic timeline: the RPU (1965–1980), the UVF (1975–2001), the *Planungsverband* (2001–2011) and the *Regionalverband* (since 2011) (Lackowska 2011a; Zimmermann and Heinelt 2002). In the historical context, these institutions represent the metropolitan governance bodies for collaboration in FRM. Their main duties were the establishment and preparation of a common land use plan of the region, the joint implementation of a regional traffic system, and the economic development of the whole region. Today, the OECD recognizes the *Regionalverband* as the official metropolitan governance body of FRM (Ahrend et al. 2014: 27). Therefore, these regional organizations seem *ex ante* to be important for the regional topic of the extension of Frankfurt Airport.

*Regionale Planungsgesellschaft Untermain (1965–1980)*

After World War II and due to troubles in regional planning and spatial development, German municipalities were provided with planning jurisdictions in the 1960s. After only two years the Federal State of Hesse introduced its plans for a regional planning body. Subsequently, in 1965, the RPU was established as the special purpose association, to support the regional land use planning in accordance with the Hessian Spatial Planning Act. The RPU was organized by the executive committee and the general meeting, which itself constituted planning committees (ISG 2014a). Accordingly, FRM was one of the first metropolitan regions to set up a metropolitan governance body and institutionalizing regional cooperation (Ahrend et al. 2014: 11).

Nonetheless, the RPU was limited in its actions, while its catchment area was also limited. Not even all neighboring municipalities of Frankfurt am Main were covered.[17] Owing to misinterpretations of the member cities and the regional development – especially regarding an expected population explosion – as well as due to internal and external conflicts of the RPU, the institution was unable to act effectively. In the eyes of the administration of Frankfurt am Main, the other municipalities had too much power and the RPU itself struggled for the increase of competencies with the Federal State of Hesse. Although the concept of the *Regionalstadt* was refused in 1971, which further contributed to the weakening of RPU (Lackowska 2011a: 82f). The RPU operated until 1980 (ISG 2014a).

*Umlandverband Frankfurt (1975–2001)*

Five years before the RPU was officially abolished and based on the *Gesetz über den Umlandverband Frankfurt* (GVBl. I S. 427), the UVF was founded on January 1, 1975. The association came closer to the definition of a metropolitan-wide institution for regional cooperation. Its duties were the coordination of spatial planning, the support of the development and implementation of regional systems and grids, such as the regional public transport, and the freshwater and wastewater system. Further topics were the advancement of energy, economic development, healthcare, and the environment.[18]

Accordingly, the UVF was *de jure* an important metropolitan governance institution for FRM and at this time the most innovative metropolitan body in Germany. Its main duty was the development and preparation of the land use plan.

Despite its institutional position, the UVF was not utilized as a common organization to improve joint discussions of regional topics; rather, its member municipalities only presented local self-interests and hampered a common metropolitan outlook. This was accompanied by a lack of public interest, although the council was elected directly at the same time as general local elections took place. Aside from some leisure facilities, the UVF was almost unknown in civil society. Due to the central position of

---

17 Members of RPU: independent cities of Frankfurt am Main, Hanau, and Offenbach am Main, as well as the counties of Büdingen, Dieburg (partly), Friedberg, Gelnhausen, Hanau, Main-Taunus (partly), Obertaunus, Offenbach, Schlüchtern, and Usingen (ISG 2014a).

18 According to §3 GVBl. I S. 427.

the city of Frankfurt am Main, the UVF failed to promote and implement larger joint regional projects.

The weakness of the UVF was further escalated by the establishment of the *Frankfurter Verkehrs- und Tarifverbund.*[19] It took over the development and improvement of the public transport system, which was previously an important task of the UVF (Lackowska 2011a: 84–86).

*Planungsverband Ballungsraum Frankfurt/Rhein-Main (2001–2011)*

The UVF was abolished by the *Gesetz zur Stärkung der kommunalen Zusammenarbeit im Ballungsraum Frankfurt/Rhein-Main* (BallrG). Based on the *Gesetz über den Planungsverband Ballungsraum Frankfurt/Rhein-Main* (PlanvG), the Planungsverband was founded and implemented on April 1, 2001. Its main task was also the development and preparation of the land use plan for the region.[20] The *Planungsverband* was structured by an association chamber and its board. The catchment area – with 75 municipalities – was larger than those of its predecessors, although as a single-purpose institution the *Planungsverband* had fewer competencies.[21]

Given that the tasks of the *Planungsverband* had to be coordinated and agreed with the politicians and administrations of the member municipalities, it became *de facto* a political body. Also, contrary to the former approach, the *Planungsverband* was seated by representatives of the member municipalities, not directly elected. Consequently, the *Planungsverband* also did not overcome the challenge, that the single municipalities only expressed their self-interests instead of establishing and fostering a coherent common regional approach. In this context, its major success was the development of a legally-non-binding strategic vision for the region (Lackowska 2011a: 95f).

*Regionalverband FrankfurtRheinMain (since 2011)*

As the contemporary metropolitan governance body of FRM, the *Regionalverband* replaced the *Planungsverband* with the MetropolG. As a single-purpose institution, the *Regionalverband* took over the duties of

---

19 Today *Rhein-Main-Verkehrsverbund.*

20 According to §2 PlanvG.

21 Basically the UVF just prepared the land use plan, but according to §3 GVBI. I S. 427 it had more responsibilities regarding regional topics *de jure.*

the *Planungsverband* for regional cooperation.[22] Its main duty is also the development, change, or annulment of the land use plan for the metropolitan region.[23] Nevertheless, the *Regionalverband* – following permission of the executive committee – is allowed to engage in each topic with major regional importance.[24] In all interests, the municipalities of the metropolitan region are encouraged to coordinate and promote joint regional cooperation within the fields of traffic planning and management, the establishment and operation of the *Regionalpark RheinMain*,[25] the marketing for and the promotion of the regional economy, as well as the promotion of cultural installations of supra-regional interest.[26]

The dispatch of representatives instead of the direct election was retained.[27] Besides the municipalities, envoys of employers' associations, unions, and nature conservation organizations are consulting members of the *Regionalverband*. Furthermore, civil society actors from the economy and the scientific community are involved as invited "advisory guests" (Regionalverband 2013: 6).

#### *4.2.2.2 Private institutions*

For the presented analysis, private institutions hold interest regarding their involvement in the topic of Frankfurt Airport. Private metropolitan institutions hold interest based on their task of improving the metropolitan region of FRM. The relevance of these private metropolitan institutions is based on their duties regarding the economic development of the region and environmental or societal issues, which both have to be in close relation to the extension process of Frankfurt Airport. Moreover, there are private metropolitan institutions identifiable, of which the corporate purpose is the improvement of logistics and transport sectors in the region of FRM. Economic actors excel regarding their business position for

---

22 According to §5 MetropolG.
23 According to §8 MetropolG.
24 According to §8 I 3 MetropolG.
25 The *Regionalpark RheinMain* was designed to secure remaining open areas in FRM for the relaxation and leisure of the people of the region (Regionalpark 2014).
26 According to §1 MetropolG.
27 According to §11 MetropolG.

Frankfurt Airport. This might be especially due to shares of the operating company or the corporate purpose as an operating company.

### Metropolitan logistics actors

*Logistik RheinMain. Die Wissensinitiative (2007–2014)*

*Logistik RheinMain. Die Wissensinitiative* (LRM) was initiated in 2007. Funded by the EU, its main duty was to establish the House of Logistics and Mobility (HOLM), as the main cooperation platform for logistics and logistics-related issues in FRM. Moreover, it was responsible for the promotion of regional logistics topics at the regional and national levels. Since 2009, LRM has been supported by the newly-founded *Gründungsinitiative Frankfurt HoLM (House of Logistics and Mobility)*,[28] which has the duties to integrate and promote logistics and mobility topics as well as actors to foster cooperation between economic, research and political institutions (HOLM 2015d). With the opening of the HOLM office building in 2014 and the phasing-out of the EU funding, LRM was integrated into the HOLM structure (LRM 2015).

*House of Logistics & Mobility (HOLM) GmbH (since 2010)*

The HOLM was established in 2010 as a corporation. The shares of the corporation are held by the Federal State of Hesse (86.5 percent), the city of Frankfurt am Main (12.5 percent), and the HOLM e.V. (1 percent). Additionally, it is supported by the economy,[29] the scientific community,[30] and the Regionalverband (HOLM 2015c). Nevertheless, non-public actors are the major shareholders (99 percent).

Besides LRM working on the logistics pillar, *Mobilität RheinMain* (MRM) forms the mobility pillar and Hessen Aviation[31] as the third pillar, promotes the aviation cluster (HOLM 2015a). Therefore, the HOLM is engaged in all relevant transport topics according to traffic carriers.

---

28 Today HOLM (House of Logistics and Mobility) e.V. (HOLM e.V.) (Preamble HOLM statutes).

29 The DLH, the Fraport AG, and other manly logistics firms.

30 Goethe University Frankfurt, Technische Universität Darmstadt, and other universities, mainly located in FRM.

31 Before HESSENMETALL Aerospace-Cluster, 2009 to 2013.

With the integration of LRM, the HOLM is also supported by HOLM e.V. and its 235 members[32] in the aforementioned issues (HOLM 2015b).

*CargoCity Frankfurt Task Force (2012–2014)*

Founded in 2012, the CargoCity Frankfurt Task Force (CCFTF) had the duty to promote Frankfurt Airport to advance the airport to Europe's number one air cargo hub. CCFTF concentrated on interests of air cargo companies in FRM, to advance the location of Frankfurt Airport. Its members were major logistics interest groups, as well as Fraport AG and DLH (Cargo) (Fraport 2011; LFU 2012).

*Air Cargo Community Frankfurt e.V. (since 2014)*

The Air Cargo Community Frankfurt e.V. (ACCF) was founded in 2014 as the successor of CCFTF. Nonetheless, the paramount goal is the advancement of Frankfurt Airport to the number one air cargo hub in Europe. Beside the major logistics interest groups, fifteen companies located in FRM region also replenish the members (BARIG 2014; Fraport 2014).

**Economic actors**

*Fraport AG Frankfurt Airport services worldwide*

As the operating company of Frankfurt Airport, Fraport AG is the owner of the airport. Its main field of engagement is the development and operation of airports, mainly in Europe (6), as well as Africa (1), Asia (3), and South America (1) (Fraport 2015a). It is organized as a public stock corporation with stock capital valued at EUR 922.9 million. The main shareholders are the Federal State of Hesse (31.4 percent), the Stadtwerke Frankfurt am Main Holding GmbH (20 percent),[33] the DLH (8.5 percent), Lazard Asset Management (3.2 percent), and RARE Infrastructure Limited (3.1 percent). The last 33.4 percent of the share is below the reporting threshold (Fraport 2014: 5). Based on shares the Federal State of Hesse and the city of Frankfurt am Main have the absolute majority (51.4 percent). Therefore, non-public actors are the most important shareholders of Frankfurt Airport.

32 As of February 26, 2015.

33 The city of Frankfurt am Frankfurt is the only shareholder of the close corporation (Frankfurt 2015a).

*Deutsche Lufthansa AG*

DLH can also be seen as an important shareholder. It only holds 8.5 percent of the shares, although DLH is the world's 7th largest airline and Frankfurt Airport is its home base (Statista 2015). This gives DLH a prominent position at the Frankfurt Airport. Furthermore, it is the only air transport-related shareholder above the reporting threshold.

**Further institutions**

There are also several further institutions identifiable. In the metropolitan context, Lackowska (2011a) and Zimmermann and Heinelt (2002) also identified institutions which might be relevant to the presented research. In particular these are *FrankfurtRheinMain: Verein zur Förderung der Standortentwicklung e.V.*, *FrankfurtRheinMain GmbH (FrankfurtRheinMain International Marketing of the Region)*, *Hessen-Agentur*, *IHK Forum Rhein-Main*, the *Metropolitana*, the *Wirtschaftsförderung Region Frankfurt RheinMain e.V.*, and *Wirtschaftsinitiative FrankfurtRheinMain e.V.* They all have the corporate purpose or goal to promote the economic development of FRM. Moreover, they are all engaging in societal issues with a metropolitan character (Lackowska 2011a; Zimmermann and Heinelt 2002).[34] According to BallrG, at least *FrankfurtRheinMain GmbH* is also officially recognized as an institution with a metropolitan character for FRM (HMIS 2015).[35] However, strong involvement in airport topics is not evident upfront. Furthermore, Lackowska (2011a) and Zimmermann and Heinelt (2002) identified several additional institutions, although these are neither airport-affine nor do they have an identifiable focus on airport issues. Moreover, the additional regional cooperation platforms listed by the Hessian Cluster Portal and the Regionalverband are not in the focus of the presented research question (Annex A.2).

*Ex ante*, the further identification of possible relevant institutions is not possible. Regarding the extension process of Frankfurt Airport, it is assumed that the presented institutions represent the major relevant actors

---

34 According to their websites or statutes.

35 According to §1 BallrG.

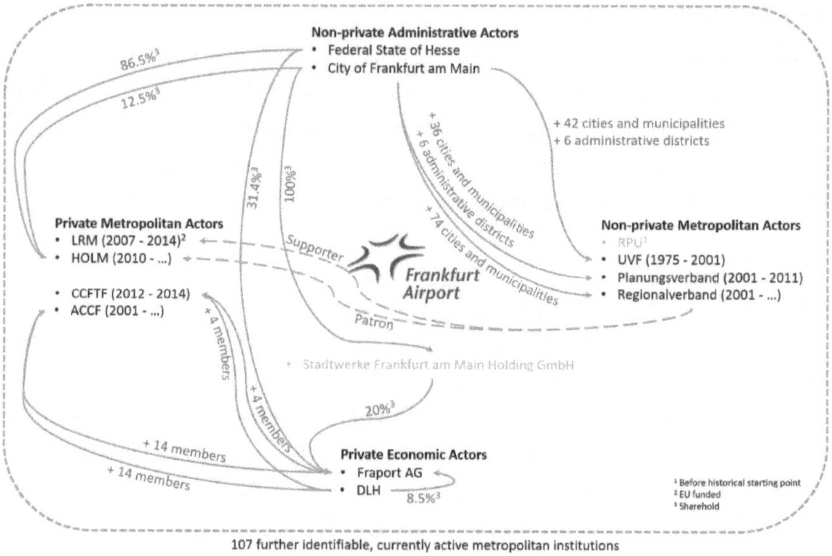


*Figure 4.5: Scheme of the institutional arrangement of Frankfurt Airport, own depiction.*

in the institutional arrangement of FRM.[36] Accordingly, the following schematic overall institutional arrangement can be drawn.

## 4.3 Frankfurt Airport and the governance of the extension process

As an example of the development of an airport, a brief historical introduction to the development of Frankfurt Airport is presented in

36 There is no exact number available on how many NGOs or local citizens' groups exist in Germany or even FRM. Surveys from the 1970s oscillate between 1,000 and 50,000 organizations in Germany (BPB 2015) and the *Bundesministerium für wirtschaftliche Zusammenarbeit und Entwicklung* (BMZ) counts approximately 2,000 NGOs in Germany only in the field of international cooperation and development (BMZ 2015). The research of Geis (2005) shows, that with the publication of the plans to extend the Frankfurt Airport an unfinished list of local citizens' group appeared (Geis 2005: 125–130).

the following. For the subsequent analysis, 1997 and the initiation of the mediation process by the Gesprächskreis Flughafen is the historical starting point. At that point in history mediation came into fore, as an experimental process of municipal policy-making, came into the fore and in Germany the *Gesetz zur Förderung der außergerichtlichen Streitbeilegung* came into force (Geis 2005: 17; Schlieffen 2010: 1; Strempel 2002: 129–132). Accordingly, the mediation process of the extension of the Frankfurt Airport can be seen as the initial spark of mediation for political participation in Germany (Geis 2005: 15f).

The analysis of Frankfurt Airport follows six major periods: the airport extension between the 1960s–1980s (Sec. 4.3.3, the mediation process since 1998 (Sec. 4.3.4), the RPP and the RPD of 2000–2002 (Sec. 4.3.5), the RSE and the AoP A380 Shipyard of 2002–2004 (Sec. 4.3.6), the RSE and the AoP Runway North-West/Terminal 3 of 2004–2007 (Sec. 4.3.7), and the Adjustment of the Plans from 2007 and beyond (Sec. 4.3.8).

The results presented in the interim findings of the respective period are two folded. First, by reconstructing and challenging ERG and the decision-making process, the research question concerning whether ERG is an appropriate approach for policy analysis to analyze decision-making processes in geographic-functional regions is answered. Additionally, the presented research-guiding hypotheses are confirmed. Second, the analysis presents the institutional arrangement surrounding the particular decision-making process. With a focus on non-private metropolitan governance bodies and private metropolitan logistics actors, this identifies the relevant stakeholders in the decision-making process.

Finally, these interim findings are summarized and discussed at the end of the chapter to present the overall context. This shows that ERG – as an analytical tool in policy analysis – is not restricted to a single phenomenon. Based on its recursive character, ERG allows setting single periods in a historic connection. Summarizing and discussing the findings, it is also analyzed whether the decision-making processes improved and if governance learned during the overall process of decision-making.

### 4.3.1 Frankfurt Airport

After the foundation of the *Frankfurter Verein für Luftschiffahrt* on November 14, 1908, the first airship landed on July 31, 1909, in the *Rebstock* area. With the foundation of the *Deutsche Luftschiffahrts AG* on November 16, 1909, and the *Internationale Luftschiffahrt Ausstellung* from July 10, to October 17, 1909, Frankfurt am Main became one of the major aviation centers in Germany.

The *Rebstock* area was reconstructed after it was destroyed in World War I and in 1924 the first scheduled flights took off again. The expansion of flight operations resulted in the foundation of *Frankfurter Flughafen GmbH* on April 1, 1925. After the first enlargement to 100 hectares, the *Flugplatz Rebstock* gained its operating license as an Airport of First Order on August 16, 1926.[37] With growing air traffic, the need for a bigger airport was formulated. The construction of a new airport was started in 1934. In July 1936, the whole air traffic was taken to the new area of *Flug- und Luftschiffhafen Frankfurt Rhein-Main*, today's airport area. Participating companies were *Südwestdeutsche Luftverkehrs AG* and the *Rhein-Main Luftschiffhafen GmbH*. In 1938 and 1942, the first and second general expansion plans were introduced, respectively. During the time of World War II, the airport was used as an air force base. The airport was nearly destroyed completely at the end of the war. Nonetheless, an air force squadron of the US Air Force was based at Frankfurt Airport in 1945.

During the following year, Frankfurt Airport was appointed and established as the European air force hub of the US Air Force. Already in 1946, public air transport was re-established. Up to this point, the airport's historically most significant role emerged with the advent of the Cold War and the establishment of the Berlin Airlift to supply Western Berlin. In those years, the airport was also extended and the Runway North was constructed (Treibel 1992: 165–175).

On May 28, 1954, the operating company *Verkehrsaktiengesellschaft Rhein-Main* – founded in 1947 – was renamed as FAG (FAG 1996: 42).

---

37 This priority was regularly confirmed in the following years (HMWVL 2004: 74).

After the airport was partly released from military control of the US, Lufthansa entered back into service (March 1, 1955), and air sovereignty was transferred back to Germany (May 5, 1955), whereby the airport was extended to an airport for jets. The scheduled jet traffic started on April 1, 1959.

This was also the time in which it became clear, that the capacities of the airport will be exhausted in the coming years. In 1958 over one million passengers (1,212,533) had to be checked in for the first time. Accordingly, the Frankfurt Airport reached the estimated number of assessed passengers, two years before they expected (DIL 1955: 14). Previous extension plans were checked for their enforceability and on June 16, 1965, the foundation stone of the Empfangshalle Mitte (today Terminal 1) was laid. On August 23, 1966, further extension plans were authorized by the then-Hessian Minister for Economy and Traffic Rudi Arendt. On March 23, 1971, the second AoP Runway 18-West (HMWT 1971) was decided (Treibel 1992: 176–195).[38] One year later, the *Terminal Mitte* was opened on March 14, 1972 (Müller and Skaryd 1972: XVI).

Already in those times, further demands of Frankfurt Airport and its interrelations to the region were identified. The development of the national and international air traffic system as well as its site optimization and enlargement regarding runways, cargo centers, general traffic, its multi-modal connection, and other minor topics held major interest (FAG 1971; FAG 1978). The main challenge was (and still is) to fit any short-term concept into an overall long-term framework, recognizing issues of the future, not only technical but also regarding political and societal demands (FAG 1978: 32; Thießen 2012).

Although several rescissory actions were still not completely settled, the extension of the airport started on October 8, 1981. Bringing the runway into service on April 12, 1984, Frankfurt Airport was enlarged by 200 hectares (Treibel 1992: 187–189). During this time, new technological possibilities – especially regarding wide-bodied aircraft – made it necessary to continuously adjust the enlargement plans to meet the technical needs

38 The AoP was finally confirmed by the court decision of the *Hessischer Verwaltungsgerichtshof* (HessVGH) with the decision II OE 205/78 on October 21, 1980.

and the demands regarding increasing passenger numbers (Müller and Skaryd 1972: 177; Treibel 1992: 192). In 1985, the general expansion plan set the framework for the next stages of expansion (Treibel 1992: 189). This general expansion plan included the construction of the second terminal (Terminal 2), which was opened on October 24, 1994. It was the anchor point for the Airport 2000 Expansion Program to meet the demands of the upcoming century by the enlargement of Terminal 1 and further cargo buildings and areas, such as Cargo City Süd (FAG 1990; FAG 1996: 44).[39]

In the following years, Cargo City Süd (1997), Gate D (1997), and the ICE train station in 1999 were opened and entered into service (Fraport 2014: 4f). In those times, the first discussions of the further extension of Frankfurt Airport started. In 1998, the so-called *Mediationsverfahren* was initiated and it ended in 2000.[40]

In June 2000, a basic political decision in favor of the extension plans was made. The majority of the Hessian State Parliament[41] voted for the extension of the airport and in September 2000. The then-FAG decided to apply for the extension and to initiate the RSE (Amann 2012: 2; FFR 2014).

On June 11, 2001, FAG went public and was renamed. Since then, its official name is Fraport AG Frankfurt Airport Service Worldwide (Amann 2012: 2; Fraport 2014: 5).

In the same year, the RPP was initiated to prove the spatial compatibility according to the regional spatial regulation.[42] With the RPD on June 10, 2002, the *Regierungspräsidium Darmstadt* (RP Darmstadt) asserted the accordance of the extension plans with the demands of the urban and regional development planning (RPDA 2002b). After the public hearings, the AoP A380 Shipyard – as a part of the expansion plans – was published

39 Already in 1989 the number of passengers was higher than assessed in 1985 for the year 2000 (FAG 1990: 3).

40 Initially continued by the *Regionales Dialogforum Flughafen* (RDF) from 2000 to 2008, the process is now continued by the *Forum Flughafen & region* (FFR).

41 Constitution 1999–2003: leading parties: 43.4 percent CDU and 5.1 percent *Freie Demokratische Partei* (FDP); opposition: 39.4 percent SPD, 7.2 percent *Bündnis90/Die Grünen*, and 4.9 percent Others.

42 According to §1 ROV.

on November 26, 2004 (HMWVL 2004). Three years later in 2007, after the public hearing from September 2005, to March 2006, the RSE of Runway North-West/Terminal 3 was evaluated. The HMWVL presented the AoP Runway North-West/Terminal 3 on December 18, 2007, with the construction of the Runway North-West as its core (HMWVL 2007). In the same year, the next step to establish an airport city was made by starting the construction of The Squaire. In 2011, both the Runway North-West and The Squaire were opened. During the following year, in October 2012 the enlarged Gate A (Gate A-Plus) entered into service. In September 2013, the Fraport AG submitted the building permit application for Terminal 3, which was already part of the RPP in 2002 and the AoP in 2007 (FRA 2014).

Having presented the history of the extension of Frankfurt Airport, the following paragraphs will analyze the decision-making process of the extension since 1997 in further detail. First, the mediation process will be examined. Second, the RPP is analyzed. Third, the RSE A380 Shipyard is evaluated. Fourth, this is followed by the analysis of the RSE Runway North-West/Terminal 3. Finally, the decision-making process of supplementary adjustments is analyzed.

### 4.3.2 Airport extension: the official proceedings

However the consecutive proceeding of an RPP followed by an RSE is not applying for the extension regarding Runway 18-West, it will be coherently presented in the following.[43] For the RSE Runway 18-West different institutions were requested to execute expert reports, e.g. the *Deutsche Institut für Luftverkehrsstatistik e.V.* commissioned by the then-Hessian Minister for Economic and Traffic (DIL 1965). The construction of Runway 18-West only included an RSE.

Approval procedures for infrastructure projects in Germany are highly complex. Thus, many institutions have to be involved. To reduce the complexity of the overall planning proceedings German law divides it into different but dependent parts. The RPP is assessing different solutions and

43 Only with the decision of the HessVGH on October 21, 1980, the need of a planning requirement was approved (Ladeur (1983): 110).

| Year | Procedure |
|---|---|
| 1954 | General extension study for optimal positioning of a further runway (FAG) |
| 1961, September | FAG requested surrounding cities of Frankfurt Airport to align spatial planning |
| 1965, December | FAG officially requested the extension and the construction of Runway 18-West |
| 1966, May | Hessian state government approved the extension subject to defined requirements |
| 1968, March | First AoP issued, rejected by Staatsgerichtshof des Landes Hessen |
| 1970, June - July | 2nd Public display of planning documents |
| 1970, September | Public hearings of the extension |
| 1971, March | Publication of 2nd AoP by HMWT |
| 1978, July | Confirmation of 2nd AoP by BVerwG |
| 1979, March | FAG started extension of existing runways |
| 1980, October | Final confirmation of construction Runway 18-West by HessVGH |
| 1984, April | Runway 18-West was opened and set into operation |
| 1997 | First official discussion in regard to the further airport extension |
| 1998 - 2000 | Mediation process |
| 2000, June | Basic policy decision in favor of the airport expansion |
| 2000, September | Final decision of Fraport AGs' supervisory board for the airport extension |
| 2001 | Start of regional planning procedure |
| 2002, June | Regional planning decision by RPD, conclusion of regional planning procedure |
| 2003, April | Scoping of road scheme evaluation |
| 2002, May | Start of road scheme evaluation for A380 Shipyard |
| 2004, January - March | Public hearings A380 Shipyard |
| 2004, November | Regional planning decision by HMWVL, conclusion of regional planning procedure A380 Shipyard |
| 2004, November | Start of 1st Road scheme evaluation Runway North-West |
| 2005, September - 2006, March | Public hearings Runway North-West and Terminal 3 |
| 2006, November | Start of construction The Squaire (Airrail Center) |
| 2007, February | Start of 2nd Road scheme evaluation Runway North-West and Terminal 3 |
| 2007, December | Regional planning decision by HMWVL, conclusion of regional planning procedure Runway North-West and Terminal 3 |
| 2008, January | Opening A380 Shipyard |

*Figure 4.6: Historical overview of the extension of Frankfurt Airport, from 1960s Pt. I, own depiction, based on Amann (2012), Fraport (2015b), Fraport (2014), and Johnsen (1996).*

their spatial complexity. It is followed by the RSE to specify the plans and officially approve them (Scholles 2008: 105).[44] However, a standardized

44 An RPP can be followed by several RSEs, if the RPD has covered the topics already. This is the case for the extension of the Frankfurt Airport, where the PRD covers the A380 Shipyard, the Runway North-West, and Terminal 3.

| Year | Procedure |
| --- | --- |
| 2009, August | Decision of VGH Kassel (Runway North-West) |
| 2011, October | Opening Runway North-West |
| 2011, October | Opening The Squaire (Airrail Center) |
| 2012, October | Opening of Gate A-Plus (enlargement of Gate-A) |
| 2013, April | 1st amending order Runway North-West (Side commercial use) |
| 2013, May | 2nd amending order Runway North-West (Vortex I) |
| 2013, September | Submission of building permit application Terminal 3 |
| 2014, May | 3rd amending order Runway North-West (Vortex II) |
| 2014, August | Approval of first stage of construction Terminal 3 |
| *2015, planned* | *Start of construction Terminal 3* |
| *2017, planned* | *Gradual opening of Terminal 3* |
| *2021, planned* | *Final opening Terminal 3* |

*Figure 4.7: Historical overview of the extension of Frankfurt Airport, from 1960s Pt. II, own depiction, based on Amann (2012), Fraport (2015b), Fraport (2014), and Johnsen (1996).*

airport planning procedure and decision-making process at the federal state level are not available (Hösch 2009: 75).

The RPP is an instrument to realize spatial planning. It is one of the major coordination processes for approval proceedings of infrastructure planning in Germany. The final RPD represents its conclusion (Höhnberg 2005: 884). The goal is to coordinate spatial planning with supra-regional importance.[45] In this context, the RPP clarifies upfront the final approval of the actual project, the compatibility relating to regional planning policy and environmental issues. It identifies the most feasible possibilities for the planned infrastructure project (Höhnberg 2005: 885; Scholles 2008: 105).[46] Especially social, infrastructural, economical, ecological and cultural aspects have to be taken into account.[47] Sustainable economic development, as well as sustainable services for the public, promotion of innovation, and natural resource management are key indicators for the approval. Each and every development plan assessed by an RPP has to be

45 According to §1 I (1) ROG.
46 BVerwGE 68, 311/318.
47 According to §2 II (1) ROG.

compatible with the goals and principles of the ROG (Gohl and Meister 2012: 95).

Even though the RPP is an internal administrative proceeding, public participation is intended. For projects with a supra-national character, such as the extension of an airport, public participation is obligatory.[48] Public participation is widely understood and embraces everyone, with the exception of public bodies (Spannowsky, Runkel, and Goppel 2010: §10 recital 42).

Thus it can be said, that the RPP follows the understanding of Geißel and her understanding of cooperative governance. Decision-making in the RPP is negotiated between private and non-private institutions as well as citizens. The lower levels are empowered and power is shared between private and non-private institutions (Geißel 2013: 10f; Geißel 2012b: 165). Accordingly, the RPP can be described as a deliberative, dialog-oriented approach of decision-making.

An RPP has to be initiated if a project is of supra-regional importance.[49] The specific embodiment is undertaken by national law and federal state law (Höhnberg 2005: 885).[50] The extension of Frankfurt Airport represents such a project with supra-regional importance.[51]

After the RPP, the RSE is the second step of the formal proceeding for the approval of infrastructure projects with supra-regional impacts (Scholles 2008: 100).[52] The goal of the RSE is a decision on the authorization of the actual project plan (Stelkens, Bonk, and Sachs 2014: §72 recital 3f). The responsible authority assesses the plan of the project proponent by trading-off private and public interests. In this trade-off, all relevant and involved institutions – meaning affected municipalities, recognized associations[53] as well as the civil society – have to be involved in the decision-making process (Scholles 2008: 100–102). Like the RPP, the RSE is also characterized by the understanding

48 According to §§10, 11 III, and §15.3 ROG.

49 According to §1 I (1) ROV.

50 According to §15 ROG, ROV and for the case of Frankfurt Airport the *Hessischen Landesplanungsgesetzes* (HLPG).

51 According to §8 LuftVG.

52 According to §§72-78 *Verwaltungsverfahrensgesetz*.

53 According to §58 *Bundesnaturschutzgesetz* (BNatSchG).

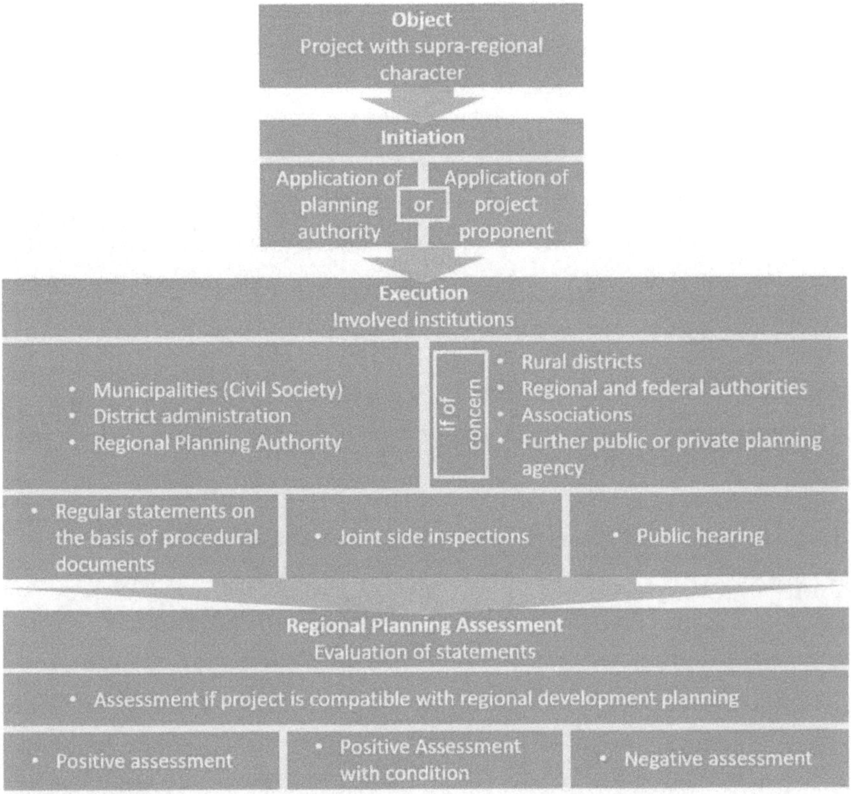


*Figure 4.8: Scheme of regional planning procedure, own depiction, based on Höhnberg (2005): 886.*

of cooperative governance presented by Geißel. Enforced by law, the decision-making in the RSE is negotiated between private and non-private institutions. Citizens' interests are especially included by local citizens' groups. During the hearing, the lower levels are integrated into the decision-making process. Under these circumstances, they are empowered and power is shared between private and non-private institutions, as is the case for cooperative governance (Geißel 2012b: 165; Geißel 2013: 10f). Accordingly, the RSE can be described as a deliberative, dialog-oriented approach of decision-making.

The RPP and the RSEs of Frankfurt Airport seem to be a good example to analyze the regulatory structure of governance in geographic-functional

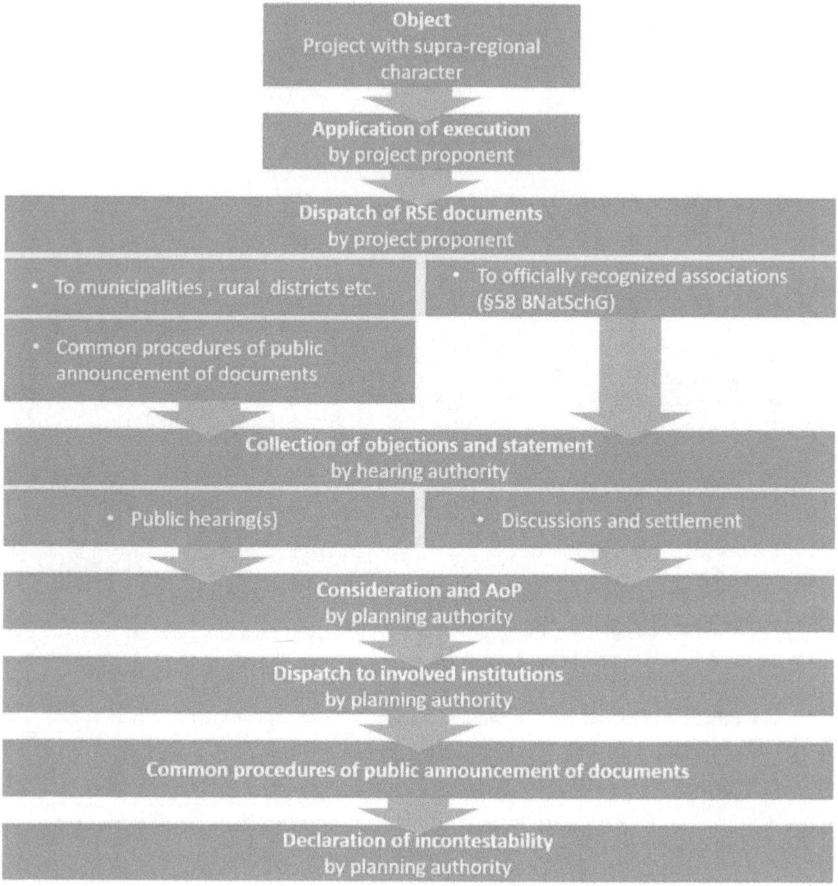


*Figure 4.9: Scheme of road scheme evaluation, own depiction, based on Scholles (2008: 107).*

regions, the decision-making processes, and institutional arrangements as well as to identify governance learning. The RSE for the Runway 18-West – again the RPP as an official proceeding was not applying to the Runway 18-West extension – as well as the RPP and RSEs for Cargo City Süd, further maintenance shipyards and construct Terminal 3 are analyzed following the theoretical derived concept of ERG, presented in this research.

First, the regulatory structure of governance and its decision-making processes are reconstructed to test the concept of ERG and its five aspects. Second, a clear picture of the institutional arrangements of the official proceedings is drawn. With a focus on non-private metropolitan governance bodies and private metropolitan logistics actors, the institutional arrangements and relevant institutions of the RPPs and RSEs are outlined. Third, the research question and the minor research questions are answered. Finally, the research-guiding hypotheses are confirmed for the regulatory structure of governance, its decision-making processes, and institutional arrangements as well as to identify governance learning.

### 4.3.3 Airport extension, the 1960s–1980s

Today, Runway 18-West is one of the most used runways at Frankfurt Airport. It was planned and built between 1962 and 1984. At the beginning of the 1980s, its construction became the culmination of conflicts between the state power and a resistance movement.

The demand to extend Frankfurt Airport emerged from different prospects. First, a strong increase in air traffic volumes affecting the operating procedures of the airport was expected. Second, DLH decided in favor of Frankfurt Airport for its home base for flying personnel and aircraft maintenance. Finally, the plan for the extension was linked to an official order by the German Government to make Frankfurt Airport to the first German airport meeting the requirements of intercontinental scheduled jet aircraft traffic for which the airport did not yet offer optimal conditions for the operation (Arndt 1975: 327f; Treibel 1992: 176–195).

Already in 1954, the FAG examined with a general extension study the optimal positioning of a further runway. These results were confirmed by a further study presented in February 1958 by the FAG (HMWV 1966: 2). Based on topographical and climatological characteristics, six different layouts of a future runway system at Frankfurt Airport have been examined. After around three years of planning, the FAG applied for the airport extension to the then-Hessian Minister of Economics and Traffic, Rudi Arndt on December 28, 1965. The plan embraced the extension of Frankfurt Airport's existing runway system by expanding the capacities of the two existing runways and building an additional pure runway at the western end of the old runways, the Runway 18-West. All had spatial-relevant impacts (Arndt 1975: 327f; Treibel 1992: 176–195).

Regarding efficiency, a parallel runway to the existing runways would have been the optimal solution. With little impact on the adjoining forest and the relocation of several US military installations, this solution could have been realized on the existing grounds of Frankfurt Airport. However, that option would have had a serious impact on the bordering cites regarding noise emission. Further research showed, that the sole installation of technical improvements for the existing runways would not have lead to the needed capacity increase, necessary for proper operation and maintenance.[54] Finally a compromise solution between social and environmental impacts was chosen (Arndt 1975: 327f; Johnsen 1996: 160). On the one hand, it was intended to avoid additional impairment of the adjacent resident population. On the other hand, it was intended to cut down parts of an intact forest area, a considerable part of the so-called avalanche forest (Arndt 1975: 327; HLGL ny).

#### *4.3.3.1 Runway 18-West*

Before the official proceedings in September 1961, the FAG requested the surrounding cities of Frankfurt Airport to align spatial planning with the land development plan for the airport. During the so-called argumentative phase (Jahn 1984: 319), FAG presented the plans to the bordering cities and municipalities. It demanded the coordination of the urban land use planning accordingly to the area required for the expansion. All 32 affected communities rejected FAG's application. As a result, not only the plans of the FAG to extend the airport – supported by the federal government, the state, and the city of Frankfurt am Main – were impeded. Also, the land use planning processes of the municipalities were hampered. On the one hand, this situation led to extensive discussions in the community councils and city parliaments. On the other hand, it led to a broad exchange and publication of personal and open letters, e.g. between Kurt Oeser, the then-protestant pastor in Mörfelden and later mediator

54 According to §45 *Luftverkehr-Zulassungs-Ordnung.*

of the extension process and the then-Hessian Minister-president Georg-August Zinn. Also, additional statements, petitions as well as proposals and counterproposals to solve challenges of increasing air traffic and the planned extension have been published – e.g. Königsstein (1961). Provided by the media, further information on extending Frankfurt Airport triggered waves of protest (Jahn 1984: 319; Johnsen 1996: 27; Rucht and Sauer 1984: 212).

During further years, additional insights regarding the operations of propeller turbine aircraft and jet aircraft have been included in the planning. This understanding showed that a parallel runway would not be feasible due to the noise emission of jet aircraft. With this additional information, the official application for approval was prepared (Arndt 1975: 328; HMWV 1966: 2f).

On December 28, 1965, the FAG officially requested the extension and the construction of Runway 18-West. The plans included the extension of the parallel runway system to the west and the construction of the runway to the south (Arndt 1975: 327). Already before – triggered by plans to extend the airport – the first local citizens' group *Interessengemeinschaft zur Bekämpfung des Fluglärm*, dealing with the airport extension, was founded by Kurt Oeser and further citizens from the municipalities of Egelsbach, Groß-Gerau, Kelsterbach, Walldorf, and Mörfelden, on April 27, 1965. Based on the results of further studies[55] the Hessian State Government approved the extension subject to the defined requirements on May 10, 1966.[56] The same day, the then-Minister for Economy and Traffic informed the affected cities and municipalities during the cabinet meeting about the extension (Arndt 1975: 327).

In fall 1967, members of the local citizens' group *Interessengemeinschaft zur Bekämpfung des Fluglärm* initiated the nationwide *Bundesvereinigung gegen Fluglärm*. It became the core of the *Rhein-Main-Aktion gegen Umweltzerstörung* in 1970 (Jahn 1984: 27).

The RSE for the Runway 18-West extension was initiated to combine the official proceedings with the full participation of the citizens affected. Around 4,000 objections were made in the official proceeding, particularly

55 See e.g. DIL (1965), DW (1965), HMWV (1966), and Rössger (1965).
56 According to §6 LuftVG.

concerning the effects of aircraft noise on neighboring communities and damage to forests. Objections raised were discussed together with the comments of the authorities responsible for the procedure in four meetings in fall 1967. Based on the further insights, on March 26, 1968, the then-responsible Hessian Ministry of Economics and Transport Rudi Arndt, issued the first AoP. Municipalities and private individuals filed 42 lawsuits.

Two major issues led to the AoP's rejection. The proceedings failed to include the city of Offenbach am Main in the procedure and the order issued by Rudi Arndt was not published according to the official standards (HLGL ny; Johnsen 1996: 28).[57] Jahn (1984) describes that phase as the "juridical phase" of the extension (Jahn 1984: 319). Characterized by litigations in that period, the FAG faded into the background and the Hessian State Government became the major supporter of the airport extension (Jahn 1984: 319).

During the further official proceeding, the responsible RP Darmstadt requested for objections.[58] The publication of the planning documents was announced in the then-standard media. The planning documents were publicly displayed between June 1, and July 22, 1970, in all affected cities and municipalities (HMWT 1971: 754).

The official proceedings of the RSE Runway 18-West included only non-private administrative institutions (HMWT 1971: 754). With 40 institutions, the majority came from the sub-federal state level. No non-private metropolitan institution was founded yet (Fig. 4.10).[59]

Six of the involved institutions did not object against the extension plans, five institutions gave input, which had to be handled outside the official proceedings. Overall, 8,900 objections have been submitted by the involved institutions and the general public (HMWT (1971): 754).

During the five hearings in September 1970, the objections have been discussed. Around 400 supporters and objectors attended the hearings. On March 23, 1971, the second AoP – identical to the previous version of March 26, 1968 – was published by the *Hessisches Ministerium für*

57 File number P.St. 569, *Staatsgerichtshof des Landes Hessen.*
58 According to §10 II LuftVG.
59 See Annex A.3 for more detail.

| # | Institutional form | | Institution | Delegates |
|---|---|---|---|---|
| | Non-private | | | |
| | | Administrative | ------------------------------------------------------------ | 49 |
| 1 | | | Federal Republic of Germany | 5 |
| 2 | | | Federal State of Hesse | 4 |
| 3 | | | Sub-federal state level | 40 |
| | | Metropolitan | ------------------------------------------------------------ | 0 |
| | | | - | |
| | Private | | | |
| | | Metropolitan | ------------------------------------------------------------ | 0 |
| | | | - | |
| | | Economic | ------------------------------------------------------------ | 0 |
| | | | - | |
| | Further | | ------------------------------------------------------------ | 0 |

*Figure 4.10: Institutional aggregation of RSE Runway 18-West extension, own depiction, based on HMWT (1971: 754).*

*Wirtschaft und Technik* (HMWT) (HMWT 1971). However, due to the objections could not be eliminated. As a result, the AoP was refused. The HessVGH annulled the AoP. The major argument was again the improper publication.

In the following years, further expert opinions and counter-opinions[60] and several lawsuits were brought to the HessVGH and *Bundesverwaltungsgericht* (BVerwG). In March 1974, the BVerwG objected to the decision of the HessVGH. As a result of that decision, the HessVGH decided in favor of the extension of the existing runways, with obligations of a night-flight ban. Still, the construction of Runway 18-West was not allowed. Regarding that, the FAG and the State of Hesse approached the BVerwG, which declare that an appeal against that judgment is currently admissible.

Until 1976 no Hessian administrative court decided in favor of the extension. During that time, among other institutions, the *Deutsche Angestellten-Gewerkschaft* (DAG) criticized these decisions and argued for the extension (DAG 1976). On July 7, 1978, the BVerwG reversed the

60 See e.g. DW (1965) and Eschner (1971).

decision of the Hessian State Court from September 10, 1976, and restored the immediate enforceability of the runway relocation and expansion.[61] The court decided that the authority of the HMWT includes the order of an AoP and that there are no legal concerns regarding jurisdiction (HLGL ny; Johnsen 1996: 28, 52). Further, the court confirmed the extension of the existing runways. For the construction of Runway 18-West, additional expert evidence was requested. It was the turn of regarding the decisions made before (Johnsen 1996: 52). With the decisions of the BVerwG and the HessVGH, the FAG started the constructions in March 1979 for the extension of the existing runways (Jahn 1984: 321; Johnsen 1996: 94).

After BVerwG gave back the case to the HessVGH, the court reviewed the circumstances of the RSE of Runway 18-West again. Further expert opinions were requested and recognized in the proceedings, e.g. Dornier (1980), Hooton (1980), and DW (1980).[62] Already before the final decision of the courts, the then-Hessian Economic Minister Heinz-Herbert Karry decided the immediate execution of the AoP of 1971 on July 22, 1980 (Johnsen 1996: 200).

Then, on October 21, 1980, based on the further legal proceedings[63] and requesting additional expert evidence, the HessVGH confirmed the validity of the official proceedings of the RSE and allowed the construction of Runway 18-West (HLGL ny; Johnsen 1996: 28, 52). The decision of the HessVGH abolished the last lawsuit of the city of Mörfelden-Walldorf.[64] It confirmed the AoP of March 23, 1971. Any official appeal was no longer admitted. Opposers of the planned extension reacted with a constitutional complaint, however, without success. Initial demonstrations against the extensions followed right after. Furthermore, the extension was continually discussed in the public and media, e.g. HR1 (1980).

On February 2–5, 1981 an additional hearing in the Hessian Parliament was conducted. On the one hand, beside the already included administrative institutions, further nonadministrative stakeholders and experts were

---

61 Decision 4 C 79.76.

62 See also e.g. ATSC (1980), DLH (1980), Lahmann and Prescher (1979), RB (1980), RB (1979), and Mock and Gonsowski (1981).

63 See decision II R 75/80.

64 Decision II OE 205/78 HessVGH.

invited to discuss the airport extension.[65] On the other hand, evaluators suggested by the affected municipalities were not invited (KAF 1981). The general public was included with the invitation of local citizens' groups, whose members requested the presentation of their catalog of questions (Bi-Plenum 1981). Former individual leaders of the discussion, like Kurt Oeser, were also requested to participate in the hearing (Oeser 1981). In addition, other private individuals independently requested their participation in the hearings.[66] It is also argued that the hearings were just initiated to unite the then-ruling SPD again and that it was more a farce in which only well-known things were repeated (Geis 2005: 107; Ladeur 1983: 111).

Besides the juridical achievement of the FAG and the support of DLH, the federal government, the state, and the city of Frankfurt am Main, and further institutions such as the DAG,[67] more regional institutions such as the *Deutscher Gewerkschaftsbund* (DGB), but with conditions (DGB 1981), the *Hessischer Handwerkstag* (Handwerkstag 1981), the *Industrie und Handelskammer* (IHK) (IHK 1981), and the *Vereinigung Cockpit e.V.* (Cockpit 1981) officially supported the extension plans from the beginning of the year 1980 (Johnsen 1996: 78f; Rucht and Sauer 1984: 209–214). Amongst others, the *Bundesanstalt für Flugsicherung* (BfF 1981) and the *Dekanat Rüsselsheim* (Rüssesheim 1981) criticized and opposed the plans.

In the same year, the local citizens' group against the airport extension in Mörfelden-Waldorf was founded. Also, the first major demonstration on the premises of Frankfurt Airport was organized.

Until 1982, 25 local citizens' groups against the extension were founded, e.g. in Flörsheim, Frankfurt am Main, Groß-Gerau, Kelsterbach, Offenbach am Main, Nauheim, Rüsselsheim, and Wiesbaden (Johnsen 1996). The demonstrations and actions against the extension were continued and intensified after the decision of the HessVGH. Jahn (1984)

65 E.g. the Board of Airline Representatives in Germany (BARIG) (BARIG 1981), the *Bundesanstalt für Flugsicherung* (BfF 1981), and the DLH (DLH 1981) as well as fight plan coordinator Hein Gering (Gering 1981) and the evaluator Ted Hooton (Hooton 1981).

66 See e.g. Schmidt-Schaun (1981).

67 The DAG repeated its support for the Runway 18-West during the years (DAG 1979); DAG 1980.

describes this phase as the "political mobilization phase" (Jahn 1984: 321). A "culture of 'lived everyday resistance' was created" (HLGL ny). Not only the "slobs" demonstrated (Wilk quoted by SPIEGEL (2008)). Suddenly ordinary citizens engaged in the protest (FAZ 2010; FR 2004d). People that before collecting signatures, started to actively engage in the demonstrations on-site (FR 2011). "It was the uprising of the long- and the grey-haired" (Wilk quoted by SPIEGEL (2008)).[68] Objectors made use of conventional and unconventional instruments to prevent the extension. Actions varied from the commissioning and publishing of expertise on environmental matters thought occupations of the forest. Protestors constructed the *BI-Hütte*, an information booth built in May 1980 by local citizens' groups, and initiated the "Freie Universität Startbahn West or Walduniversität Mörfelden-Walldorf" (Jahn 1984: 321). Finally, a referendum to stop the extension was started.

The protestors were supported by the church and parts of the Hessian SPD as well as local media and authors (Jahn 1984; Johnsen 1996). The momentum of this resistance – supported by ordinary citizens – got the then-Minister-president Holger Börner (SPD) and its ruling party in the State of Hesse, the SPD into trouble. On the one hand, supporters of the extension like Holger Börner himself argued that the extension and the Runway 18-West are "economically unavoidable" (Börner quoted by SWR (2013)).[69] On the other hand, opposers like Frank Breucker (SPD), then member of the Hessian State Parliament argued that "[t]he runway is no longer politically feasible" (Breuker quoted by SPIEGEL 2008; Johnsen 1996).[70] To counteract the fragmentation of the state government and the SPD, Holger Börner had SPD chairman Willy Brandt flown in "to bring his faction back into line" (SPIEGEL 2008).[71] This phase ended with the decision of the HessVGH (Jahn 1984: 321).

The following phase Jahn (1984) characterizes as the "action phase" (Jahn 1984: 321). It was dominated by the occupation and clearance of the protest areas. Peaceful actions were now accompanied by civil

68 Translated by the author.
69 Translated by the author.
70 Translated by the author.
71 Translated by the author.

disobedience, partly culminating in attacks on tangible objects, targeted violence against police forces and riots as well as police violence against squatters. Besides the decision of the HessVGH[72] and the immediate clearance of woodland on October 21, 1980, this period is further characterized by the first occupation of land on October 28, 1980, and its partly violent disbandment (Jahn 1984: 321f). One year later, starting with the disbandment of a seven-hectare area and its fencing on October 6, 1981, the first serious clashes and protests took place (FR 2004c; Jahn 1984: 323). On November 2, 1981, the so-called *Hüttendorf*, the then-center of the organized protests, which evolved around the *BI-Hütte* on the area of the planned Runway 18-West, was demolished and the building land was fenced. The highlight of the peaceful actions was the demonstration of November 14, 1981, in Wiesbaden with 120,000 activists and the handover of a petition for a referendum to the State Election Commissioner, counting over 220,000 supporters (Jahn 1984: 323; Johnsen 1996: 86; SPIEGEL 2008). However, the petition, which was also rejected later by the *Hessischer Staatsgerichtshof* on January 15, 1982, did not change the mind of the then-Minister-president Holger Börner (SPIEGEL 2008). The "inspection" of the airport terminal and the area of Frankfurt Airport of approximately 12,000 activities the next day, on November 15, 1981, led to blocking of the road traffic around Frankfurt Airport, riots, and a major police operation with several hundreds of police forces (FR 2004c; Jahn 1984: 323; SWR 2013). According to Jahn (1984) the failed "inspection," the major violence during the protest, and the rejection of the petition for a referendum against the construction of Runway 18-West "broke the [political-moral] neck" of the protest movement (Jahn 1984: 323; Johnsen 1996: 86).[73] Local citizens' groups faced internal disputes initiating their collapse, but not the reappraisal of the violent riots. The political core of the local citizens' groups vaporized. Within the protest movement, environmental organizations, activists criticizing the general political system, and violent fringe groups came into fore (HLGL ny; Jahn 1984: 321–324). How

72 The AoP was finally confirmed by the court decision of the HessVGH with the decision II OE 205/78 on October 21, 1980.

73 Translated by the author.

tense the situation was is not only shown by actual riots. Also, comments by politicians like Holger Börner show the confrontational atmosphere during the "action phase."

> "I regret that my high state office forbids me to punch the guys myself in the face. Back in the day, in construction, you did sort these things with a roof bar." (Börner quoted by Bunte (1982); see also SPIEGEL 1982)[74]

During the following period, the so-called "palamental phase", further militant actions took place (Jahn 1984: 324). However, that period was also characterized by several lawsuits against activists, e.g. the case filed against Frankfurt's magistrate Alexander Schubart (Jahn 1984: 324; taz 1991; Zeit 1982). In February 1982, the conflicts and discussions moved back to the political arena. Nevertheless, due to the dispute about the violence and divergent opinions within the bordering cities of Frankfurt Airport, the political actions of objectors were characterized by resignation. Opposers of the extension "have reluctantly resigned themselves to the verdicts of the highest courts and the steadfastness of the state government and parliament" (Jahn 1984: 325).[75] Above all, the alternative perspectives regarding the extension, the operation of Runway 18-West, or reforestation had not met with any significant response. The communal consensus of the political parties broke. The further resistance was now only promoted by the *Bündnis90/Die Grünen* and smaller left-wing fringe group (Jahn 1984: 325). On April 12, 1984, the Runway 18-West was opened and set into operation. Due to the previous protest and the extensive conflicts, the opening was not accompanied by bigger celebrations (FR 2004d; Johnsen 1996: 204, 213).

After the opening of the runway, the protests continued, e.g. with the so-called *Sonntagsspaziergängen.* A broad spectrum of opponents participated in these walks around the airport, but the previously strongly represented mass organizations, like unions, environmental protection organizations, parties, or churches were not among them (FR 2004c; HLGL ny). Protesters continued trying to come onto the premises of Frankfurt Airport to impede the construction (FR 2004c).

---

74 Translated by the author.
75 Translated by the author.

The larger, organized protests ended abruptly on November 2, 1987. It marks the sixth anniversary of the clearing of the *Hüttendorf*. During the demonstrations on November 2, 1987, the two police officers Klaus Eichhöfer and Thorsten Schwalm where shoot death (FR 2004c; RP 2004; SPIEGEL 2008).[76] From that day on the resistance shifted to individual actions (SPIEGEL 2008). Rudi Hechler, a protester of the first hour, said: "At that moment I knew that the runway movement was finally over." (FR 2004d). From now on, the continuous protest of the *Sonntagsspaziergänge* were limited to the old activists – the "Runway veterans" (FR 2011).[77]

#### *4.3.3.2 Interim findings: airport extension, the 1960s–1980s*

In the following, the airport extension in the 1960s–1980s is analyzed by the theoretical derived concept of ERG. The latent content of the five aspects of ERG is reconstructed in the regulatory governance structure and decision-making process of the airport extension during the 1960s–1980s. This analysis is followed by establishing the missing link between the structural perspective of political science and the procedural character of policy-making, which Mayntz already proclaimed in the 1980s (Mayntz 1983), by discussing the institutional arrangement of that period.

*Discussion of shared perceptions of joint challenges*

The airport extension of the 1960s–1980s was characterized by the formulated demand of the FAG and the State of Hesse to extend the airport based on economic reasons and the official order of the German Government to make Frankfurt Airport to the first German airport meeting the requirements of intercontinental scheduled jet aircraft traffic, less by shared perceptions of joint challenges regarding the extension. On the one hand, several individual expert opinions were presented to justify the extension plans or to encounter it in order to present the

---

76 If the two police officers were the first victims of the extension process is unclear. Unknown persons shot the then-Hessian Economic Minister Heinz-Herbert Karry in his home. Due to the perpetrators were never caught, it is still unclear whether the crime was connected with the conflict over the Runway 18-West (Geis 2005: 107; RP 2004).

77 Translated by the author.

best available techniques. Still, these individual expert opinions were not just limited to facts and figures as well as best practices to promote a common understanding of the challenges. Rather, the assessments also included opinion-forming statements alleging non-academic approaches (Eschner 1971) or allegation of falsification and manipulation of data, leading to criminal complaints (Haselberg 1981). Also, the actions of the Hessian State Government were more characterized by order than by discussion and heterarchical exchange. This holds especially true regarding the escalating violence between protesters and police forces during the "action phase" and "parlamental phase," ending with the death of the two police officers Klaus Eichhöfer and Thorsten Schwalm on November 2, 1987. The institutions already in place focused on their initial agenda to extend the airport, rather supporting the recursive policy-making and learning from the experience of lower levels. On the other hand, there have been additional hearings on February 2–5, 1981, in the Hessian Parliament. However, since the discussions were downstream of the official proceedings, these hearings can only be interpreted as an initial structure to discuss shared perceptions of joint challenges. Upfront, neither the elements of identification of the topic and best available techniques for a joint advancement identifiable, nor the establishment of heterarchical relations beyond traditional border such as parliamentary session and the proactive delegation of the mandate of certain topics and the increase of ownership of different levels. Rather, local citizens' groups and individual leaders of the protest like Kurt Oeser, had to force their integration. In addition, interested individual citizens excluded from the official proceeding had to request their participation individually, e.g. Schmidt-Schaun (1981). It is not clearly identifiable that tacit knowledge and the lower-level experience was integrated into the discussions to foster the objectivity of the negotiations. Given that the additional hearings in February 1981 were downstream, they do not represent capacities to discuss joint challenges and review the goals as well as the measurement systems of the planned project. The hearings failed to create an open, transparent multi-actor discussion within a heterarchical environment. In addition, insights and information were hardly shared, exchanged through the levels, and pooled during the airport extension of the 1960s–1980s.

*Definition of goals and measurement systems*

High-level goals were set by the FAG and the Hessian State Government, mainly based on the expected increase of in air traffic volume affecting the operating procedures of the airport, DHL's decision-making Frankfurt Airport to its home airport, and the official order from the German Government to make Frankfurt Airport to the first German airport meeting the requirements of international scheduled jet air traffic. Based on topographical and climatological characteristics, six different layouts of a future runway system at Frankfurt Airport have been examined. However, these examinations and the programming of a measurement system were initially established by the FAG and the Hessian State Government. With that, it becomes evident, that the definition of goals and the measurement system during the decision-making process was rather based on political will than on joint high-level goals and common objects. Only lawsuits against the AoP – especially the court orders of the BVerwG and HessVGH – made it possible to request further expert opinions and consider these insights. Even the AoP and the planning documents can be interpreted as a practical and workable implementation plan and feasible regulations, the responsible institutions failed to discuss the results as a proposal with the addressees. The additional hearings in February 1981 could not heal that circumstance, because they were downstream of the publication of the second AoP on March 23, 1971, and the final confirmation by the HessVGH on October 21, 1980. With that, the AoP and the planning documents became the reference document with legally (non-)binding rules. In regard to that, an environment to define goals and measurement systems for the decision-making process was hardly established. Besides the AoP and court decisions, no reference documents or recommendations to elaborate legally (non-)binding, were provided to clarify and discuss the approach. Further, results of expert reports commissioned by the government and the FAG were, without broader discussion, directly integrated into the AoP.

*Implementation*

The implementation of measures taken to extend Frankfurt Airport was characterized by order, either by political control or court decision. Its operationalization was hardly transferred to lower levels. With that, lower-level units were not fully entrusted. Responsibilities and power were barley delegated to lower-level authorities or experts. Exemptions were the execution of the extension and construction of the Runway 18-West by the FAG as the airport operator and the municipalities to adapt spatial planning. However, the freedom of implementation only concerned the adoption and fulfillment of the goals, not the modeling of the goals. Tacit knowledge was hardly integrated. With that, it is not evident that the goals and the measurement systems set at the beginning of the official proceedings were adjusted regarding tacit knowledge and subsequent experience. That holds also true for the adjustment of existing rules based on tacit knowledge and subsequent experience. The overall framework was forced into being, rather than fulfilled by soft guidance. Further cooperation between the FAG and the Hessian State Government with the affected municipalities and the general public as addressees was hardly established. No imperial structure or institution was established. The hearings in February 1981 can be seen as the only effort to foster cooperation and to create such an institution. Still, these hearings were downstream. With that, lower-level experience, presented during the hearings in February 1981 and other tacit knowledge presented outside court hearings was barely taken into account. This is especially reflected in the publication of the second AoP. The AoP presented on March 23, 1971, was identical to the previous version of March 26, 1968. Its final confirmation by the HessVGH on October 21, 1980, integrated only minor adaptations, necessary for formal accuracy, e.g. the previously missed integration of the city of Offenbach am Main.

*Report and review*

During the decision-making process of the RSE, joint monitoring and evaluations as well as peer reviews and reporting hardly took place. It was mainly characterized by the possibility of reviewing the planning documents and the possibility to formally object. During the hearings in September 1970 objections have been discussed. However, these hearings were less an exchange on the issued AoP. It rather was a discussion and counteract regarding the objections made. The following decision-making process was more like a cut and thrust of politics or juridical clashes between the promoters and objectors of the extension. A dialog was hardly initiated, not to mention a constitutional and collective regulatory dialog. From the objector's side, there have been attempts, mainly characterized by the efforts of Kurt Oeser. From the proponent's side, the only effort in that regard has been the initiation of the additional hearings of February 1981, downstream the official decision-making process. However, an environment of continuous adjustment and learning was not introduced and institutionalized. With that, taken activities and performance had mainly to be justified during court hearings. Periodical assessments, audits, or peer reviews were not jointly enforced. On the one side, tacit knowledge and information were often distributed upwards by respective lawsuits. On the other side, official information downwards was mostly announced by political order or court decisions. The consultation of additional opinions or any relevant scientific issues was only initiated on a court decision. Additional expert opinions of evaluators suggested by the affected municipalities were not integrated into the hearings in February 1981.

*Review of the measurement system, the decision-making process, and the stakeholder*

However the review of the measurement system, the decision-making process, and the stakeholder was enforced by the abolishment of the first AoP, the presented research shows that this review was less characterized by reassessment, questioning, and cross-checking of the intended goals and measurement system. A recursive revision of policy objectives, procedures of political programming, legislation, and best practices was not initiated. This becomes evident mainly by two events. First, during the publication of the second AoP the same mistakes regarding the publication procedures have been made. These procedures had to be assessed and decided by court decision again. Second, which is the main aspect, is the publication of the second AoP, published on March 23, 1971, which was identical to the previous version of March 26, 1968. New developments and innovations were only taken into account regarding the operations of propeller turbine aircraft and jet aircraft, mainly regarding the technical needs and feasibility. Further stakeholders were formally only brought into the overall regulatory structure of governance and decision-making process by court decision and formal needs, e.g. the city of Offenbach am Main, which was missed to integrate into the first AoP.

### Institutional arrangement of the RSE Runway 18-West

Besides the sketch of the decision-making process, the previous paragraph also drew a clear picture of the institutional arrangement of the airport extension from the 1960s to the 1980s. The institutional arrangement was not characterized by joint heterarchical relations between private and non-private institutions as well as private individuals. The research shows that the official proceedings of the RSE Runway 18-West included only non-private administrative institutions. No non-private metropolitan institution was founded yet.

By requesting the participation of further nonadministrative stakeholders like Board of Airline Representatives in Germany (BARIG), the *Bundesanstalt für Flugsicherung*, DLH, the fight plan coordinator Hein Gering and the evaluator Ted Hooton as well as the general public by inviting local citizens' groups and individual leaders of the discussion, like Kurt Oeser, further stakeholders have been integrated. However, the interaction was downstream the official decision-making process and after confirmation of the validity of the official proceedings and approval of the construction of Runway 18-West by the HessVGH.

In front of that background, only the additional hearing after the official decision-making process included different levels of the regulatory framework of governance in the geographic-functional region of FRM.

### Confirmation of the research-guiding hypothesis for the airport extension, the 1960s–1980s

This interim conclusion shows that the theoretical derived concept of ERG is a suitable analytical framework for analyzing the regulatory structure of governance, the decision-making processes, and the institutional arrangements in the case of the airport extension from the 1960s to the 1980s and the RSE Runway 18-West. With that, the research-guiding hypothesis can be confirmed. It also shows that the concept of ERG counterfeits the challenge of a missing link between the structural perspective of political science and the procedural character of policy-making, which Mayntz already proclaimed in the 1980s (Mayntz 1983), by including the analysis of the institutional arrangement of that period.

### 4.3.4 Mediation process, since 1997

This paragraph initially introduces the general idea of mediation in decision-making processes. Then, the research follows the general approach as presented (Ch. 4).

In the globalized world of today, conflicts are linked beyond traditional borders and levels. Economic, political, environmental, and social challenges are dependent. Especially the governance of major projects, like the construction or extension of an international airport, are "games" between the levels, where different opinions, goals, and approaches have to be discussed and mediated (Gohl and Meister 2012: 1).

Coming mainly from the field of jurisprudence, mediation in German court ruling can be traced back to the "Civilprocessordnung" as of January 30, 1877.[78] Solving disputes between opposing parties by amicable settlements – as an alternative dispute resolution – was the goal (Haft 2009: 70; Strempel 2002: 108). Today, mediation is applied to many fields of law, e.g. family law, industrial law, and economic law (Haft and Schlieffen 2009). From these legal-procedural perspectives, the focus is to provide assistance with the negotiations by an imperial supporter (Haft 2009: 70).

This definition also holds true for a political science perspective (Gohl and Meister 2012: 3f). Here, mediation is characterized as a form of cooperative democracy (Geis 2002: 269). Mediation as an administrative approach is particularly utilized in situations where common political proceedings such as steering fail. This is especially the case where overall complexity is very high. The goal is to foster administrative efficiency (Troost 2001: 248f). Today, following Geißel, mediation can be described as a deliberative, dialog-oriented approach of decision-making (Geißel 2013: 11f; Geißel et al. 2014a: 17f).

The interest in mediation – as an experimental form of policy-making – was especially enforced by the mediation process of the extension of Frankfurt Airport. Although the interest of politics and policy-making are still stepping behind the academic interest, the "Frankfurter Verfahren"

78 §§268 and 471.

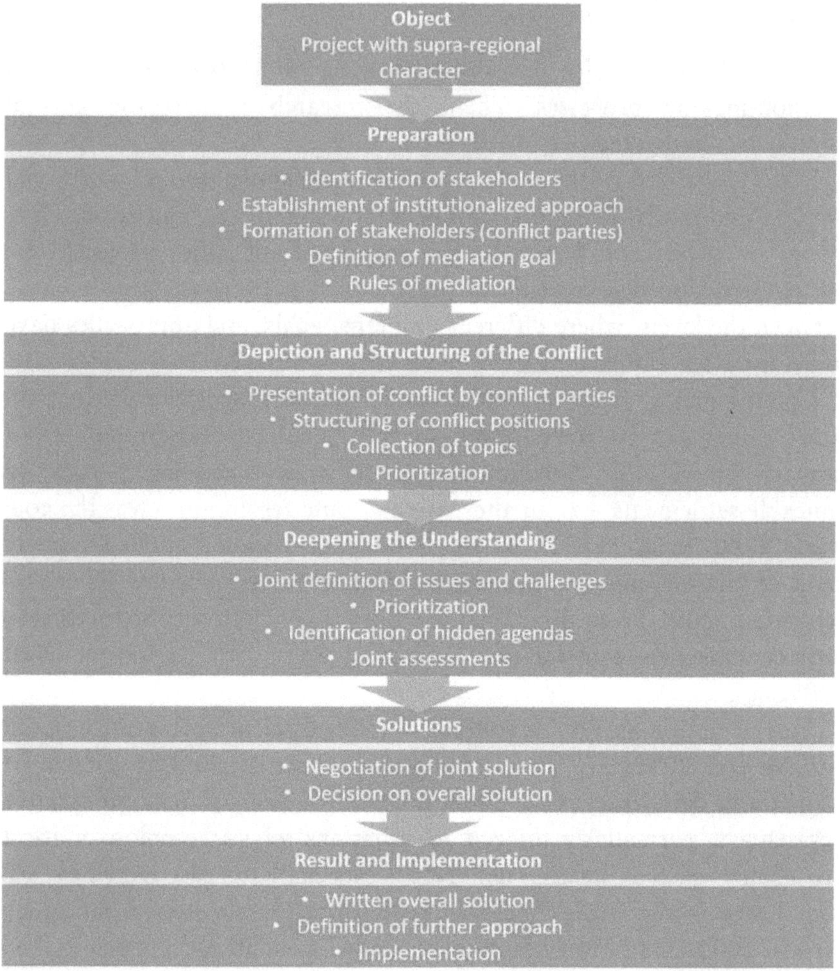


*Figure 4.11: Scheme of mediation process, own depiction, based on Gohl and Meister (2012: 31f).*

(Troost 2001: 252) received more attention than any other meditation process in Germany.

Internationally the mediation process at Seattle-Tacoma International Airport and the signing of an agreement of noise reduction went ahead of the extension and mediation process of Airport Frankfurt. In the 1970s, residents of the surrounding area of Seattle-Tacoma Airport filed lawsuits against the extension. Due to the extensions, affected stakeholders feared increased noise and smoke pollution, vibrations, and other challenges. The challenges were addressed in the official proceedings and the spatial planning, finally adopted in the Sea-Tac Communities Plan of 1976. In 1979 the Congress initiated a noise-compatibility program.[79] Based on that, in 1985, Seattle-Tacoma Airport became one of the first airports in the US to conduct a so-called Part 150 study (PoS 2013b: 31). The study is voluntary planning to identify current and future noise impacts, defined by the Federal Aviation Administration and in line with the Aviation Safety and Noise Abatement Act of 1979 (PoS 2013b: 5).[80]. The last Part 150 study update was completed in 2014 (PoS 2013b). It addressed 15 measures to continue, such as the exploration to limit the rescheduling of nighttime flights, expansion of the Fly Quiet Program, insulation of schools or property advisory services, four new measures such as minimize run-up noise and further sound insulation, and three program management measures (PoS 2013a).[81]

#### *4.3.4.1 Gesprächskreis Flughafen, 1997–1998*

After the first official discussions of Fraport AG, DLH, and BARIG to extend Frankfurt Airport, the frontlines between the proponents and the opponents of the extension were reflected immediately. Above all, the Fraport AG and the airport companies supported the extension. The regional association of the CDU and the FDP also argued in favor of

79 H.R.3547 – 96th Congress (1979–1980).

80 Order 1050.11A, issued on January 13, 1986, gives the Federal Aviation Administration "[...] responsibilities in relation to the review of airport noise control plans and programs [...] " (Order 1050.11A: 1.) The Aviation Safety and Noise Abatement Act was issued to "[...]provide and carry out noise compatibility programs, to improve assistance to assure continued safety in aviation and for other purposes.", Public Law 96–193 (PoS 2013b: 1).

81 Until today, with the last update of 2014, the Seattle-Tacoma Airport spent approximately 400 million USD for over 9,400 single-family homes to be sound insulated, five condominium complexes to be sound insulated with 246 individual units, 22 schools and colleges, and around 360 mobile home units were acquired and residents relocated (PoS 2013c).

the extension. Opponents were environmental associations, local citizens' groups, and the affected municipalities, as well as the *BÜNDNIS 90/DIE GRÜNEN*. The SPD of Hesse was torn (Becker-Ritterspach 2015; Busch 2000: 17–22).

Based on this mixture, the then-Minister-president Hans Eichel initiated the *Gesprächskreis Flughafen* to jointly discuss the extension plans in civil society (Gohl and Meister 2012: 36; Hochfeld et al. 2000: 11). The idea was to counterfeit the Sword of Damocles of the riots regarding Runway 18-West from the beginning (Busch 2000: 22). From a constructive perspective, this political dilemma leads to a change in the governance approach and the implementation of the mediation process (Koch 2013: 401–405). This political situation in the Hessian State Government, the avoidance strategy to actively discuss the extension in the political arena of party politics and the experiences from the planning process of Berlin-Brandenburg Airport represent the background of the initiation of the *Gesprächskreis Flughafen* (Geis 2005: 130–135).

The 16 members of the *Gesprächskreis Flughafen* came from different institutional backgrounds (Annex A.4).

Members of the *Gesprächskreis Flughafen* were mainly non-private administrative institutions. Fraport AG – as the operating company of Frankfurt Airport – was the only private economic institution. The UVF –as the then-metropolitan governance body – was not involved. The *ex ante* identified private metropolitan logistics actors were not yet established. Accordingly, no metropolitan governance institution was involved. Only further institutions participated, albeit neither with a focus on metropolitan governance nor an affinity to regional logistics and transport (Fig. 4.12).[82]

The major goal of the *Gesprächskreis Flughafen* was to design and establish an objective and imperial institution, to foster an open, joint discussion (Busch 2000: 22f). After gathering the challenges regarding the extension of Frankfurt Airport, the members of the *Gesprächskreis Flughafen* decided to initiate an open and unbiased mediation process,

82 See Annex A.4 for more detail.

| # | Institutional form | | Institution | Delegates |
|---|---|---|---|---|
| | **Non-private** | | | |
| | | **Administrative** | ---------- | 9 |
| 1 | | | Federal State of Hesse | 7 |
| 2 | | | Sub-federal state level | 2 |
| | | **Metropolitan** | ---------- | 0 |
| | | | - | |
| | **Private** | | | |
| | | **Metropolitan** | ---------- | 0 |
| | | | - | |
| | | **Economic** | ---------- | 1 |
| 3 | | | Fraport AG | 1 |
| | **Further** | | ---------- | 6 |
| 4 | | Church | Protestant | 1 |
| 5 | | | Catholics | 1 |
| 6 | | Civil Society | Author | 2 |
| 7 | | Union | Deutscher Gewerkschaftsbund (Hessen) | 1 |
| 8 | | Chambers | Chamber of Commerce and Industry | 1 |

*Figure 4.12: Institutional aggregation of Gesprächskreis Flughafen members, own depiction, based on Hochfeld et al. 2000: 146.*

which was comprehensively described in the mediation framework (GKF 1998).

### *4.3.4.2 Mediationsgruppe Frankfurt Flughafen, 1998–2000*

The mediation process was intended as an additional platform of participation, besides the formal proceeding. With the initiation of the *Mediationsgruppe Frankfurt Flughafen* (*Mediationsgruppe*), the mediation process officially started in May 1998. It was never assumed or considered that the mediation process would bind the administrative institutions or even replace the official proceedings (Hänsch, Niethammer, and Oeser 2000: 11). Moreover, the *Mediationsgruppe* represented an institutionalized platform for the collective discussion on how Frankfurt Airport could improve and support the competitiveness of FRM's economy without ignoring local effects like pollution. Together with the three basic topics of traffic, ecology and the economy, this was the framework and formal goal of the mediation process, initiated by the *Gesprächskreis*

| # | Institutional form | | Institution | Delegates |
|---|---|---|---|---|
| | **Non-private** | | | |
| | | **Administrative** | ---------------------------------------------------------------- | **12** |
| 1 | | | Federal Republic of Germany | 1 |
| 2 | | | Federal State of Hesse | 2 |
| 3 | | | Sub-federal state level | 9 |
| | | **Metropolitan** | ---------------------------------------------------------------- | **0** |
| | | | - | |
| | **Private** | | | |
| | | **Metropolitan** | ---------------------------------------------------------------- | **0** |
| | | | - | |
| | | **Economic** | ---------------------------------------------------------------- | **3** |
| 4 | | | Fraport AG | 1 |
| 5 | | | DFS Deutsche Flugsicherung GmbH | 1 |
| 6 | | | DLH | 1 |
| | **Further** | | ---------------------------------------------------------------- | **6** |
| 7 | | Association | Vereinigung der hessischen Unternehmerverbände | 1 |
| 8 | | | BARIG Board of Airline Representatives in Germany e.V. | 1 |
| 9 | | | Offenbacher Vereinigung gegen den Fluglärm e.V. | 1 |
| 10 | | Union | Deutsche Angestellten-Gewerkschaft (Hesse) | 1 |
| 11 | | | ötv Hesse | 1 |
| 12 | | Chambers | Chamber of Commerce and Industry (Frankfurt) | 1 |

*Figure 4.13: Institutional aggregation of the Mediationsgruppe members, own depiction, based on Hochfeld et al. 2000: 147.*

*Flughafen*. However, the implementation and governance of the mediation process were not determined in detail. Accordingly, the *Mediationsgruppe* had the most freedom to do so in a self-organized way. In the following, it developed its own rules of procedure (RoP) (Mediationsgruppe 2000c). Additionally, the *Mediationsgruppe* and the mediators were independent of the directions, and since their first constitutional meeting, they were also the official custodian of the mediation process (Geis 2005: 141, 149). Again, involved members came from private and non-private institutions (Annex A.5).

The majority of the members were non-private administrative institutions. Besides, several private economic institutions were formally integrated into the mediation process. The then-metropolitan governance body UVF was not a member of the *Mediationsgruppe* and thus it was not actively involved in the mediation process. The private metropolitan logistics actors were not yet established. Further member institutions were represented by associations, unions, and chambers (Fig. 4.13).[83]

However, some institutions, such as local citizens' groups and environmental associations refused their participation willfully from the beginning or after the official beginning of the mediation process.[84] Nonetheless, the Hessian State Government tried to integrate their positions both formally and informally (Geis 2005: 149–151). Chaired by three mediators[85] the *Mediationsgruppe* started its work on July 16, 1998.

The major task of the *Mediationsgruppe* was to assess the economic, ecological, and social relations between Frankfurt Airport and FRM.[86] Additionally, the future impacts of the development and extension of Frankfurt Airport on metropolitan FRM should be analyzed and discussed. Regular interim reports of the status quo were dispatched to the Hessian State Government (Geis 2005: 140). Therefore, the *Mediationsgruppe* jointly discussed economic, ecological, and social goals, which should represent the guardrails of the future extension of Frankfurt Airport. A measurement system to assess the major impact factors was established. Both the goals and the measurement system were adjusted by newly developed insights throughout the whole mediation process.

The members of the *Mediationsgruppe* were personally responsible to report and give feedback to their institutions (Mediationsgruppe 2000c). As a result, the mediators were asked to secure this feedback by peer-reviewing this process with specific and regular discussions. This was

83 See Annex A.5 for more detail.

84 For the detailed analysis of that phenomena see Geis (2005: 142–152).

85 Dr. Kurt Oeser, former commissioner of environmental issues, *Evangelische Kirche Deutschland*, Dr. Frank Niethammer, President of the Chamber of Industry and Commerce (1991–2000), and Dr. Klaus Hänsch (Member of EU Parliament, 1979–2009). Usually, there is one neutral third party in a mediation process, but it was considered that due to the size of the mediation process, the short time frame, and the foreseeable workload to convene three mediators (Geis 2005: 139).

86 According to the preamble of the RoP (Mediationsgruppe 2000c).

especially necessary for more loose institutions such as local citizens' groups (Geis 2005: 137).

In the first meetings, it was recognized, that not all questions could be addressed by the *Mediationsgruppe* in detail. To discuss specific topics and enforce the goals, three task forces (*Arbeitskreise*) with the topics of "traffic," "ecology, health, and social issues," and "economy" were established. Most of the members of the task forces came from the institutions of the *Mediationsgruppe*. Nonetheless, all task forces were open to further stakeholders to join the mediation process officially. Henceforth, there was the possibility to integrate further private and non-private institutions, even those that have been unable to participate in the official mediation group.[87] Also, the work of the task forces was accompanied by reviews and the coordination of the results. Furthermore, they were guided by external assessments, hearings of experts, and statements to take better results and tacit knowledge into account (Gohl and Meister 2012: 35–39; Hochfeld et al. 2000: 11–23).

*Arbeitskreis Verkehr*: To discuss common challenges with the focus on traffic, the *Arbeitskreis Verkehr* started in December 1998. It was moderated by one mediator.[88] A further institution supported it with a scientific assistant.[89] During the discussions and the ongoing process, it became clear that for some topics deeper insights wee needed. Therefore, the task force established sub-task forces (air traffic control and navigation; external risk) to jointly discuss and work on topics and challenges specifically regarding these fields.

Overall, the task force met twelve times. In November 1999, it handed over the conducted results to the *Mediationsgruppe* (Hochfeld et al. 2000: 24f). With the finalization, fifteen assessments were executed and the

87 E.g., it was not intended by the *Gesprächskreis Flughafen* to integrate the cities of Offenbach am Main and Darmstadt into the mediation group. Nevertheless, delegates tried to attend the first constitutional meeting, but concerning the guideline of the Gesprächskreis Flughafen and work economically aspects the mediation group refused their request until the final refusal of the environmental associations to participate (Geis 2005: 153, 157). For other examples, see Geis (2005: 153–156).

88 Dr. Frank Niethammer.

89 *Öko-Institut e.V.*

results were pooled and made available electronically (Mediationsgruppe 2000a).

Members of the task force came from private and non-private institutions. Non-private administrative institutions represented the majority. No metropolitan governance body participated in that task force and its sub-task forces (Annex A.6).

*Arbeitskreis Ökologie, Gesundheit und Soziales*: Started as the *Arbeitskreis Lärm*, the discussions were moderated by one mediator[90] and began in November 1998 focusing on aircraft noise. In March 1999, the field of the topic was enlarged by ecology, health, and social issues. Since then, the task force was named *Arbeitskeits Ökologie, Gesundheit und Soziales*, whereas aircraft noise was still the most important issue. To gather and utilize further tacit knowledge, two sub-task forces were initiated (housing and social issues; aircraft noise). Further knowledge and opinions were integrated by specific assessments, hearings, and meta-analysis of text-based material.

During the mediation process, the task force met thirteen times. In January 2000, it passed its results to the *Mediationsgruppe* (Hochfeld et al. 2000: 27f). Until the last meeting, 20 assessments were executed. The results were pooled and made available electronically (Mediationsgruppe 2000a).

Members of the task force came from private and non-private institutions. Again, non-private administrative institutions represented the majority. In the sub-task force, dealing with housing and social issues, the UVF – as the then-metropolitan governance body – was engaged by one representative (Annex A.6).

*Arbeitskreis Ökonomie*: Under the direction of one moderator[91] and with the assistance of two further institutions[92] the task force was launched in November 1998. It mainly collected further opinions of income and employment effects, by several external assessments and statements.

By the end of the 10th meeting in November 1999, the task force handed its results to the *Mediationsgruppe* (Hochfeld et al. 2000: 28f).

90 Dr. Kurt Oeser.

91 Dr. Frank Niethammer.

92 *Gesellschaft für Forschung Planung Entwicklung mbH* and *Öko-Institut e.V.*

Eight assessments were executed until the last session. The results were pooled and made available electronically (Mediationsgruppe 2000a).

As with the two other task forces, members came from private and non-private institutions, whereas here non-private administrative institutions also represented the majority. No metropolitan governance body participated in that task force and its sub-task forces (Annex A.6).

Additionally, the *Mediationsgruppe* itself worked on several topics. It benchmarked other European airports and airport regions with Frankfurt Airport and FRM to analyze the competitiveness based on best practices and learn from other European examples (Hochfeld et al. 2000: 51).

After the *Mediationsgruppe* received the first results from the task forces, the mediators started to formulate the final report. Starting in fall 1999, they presented interim results to the members of the *Mediationsgruppe* and discussed the findings collectively (Hochfeld et al. 2000: 138). Besides the preparation of the final report (Hänsch, Niethammer, and Oeser 2000),[93] the *Mediationsgruppe* suggested to continue the joint, open discussion by establishing a dialog forum for the future discussion of common issues and challenges (Hänsch, Niethammer, and Oeser 2000: 180; Hochfeld et al. 2000: 142f). The presentation and the further discussion of the results was accompanied by public demonstrations in the affected municipalities and political discussions (Geis 2005: 217; HessL 2000a: 2820–2853; HessL 2000b). Throughout the whole mediation process, several protests, debates of the Hessian State Parliament, political hearings, and the engagement of the economic institutions above the mediation process, contributed to the overall decision-making process.

In the process, three positions in the general public are identifiable (Busch 2000; RDF 2002a: 3). First, the Fraport AG, further air transport companies, the economy, and the parties of CDU and FDP can be named as supporters of the extension plans (Busch 2000: 17). As already emphasized in the 1960s to 1980s, the main argument was that the Frankfurt Airport is a "Jobmaschine" (CDU 1998) that has to be

93 In 1998 the *Gesprächskreis Flughafen* proposed a final report with the character of a "informed public opinion" as judgment support for the administration and as the indicated result of the mediation process (Geis 2005: 142, translated by the author).

promoted by the planned extension. All supporters of the extension dropped this argument during the whole mediation process, both private and non-private (BARIG 1997; FAG 1999; FDP 1998).

Second, the repudiators – as the second group – were mainly represented by environmental associations, local citizens' groups, the affected municipalities, and the *BÜNDNIS 90/DIE GÜNEN*. This group held on the argument of the RSE of the 1960s to 1980s. It stated that after the construction of the Runway 18-West no further runway would be planned or built in the future. Nevertheless, they were willing to discuss the enlargement of the capacities within the fence (Busch 2000: 18). Affected municipalities and environmental associations questioned the efficiency of the extension and confronted it with ecological and health aspects (BUND and NABU 2000; see footnote for further details).[94] The repudiators further argued that the extension of the airport has already reached its maximum capacity (BUND 1998; GRÜNE 1998). Besides, they always referred to the words of Holger Börner, who claimed that after the construction of Runway 18-West no further tree would be cut down.[95] They requested a sustainable measurement system, reliable goals and a maximum of pollution (BUND 1998).

Finally, the third group was located between the positions of the supporters and the repudiators. Namely, it was the SPD. On the one hand, the SPD was fixed to the non-extension, set by the coalition agreement with the *BÜNDNIS 90/DIE GRÜNEN* (SPD and GRÜNE 1995).[96] The then-Minister-president Hans Eichel also estimated that the extension was not enforceable (FAZ 1997). On the other hand, the SPD promoted the economic significance of the airport for the region (FAZ 1998; SPD 1998).

After the change of government in February 1999 and the redemption of the SPD and the *BÜNDNIS 90/DIE GRÜNEN*, the mediation process

94 Kelsterbach (1998a); Kelsterbach (1998b); NABU (1998); Neu-Isenburg (1997).

95 Holger Börner was Minister-president of the Federal State of Hesse from 1976 to 1987. In the context of Runway 18-West he said in 1981: "At the Frankfurt Airport, no tree will fall again."

96 "The area of the Rhine-Main airport is limited to the current area. An additional runway, and a shift of the parallel runway system and the use of Runway 18-West as a runway [(*Landebahn*)] is excluded" (SPD and GRÜNE 1995: 87, translated by the author).

was further supported and promoted by the new Hessian State Government (Geis 2005: 213–239; Hänsch, Niethammer, and Oeser 2000: 10).

With the final meeting of the *Mediationsgruppe* and the release of the final report, the public discussion continued seamlessly. Shortly after the presentation, the DLH questioned the joint results (DLH 2000). Environmental associations called the report a "brazen violation of neutrality" (BUND 2000a).[97] They demanded a supra-regional European-wide joint discussion on air traffic (BUND 2000b).

The final report was also an important topic for the Hessian State Parliament. It held a hearing from May 10–12, 2000. The results of the mediation process were presented by the experts and jointly discussed with the Hessian Parliament.[98] Here, the future joint discussion of challenges and dialog regarding the airport was issued several times by private and non-private institutions (HessL 2000c: 15, 21; HessL 2000h: 52–66). Moreover, the integration and anticipation of the created knowledge were highlighted. The joint implementation of the findings during the mediation process and the benchmark with best practices were named as a success (e.g. HessL 2000e: 13; HessL 2000h: 24, 30). Furthermore, it was highlighted that the mediation process as such was reviewed and its methodology was adapted (HessL 2000h: 60, 64). The then-Minister-president Roland Koch said: "It has always been clear that the package will be improved" (FR 2000).[99]

On June 5, 2000, there was a specific hearing of the local authorities, during which the challenges of Frankfurt Airport were discussed. In this discussion, the point was made that the issues regarding Frankfurt Airport had to be reviewed jointly from a geographic-functional perspective (HessL 2000b: 5).

#### 4.3.4.3 *Regionales Dialogforum, 2000–2008*

Parallel to the formal proceedings, the Hessian State Chancellery formulated the idea to establish a successor of the *Mediationsgruppe* (HessL

97 Translated by the author.
98 See protocols HessL (2000c), HessL (2000d), HessL (2000e), HessL (2000f), HessL (2000g), and HessL (2000h).
99 Translated by the author.

2000h: 56–59, 62f). In June 2000, the foundation of the RDF was decided by the Hessian Cabinet (HessL 2000a: 2843).[100] The initiators legitimated the establishment by the joint request of the *Mediationsgruppe* to create such an institution (Wörner 2010: 30).

As formulated in its Codex (RDF 2000), the RDF was designed as an intermediary institution to continue the topics and the approach of the former mediation process. Besides the continuation of the mediation process, the RDF was requested to take up the issues, that were leftover by the former processes (IFOK ny). To fulfill the duty, guidelines were formulated, although the RDF had the maximum freedom possible to work on its tasks. Its character as a learning institution was also highlighted during the process (Wörner 2010: 27). According to the RDF's Codex and to continue the mediation process, the RDF was demanded to provide an environment for effective, problem-oriented, and representative communication. The idea was to advance the topics of the mediation process by assessing future developments and innovations, to reach new insights and deeper knowledge of the identified challenges. An objectified discussion should be established to foster a more fact-based debate (Gohl and Meister 2012: 43–49; Wörner 2004).

Already entrusted with the composition of the *Mediationsgruppe*, the Hessian State Chancellery was also authorized to compose the RDF. The nomination power to participate in the RDF was with the then-Minister-president Roland Koch. Based on the experience of the former mediation process, the invitation process was planned in further detail. The institutions that had been neglected in the former process were addressed more directly and with more emphasis (Wörner 2010: 33).[101] Moreover, during official proceedings that started, Roland Koch invited the local citizens' groups and environmental associations directly for cooperation. Fraport AG – as the applicant of the RPP – also welcomed this joint approach to include further stakeholders (HessL 2000c: 15, 21).

100 Decision of the Hessian Cabinet, session on June 20, 2000.

101 In contrast to the former process, the then-Minister-president Roland Koch spoke directly with delegates of the institutions which refused their participation in the former mediation process, to convince them to engage formally (Geis 2005: 242–248).

Over the years of its existence, 73 members of 27 private and non-private institutions were nominated for participation (Annex A.7).[102]

Excluding the institutions first listed as members of the RDF by mistake, most member institutions were non-private administrative.[103] With the city of Mainz, they were also not restricted to the jurisdiction of the Federal State of Hessen. However, nearly an equivalent number was further institutions, mostly local citizens' groups. Three members were private economic institutions. No non-private metropolitan governance institution was involved in the RDF. The private metropolitan institutions were still not established (Fig. 4.14).[104]

Similar to the former mediation process, some institutions refused their participation, even though they were addressed more directly and with more emphasis than in the former process. The free seats – also similar to the former case – were passed to further municipalities and other local citizens' groups whose participation was previously refused (Geis 2005: 244–248).

With the release of the RPD on June 10, 2002 (RPDA 2002b), the discussions of the RDF were enlarged to further stakeholders, also outside the formal RDF framework (Gohl and Meister 2012: 56). Furthermore, participants – both institutions and individual citizens – had the possibility to attend joint public events, the so-called *Themendialoge* (Geis 2005: 248).

According to the Codex, the RDF was informed prematurely about the single steps of the official proceedings by the project proponent – Fraport AG – and the hearing and planning authorities (RP Darmstadt and HMWVL). Reciprocally, the RDF was requested to submit proposals and suggestions for improving discussed topics (e.g. pollution or complaint management), not only to the RP Darmstadt and the HMWVL, but also to Fraport AG, the responsible federal ministry the *Bundesministerium für Verkehr, Bau- und Wohnungswesen* (BMVBW), and the *Deutsche*

102 Not all nominated members had a seat over the whole period.

103 The Hessian authorities, named as members of the RDF in the first annual report (RDF 2001: 44), actually had observer status and were formally not members of the RDF (Wörner 2010: 27). The *Öko-Institute* was, in fact, the scientific assistance partner, and the RDF itself was also not an official member.

104 See Annex A.7 for more detail.

| # | Institutional form | | Institution | Delegates |
|---|---|---|---|---|
| | **Non-private** | | | |
| | | **Administrative** | ---------------------------------------- | **25** |
| 1 | | | Federal State of Hesse* | 2 |
| 2 | | | Sub-federal state level | 23 |
| | | **Metropolitan** | ---------------------------------------- | **0** |
| | | | - | |
| | **Private** | | | |
| | | **Metropolitan** | ---------------------------------------- | **0** |
| | | | - | |
| | | **Economic** | ---------------------------------------- | **9** |
| 3 | | | Fraport AG | 3 |
| 4 | | | DFS Deutsche Flugsicherung GmbH | 2 |
| 5 | | | DLH | 4 |
| | **Further** | | ---------------------------------------- | **39** |
| 6 | | University/ Research | Technische Universität Darmstadt | 1 |
| 7 | | | Öko-Institu e.V.* | 2 |
| 8 | | Association | Vereinigung der hessischen Unternehmerverbände | 2 |
| 9 | | | Schutzbund Deutscher Wald (SDW) | 1 |
| 10 | | | Vereinigung Cockpit e.V. | 3 |
| 11 | | | BARIG Board of Airline Representatives in Germany e.V. | 1 |
| 12 | | | Fluglärmkommission | 2 |
| 13 | | | Hessischer Handwerktag | 2 |
| 14 | | | Arbeitsgemeinschaft Deutscher Verkehrsflughäfen (ADV) | 2 |
| 15 | | | Bundesvereinigung gegen Fluglärm | 1 |
| 16 | | | Arbeitsgemeinschaft Deutscher Luftfahrtsunternehmen | 1 |
| 17 | | | Regionales Dialogforum* | 4 |
| 18 | | | Fachverband Spedition und Logistik | 1 |
| 19 | | | Bürgeraktion pro Flughafen | 1 |
| 20 | | | WiDeMa e.V. | 2 |
| 21 | | | Bundesverband der Deutschen Fluggesellschaften (BDF) | 2 |
| 22 | | Church | Protestant | 1 |
| 23 | | | Catholics | 2 |
| 24 | | Chambers | Chamber of Commerce and industry (Hessen) | 3 |
| 25 | | | Landesärztekammer Hessen | 3 |
| 26 | | Unions | ver.di (former Deutscher Gewerkschaftsbund (DGB)) | 1 |
| 27 | | | ver.di (former Deutsche Angestellten Gewerkschaft (DAG)) | 1 |

*Figure 4.14: Institutional aggregation of the RDF members, own depiction (institutions were listed as members of the RDF in the first annual report of the RDF by mistake. In the subsequent process, they were no longer regarded as members.).*

*Flugsicherung GmbH* (DFS). In its work, the RDF was independent of any instructions. It had the most freedom to regulate and carry out its work by itself. Only general deadlines of the overall process held importance (RDF 2000).[105]

The main fields of engagement were structured according to the guiding principles, elaborated by the *Mediationsgruppe* upfront. Altogether five task forces worked on the topics of "ban on night flying," "Anti-Noise-Pact," "optimization," "ecology and health," and "future perspectives." In order to reach new insights and utilize tacit knowledge of the participating institutions, some topics were also discussed jointly in sub-task forces. These task forces were merged, if necessary (Mediationsgruppe 2000d). Here, according to the former mediation process, the task forces were open to further institutions, especially those not admitted for the RDF directly. Participating institutions of the task forces were mainly non-private administrative.[106] Established sub-task forces reported regularly to the task forces and the RDF (Wörner 2010: 32, 35–37). Furthermore, the whole processes were documented by minutes taken, which were dispatched up- and downwards to all participants.[107]

For the member institutions, the RDF was especially a forum to jointly discuss issues of the airport extension. Cooperation in the RDF was fostered by different forms of participation and transparency. The RDF had also the duty to involve the general public. It ensured that all affected groups – private and non-private institutions as well as private individuals – had the possibility to address their claims and to develop ideas regarding a sustainable airport extension. As a result, the RDF enlarged the possibility of public engagement in official planning proceedings, which are already chartered by law. The goal was to identify and foster topics that hold relevance but were not covered by the official proceedings. This was ensured by the implementation of a citizens' office, press information, public discussions and hearings, two websites, a quarterly newsletter, and its annual reports (Wörner 2010: 46–52). Especially the

105 For an overview see Wörner (2010: 25).

106 See participant lists in the minutes (Mediationsgruppe 2000e).

107 In all, over 198 minutes and memos were taken and were made available (Mediationsgruppe 2000e).

quarterly newsletters (*dialogbrief*) summarized relevant information of the official proceedings, executed assessments, scientific topics, and made it available to the general public (Mediationsgruppe 2000b). Furthermore, the newsletter fostered the exchange between experts and the general public through questions and answers.[108]

Without being a decision authority, the RDF supported the official proceedings and the Hessian State Government with statements on the sustainable relationship between the airport and FRM. These statements were formulated jointly by RDF members, whereas the results have been partly interpreted quite differently. In several cases, although they issued a joint statement, members insisted on their informal point of view. Nevertheless, this joint approach led to the promotion of the dialog between the general public, the administration, and the airport's economic actors (Wörner 2010: 22–32).

During the whole process, the RDF asked for assessments to gain new insights into relevant topics of the airport extension. These insights were continuously integrated into the overall process. With further statements, quality assessments, expert hearings, and trips of delegates, the topics and the process of mediation were broadly advanced. Issues and challenges were annually reviewed and the working program was adjusted. Accordingly, new topics that otherwise would have not been part of the official proceedings, were taken into account by the responsible authorities and the legislation (Gohl and Meister 2012: 49, 54f; Wörner 2010: 22–44). However, even though the RDF's self-image was neutral[109] it published several position papers on different topics.[110]

The RDF and the mediation process were widely recognized. Several stakeholders and state politicians in Germany commented upon it positively (Gohl and Meister 2012: 61–64). The *Sieben-Punkte-Programm* of the RDF (RDF 2008a) recommended the continuation of the open

108 "Citizens ask – the RDF answers" (translated by author).

109 "Are you for or against it?" [...] "Neither, the citizens office of the RDF seeks a dialogue with citizens of the region." (RDF 2002a: 3).

110 See RDF (2008c), RDF (2008d), RDF (2007b), RDF (2007c), RDF (2007d), RDF (2007e), RDF (2006b), RDF (2006c), and RDF (2005b).

*Figure 4.15: Structure of the FFR (FFR 2015).*

dialog by establishing a successor of the RDF. It should be open to the general public and monitor the further extension (Wörner 2010: 151f).

### *4.3.4.4 Forum Flughafen & region, since 2008*

After the last session of the RDF, the Hessian State Government decided in June 2008 to establish a successor.[111] As the successor, the FFR adopted the former mediation process, to further promote the dialog between airport companies and the region. Today, involved institutions are – according to the decision of the Hessian Cabinet – relevant federal ministries, noise-affected communities, environmental and conservation organizations, local citizens' groups, chambers, unions, churches, trade associations as well as representatives of the Hessian parliamentary groups (HessSt 2008: 6).

According to its RoP, the FFR unites the *Koordinierungsrat*, the *Konvent Flughafen und Region*, the *Expertengremium Aktiver Schallschutz*, and the *Umwelt- und Nachbarschaftshaus*.

The *Koordinierungsrat* must request assessments and expertise from experts to gain new insights and promote the knowledge transfer up- and downwards the levels. To further promote the common dialog, the *Konvent Flughafen und Region* has the responsibility to work on questions regarding the extension of Frankfurt Airport and the development of air

111 Decision of the Hessian Cabinet on June 13, 2008.

| # | Institutional form | | Institution | Delegates |
|---|---|---|---|---|
| | **Non-private** | | | |
| | | **Administrative** | ------------------------------------------------------------ | **44** |
| 1 | | | Federal Republic of Germany | 2 |
| 2 | | | Federal State of Hesse | 3 |
| 3 | | | Sub-federal state level | 39 |
| | | **Metropolitan** | ------------------------------------------------------------ | **1** |
| 4 | | | Metropolitan Governance Bodies | 1 |
| | **Private** | | | |
| | | **Metropolitan** | ------------------------------------------------------------ | **0** |
| | | | - | |
| | | **Economic** | ------------------------------------------------------------ | **3** |
| 5 | | | Fraport AG | 1 |
| 6 | | | Deutsche Lufthansa AG | 1 |
| | | | DFS Deutsche Flugsicherung GmbH | 1 |
| | **Further** | | ------------------------------------------------------------ | **27** |
| 7 | | Association | | 15 |
| 8 | | Union | | 2 |
| 9 | | Church | | 2 |
| 10 | | Chambers | | 2 |
| 11 | | Party | | 5 |
| 12 | | No affiliation mentioned | | 1 |

*Figure 4.16: Institutional aggregation of the FFR members, own depiction, based on FFR (2009).*

traffic in general. This includes current topics and the future development of FRM and Frankfurt Airport. It represents the main cooperation platform within the FFR. Here, both private and non-private institutions are represented. Due to the RoP of the FFR, the *Expertengremium Aktiver Schallschutz* mainly works on topics coming from the Anti-Noise-Pact (RDF 2007a) and promotes the joint discussion of active noise-canceling activities. It can also suggest further assessments for further data gathering and integration. The *Umwelt- und Nachbarschaftshaus* represents the information center for the general public (FFR 2009).

The majority of the involved institutions are non-private administrative. With the city of Mainz, the institutional environment is not restricted to the Federal State of Hesse. In addition, with the integration of the

*Planungsverband* (today *Regionalverband*), a non-private metropolitan governance institution is directly integrated into the mediation process officially for the first time (Fig. 4.16).[112]

The first major goal of the FFR is to open the relevant topics to laymen and foster the joint discussion between all stakeholders. The second goal is to monitor the process of the joint dialog. This was especially indicated by the decision of the Hessian State Government. No further restrictions or advice was made regarding the work of the FFR (HessSt 2008).

Accordingly, a solid basis for the further common discussion and exchange was brought into being (Gohl and Meister 2012: 59).

#### *4.3.4.5 Interim findings: mediation process, since 1997*

In the following, the mediation process since 1997 is analyzed by the theoretical derived concept of ERG. The latent content of the five aspects of ERG is reconstructed in the regulatory governance structure and decision-making processes. This analysis is followed by establishing the missing link between the structural perspective of political science and the procedural character of policy-making, which Mayntz already proclaimed in the 1980s (Mayntz 1983), by discussing the institutional arrangement of that period.

*Discussion of shared perceptions of joint challenges*

The mediation process, as it was also partly manifested in the respective bylaws, was characterized by the discussion of shared perceptions of joint challenges. With the extension as the identified common challenge, the *Gesprächskreis*, the *Mediationsgruppe*, the RDF, and the FFR were created as the capacities, to jointly discuss the identified challenges. These capacities were further enhanced by the implementation of the different task forces and sub-task forces in the respective processes, e.g. like the *Arbeitskreis Verkehr*, *Arbeitskreis Ökologie, Gesundheit und Soziales*, and *Arbeitskreis Ökonomie* of the *Mediationsgruppe*. This created open, transparent multi-actor discussions across hierarchical and traditional borders. The *Gesprächskreis*, the *Mediationsgruppe*, the RDF, and the FFR and their task forces and sub-task forces linked private and

112 See Annex A.8 for more detail.

non-private institutions, coming from all governance levels to exchange knowledge and opinions. Therefore, stakeholders were connected beyond the levels. Consequently, the regulatory governance structure and the decision-making formed a participatory policy-making process. Initiating the *Gesprächskreis Flughafen* and establishing the successors, an objective and imperial process was developed to foster cooperation as well as jointly discuss and define high-level goals and measurement systems. With each step, the former process was improved by lessons learned and recursive policy-making. This is particularly evident by the more direct integration of environmental associations in the RDF in 2000 to 2008.

By establishing the task forces, not only lessons learned were put into practice. Moreover, this approach entrusted lower levels with the implementation of joint actions – e.g. executing assessments needed for the overall mediation process. Further, this delegation of mandate and increase of ownership is especially obvious by nominating the *Gesprächskreis*, as the initial institutionalized mediation process, to design and establish an objective and imperial institution, to foster an open, joint dialog. Ownership of lower levels was further created by e.g. the personal responsibility of the members of *Mediationsgruppe* to report back to their organization.

After the initial period of the *Gesprächskreis* from 1997 to 1998, the existing governance structure of mediation – as institutions already in place – was requested by the responsible non-private administrative institution, the Hessian Cabinet, to support the policy-making by at least helping to establish the respective successor, and therefore to learn from previous experience and lower levels.

Finally, up until today, the FFR provides, exchanges, and pools information of the airport extension.

*Definition of goals and measurement systems*

Since the beginning of the mediation, high-level goals and legally non-binding measurement systems were discussed and defined. Especially the focused task forces and sub-task forces contributed to establishing suitable measurement systems. Based on discussions and the insight generated in the task forces and sub-task forces, proposals to solve the identified challenges of pollution, noise emission, but also economic aspects, were negotiated with the addressees, e.g. during official mediation meetings or public hearings. In addition, several assessments were executed and expert opinion was obtained. The discussions with the addressees and the results of the assessments were the basis of reference documents and recommendations, formulated during all periods of the mediation process. Especially the RPD's Anti-Noise-Pact has to be named as a success (RDF 2007a). Guiding documents such as the Anti-Noise-Pact or the final report of the *Mediationsgruppe* (Hänsch, Niethammer, and Oeser (2000)), which were made available across hierarchies, presented sets of legally non-binding rules. With that, shared goals and the relating measurement system were presented. This document is up to date cited in the ongoing discussions of the extension of Frankfurt Airport.

*Implementation*

As it was decided or written down in the RoP and bylaws, each respective mediation institution was entrusted with the implementation of the regulatory structure of governance and its own decision-making processes. In this regard, it was also decided that besides rudimentary guidelines, the stakeholders of the mediation implement the regulatory structure of governance, decision-making process, and further aspects with the most freedom. This approach was followed without exception. First, the overall regulatory governance structure of the mediation was essentially driven by the mediators themselves in cooperation with the responsible non-private administrative institution, the Hessian Cabinet. Second, the establishment of the task forces and further sub-task forces was determined by the stakeholders of the respective mediation process. The task forces and further sub-task forces have also to be named regarding the establishment of cooperation.

Nevertheless, the actual implementation of the jointly-defined goals was not the task of the stakeholders; rather, it was the task of Fraport AG, as the project applicant and the administration to consider the insights created. The implementation of the defined goals was continuously monitored by the mediators. Again, the Anti-Noise-Pact and the installation of noise reduction measures are the best examples.

*Report and review*

In all instances, the extension process was monitored outside the official proceedings by the mediators. These reviews led to the fact that a regulatory governance structure was created in which the Fraport AG and all other private and non-private institutions, arguing in favor of the extension, had to consistently justify their actions and measurement systems. During the RDF this exchange up- and downwards beyond the traditional borders of hierarchy was even manifested in the bylaws. According to the Codex, the RDF was informed prematurely about the single steps of the official proceedings by the project proponent – Fraport AG – and the hearing and planning authorities (RP Darmstadt and HMWVL). Reciprocally, the RDF was requested to submit proposals and suggestions for improving discussed topics such as pollution or complaint management not only to the RP Darmstadt and the HMWVL, but also to Fraport AG, the responsible federal ministry the *Bundesministerium für Verkehr, Bau- und Wohnungswesen* (BMVBW), and the *Deutsche Flugsicherung GmbH* (DFS). Further, in the previous period the members of the *Mediationsgruppe* were personally responsible to report and give feedback to their institutions and the mediators were asked to secure this feedback by peer-reviewing this process with specific and regular discussions. This institutionalized periodical assessments, audits, and peer reviews from the beginning of the mediation. With that, especially based on the work of the focused task forces and sub-task forces, it was possible to release and discuss opinions and comments regarding implementation, goals, and measurement systems.

Besides that, the actions that have been taken, the status quo of implementation, goals, and measurement systems were discussed in public by several hearings, although they were also the matter of debates in the Hessian State Parliament. This initiated constitutional and collective regulatory dialogs and created an environment of continuous adjustment and learning.

*Review of the measurement system, the decision-making process, and the stakeholder*

By discussing the extension, the goals, and the measurement systems as well as monitoring the implementation, the mediation contributed to a continuous reassessment and cross-checking. However, not only the project planning and execution of Fraport AG were peer-reviewed. In addition, the procedures of political programming of the national and the Hessian State Government as well as the policy-making at the sub-federal state level were in the focus of the mediation process. Contributing factors were the manifested exchange beyond the traditional borders of hierarchy, partly proclaimed in the bylaws of the respective mediation institutions such as the RDF. The review of the decision-making process was enforced by the constant information of the broader public with different formats. During the whole mediation process, information was dispatched beyond the levels by press information, newsletters, citizens' advice offices, and further public events. Minutes and protocols of the meetings and hearings were made available. Especially the mediators themselves and the local citizens' groups have processed information up- and downwards the governance levels.

After each period, the gathered information was at least pooled in a comprehensive final report, which was also made available to the general public (GKF 1998; Mediationsgruppe 2000a; Wörner 2010). In particular, the report of the RDF stands out in this context. With the release of the final report, all information was made available on CD-ROM, which is still available.[113] Additionally to the officially involved institutions, especially further local citizens' groups have to be named in terms of improving the pooling and providing of information. They gather up to today information regarding the extension of Frankfurt Airport and make the information available on their websites for the general public.

Throughout the whole mediation, the involvement of official stakeholders was reviewed. If necessary, further stakeholders were integrated or replaced. The most evidence is represented by the more direct approach of the local citizens' groups and environmental associations to participate in the RDF.

### Institutional arrangement of the mediation process of Frankfurt Airport

Besides the depiction of the decision-making process, the previous paragraph also drew a detailed picture of the institutional arrangement

113 E.g. at the library at Goethe University (Mediationsgruppe 2000a).

of the mediation since 1997. The mediation was initially marked by individual representatives of the institutions involved or persons of public interest, rather the institution. Still, the integration of several private and non-private institutions during the different periods of the mediation created a multi-actor environment across hierarchical and traditional borders. This effect was enforced by the implementation of the task forces and sub-task forces, integrating many different stakeholders of different institutions. These institutions became linked by heterarchical relations throughout the traditional hierarchical levels, from national to local, from non-private administrative to private economic.[114]

Until the last period, metropolitan governance bodies have played no or only a very minor role. The UVF was not involved in the *Gesprächskreis* and private metropolitan governance bodies were not yet established. In the RDF the UVF was only represented in one sub-task force, the *Arbeitskreis Ökologie, Gesundheit und Soziales*. With that, the metropolitan perspective was hardly included in the main mediation process. Only with the initiation of the contemporary last mediation period, the *Planungsverband* (today *Regionalverband*) became officially involved. For the private metropolitan logistics actors, it has to be mentioned, that they were not established before 2007. However, there is no integration applicable to the contemporary mediation process.

### Confirmation of the research-guiding hypothesis for the mediation process, since 1997

This interim conclusion of the mediation process since 1997 shows that the theoretical derived concept of ERG is a suitable analytical framework for analyzing the regulatory structure of governance, the decision-making processes, and the institutional arrangements. With that, the research-guiding hypothesis can be confirmed in the case of the mediation process since 1997. It also shows that the concept of ERG counterfeits the challenge of a missing link between the structural perspective of political science and the procedural character of policy-making, which Mayntz

114 Woolcock (2001) calls this linking social capital. With that, he describes the promotion of connections and social relations (links) between individuals and institutions from different hierarchical governance levels (Sorg 2009: 66).

already proclaimed in the 1980s (Mayntz 1983), by including the analysis of the institutional arrangement of that period.

### 4.3.5 Regional planning procedure and regional planning decision, 2000–2002

The planned extension of Frankfurt Airport was an infrastructure project with spatial-relevant impacts, having a supra-regional character.[115] Besides the construction of a further runway, Fraport AG planned to restructure the former US Air Base. It was planned to establish Cargo City Süd, further maintenance shipyards and construct Terminal 3 (RPDA 2002b: 11–14).

The RP Darmstadt – as responsible authority – had the duty to consider all relevant environmental impacts.[116] This included all impacts affecting humans, animals, plants, soil, air, climate, and the countryside.[117] Local effects and impacts on the meaning of civil law were not part of the RPP. Thus, the RPP assesses principle questions of spatial planning, not the legal procedures of an infrastructure project with spatial-relevant impacts (RPDA 2002b: 29f).

#### *4.3.5.1 RPP Frankfurt Airport*

After the conflicts and riots of the 1970s/80s and the killings of two police officers in 1987, nobody imagined that Frankfurt Airport would ever grow "beyond the fence."[118] This is also reflected in the AoP as of March 23, 1971. "The fear that a further runway – e.g. parallel to Runway 18-West

115 According to §13 HLPG.

116 According to §16 II *Regionale Planungsgesellschaft UntermainGesetz über die Umweltverträglichkeitsprüfung* (UVPG), together with §13 II HLPG.

117 According to the *Regionalplan/Südhessen2000* (RPS2000) (RPDA 1999), the *Landesentwicklungsplan Hessen 2000* (LEPH2000) (HMWVL 2000), §§2 and 3 HLPG, and §2 ROG.

118 The then-mayor of the city of Mörfelden-Waldorf, Bernhard Brehl, argued that a "state government that has the courage to engage in that, has still to be found ", and the future mediator Kurt Oeser said: "The fence is the border. I cannot imagine that future state governments will change that" (Geis 2005: 120). Similar argued Johnsen (1996) when he wrote, that further forest clearance is politically not enforceable (Johnsen 1996: 24). All translations by the author.

– could be built, is without any foundation. By any circumstances, the approval of such a measure will not be granted" (HMWT 1971: 755).[119]

However, in the days when Fraport AG, DLH, and BARIG came up with the topic of extension, the general political environment was caught by the fear of mass job reductions. This fear was boosted by DLH arguing to change its home base if the extension failed to proceed (Rausch 1999). Furthermore, subsequent mass job reductions and a domino effect to other sectors were anticipated.[120] Therefore, the Hessian State Government was under high pressure to act (Geis 2002: 275f; Geis 2005: 119–124, 143f; RPDA 2002b: 10f; Wörner 2010: 14–16). The fact that other European airports such as London Heathrow, Amsterdam Schiphol, and Paris Charles-de-Gaulle were enlarged in the 1990s placed more pressure not only on Hessians but also on the national level politicians (Becker-Ritterspach 2015: 94).

With its letter of February 8, 2000, the Fraport AG announced the planning of extension to the HMWVL.[121] In response, the HMWVL informed Fraport AG that an RPP is an obligation for the planned project. The RP Darmstadt was nominated as the responsible hearing authority for the planned project.[122] On August 7, 2000, the RP Darmstadt dispatched the documents – provided by Fraport AG,[123] – to the affected state counties and municipalities as well as the involved public agencies and the environmental associations According to the *Bundesnaturschutzgesetz* (BNatSchG), these are the relevant institutions that have to be included in the proceeding in regard to the participation process.[124] The obligatory hearings – focusing on the environmental

---

119 Translated by the author.

120 The argument of giving up the Frankfurt Airport as the home base was and still is always cited (HMWVL 2004: 50f; RPDA 2004i: 26).

121 According to §45 II (1) *Luftverkehrs-Zulassungs-Ordnung*.

122 According to §16 II 2 HLPG; Replies of the HMWVL as of April 10, 2000, and September 20, 2000.

123 The 16 files included mainly the description of the planned extension, a spatial planning assessment, an environmental impact assessment, a flora-fauna-compatibility study, and several expert reports (Fraport 2001).

124 According to §29 BNatSchG.

impact assessment – took place between November 2–4, 2000.[125] Major questions regarding the content, circumference, methods, and further issues of the environmental assessment were jointly discussed. The hearing initiated further discussions, especially regarding the forest surrounding Frankfurt Airport (BUND 2001). After the construction of Runway 18-West and with the decision as of July 6, 1993, this forest gained the status of protected forest (*Bannwald*) (HMLWLFN 1993). This protection was now in conflict with the planned extension. For this reason, the decision of protection was assessed, discussed, and ultimately abolished (RPDA 2002a).[126] The final circumference of the assessment was determined by the letter as of February 14, 2001 (RPDA 2002b: 17f). With the announcement as of October 22, 2001, the RP Darmstadt initiated the actual RPP (RPDA 2001: 3693).

Between October 22, and November 2, 2001, the RPP documents were dispatched to the relevant and involved institutions. Until January 4, 2002, the documents were additionally disclosed to the public by 105 non-private institutions, mostly state counties and municipalities. Both aspects enabled all relevant and affected institutions as well as all private individuals to take notes and engage in the joint discussion. In addition, the documents were sent to 258 private and non-private institutions (Fig. 4.17)[127] to integrate them officially into the RPP (RPDA 2002b: 24).

The majority of the involved institutions were non-private administrative. Requesting statements, four non-private metropolitan governance bodies were also integrated into the RPP. The *Planungsverband* – as the major metropolitan governance body of FRM – was involved in the official proceedings. Furthermore, institutions, as the second-largest group of the involved institutions, were mainly represented by environmental associations, local citizens' groups, and chambers. The private sector was included with 34 institutions, whereas only two of them can be seen as air transport related (Fig. 4.17).[128]

---

125 According to §5 UVPG.

126 See Roller and Steinwachs (2001) and Roller (1999) for the discussion of the challenges regarding the abolishment of the protection of protected forest.

127 See Annex A.9 for more detail.

128 See Annex A.9 for more detail.

| # | Institutional form | | Institution | Delegates |
|---|---|---|---|---|
| | **Non-private** | | | |
| | | **Administrative** | --------------------------------------------------------------- | **176** |
| 1 | | | Federal Republic of Germany | 6 |
| 2 | | | Federal State of Hesse | 36 |
| 3 | | | Federal State of Rhineland-Palatinate | 6 |
| 4 | | | RPDA | 27 |
| 5 | | | RPG | 1 |
| 6 | | | Sub-federal state level | 100 |
| | | **Metropolitan** | --------------------------------------------------------------- | **4** |
| 7 | | | Metropolitan Governance Bodies | 4 |
| | **Private** | | | |
| | | **Metropolitan** | --------------------------------------------------------------- | **0** |
| | | | - | |
| | | **Economic** | --------------------------------------------------------------- | **36** |
| 8 | | | Air transport | 2 |
| 9 | | | Others | 34 |
| | **Further** | | --------------------------------------------------------------- | **42** |
| 10 | | | Association | 28 |
| 11 | | | Union | 1 |
| 12 | | | Church | 2 |
| 13 | | | Chambers | 11 |

*Figure 4.17: Institutional aggregation of stakeholders of the RPP Frankfurt Airport, own depiction, based on RPDA 2002b: 18–24.*

The general public was able to become officially involved with the examination of the documents displayed and written statements. Of the involved private and non-private institutions, 181 wrote a statement on the planned project. The majority of the affected municipalities requested at least the adjustment of the extension plans. Private individuals involved by the public disclosure raised around 48,000 suggestions and concerns (RPDA 2002b: 24).

Particularly environmental issues, the spatial compatibility, and the overall completeness of the planning documents were questioned (AZ 2002; FR 2002b; FR 2002e).[129] Also, the deviation from the results of the

129 In this time, the Fraport AG was honored with the *Umwelthammer* of the BUND Groß-Gerau (BUND 2002c).

mediation process was discussed and criticized, especially by the public and the RDF (FAZ 2002d; FNP 2002d).

In the arena of party politics, undecidedness was apparent between the parties as well as within the parties themselves (FAZ 2002f; FNP 2002b; FR 2002g). As in the mediation process, the SPD was located between supporters and repudiators with no clear vote (Geis 2002: 276). Several SPD-led municipalities expressed their defeat against the extension (FAZ 2002f; FAZ 2002g; FR 2002f). The Hessian CDU still supported the extension, although some CDU-led municipalities also raised demands against the extension plans (FAZ 2002c; FR 2002d).

The statements of the private and non-private institutions were discussed together with Fraport AG between April 8–25, 2002. During the hearings, the documents of the planned extension and the executed assessments were reviewed and debated in detail.[130] Mainly possible goals of the spatial planning and the different measurement systems regarding noise and air pollution were criticized (RPDA 2002f: 27, 54; see footnote for further details).[131] Not only conflicts of metropolitan objectives set in the RSP2000 (RPDA 1999) and the LEPH2000 (HMWVL 2000) were debated; moreover, additional divergences beyond the levels were considered.

During the RPP, the methodology of the general proceeding, the spatial impact assessment and the capacity prediction assessment were scrutinized and discussed extensively (RPDA 2002e: 34; RPDA 2002o; RPDA 2002p: 117). New insights were brought into the overall proceeding and benchmarks were discussed again (RPDA 2002d; RPDA 2002g: 59f). Furthermore, the practical integration of the results from the assessments and findings from the RDF was reviewed during the hearing (RPDA 2002d: 18; RPDA 2002e: 86f). The combination and link between the formal proceedings and the more informal process of mediation

130 See protocols RPDA (2002a); RPDA (2002c), RPDA (2002d), RPDA (2002e), RPDA (2002f), RPDA (2002g), RPDA (2002h), RPDA (2002i), RPDA (2002j), RPDA (2002k), RPDA (2002l), RPDA (2002m), RPDA (2002n), RPDA (2002o), and RPDA (2002p).

131 RPDA (2002g: 28); RPDA (2002h: 6, 29–31, 78f); RPDA (2002i: 10, 93, 96, 104, 128); RPDA (2002j: 62–66); RPDA (2002l: 8–11, 100, 156–160); RPDA (2002m: 105f, 116); RPDA (2002n: 82f).

were highlighted in these discussions. To secure this link, the RDF monitored and reviewed the RPP. It gave non-binding statements and recommendations (RPDA 2002e: 84–87). The hearing made it possible to include further tacit knowledge and benchmark the planned extension of Frankfurt Airport more objectively (RPDA 2002m: 36).[132]

After the hearing, based on the documents provided by Fraport AG and the results of the hearing, the RP Darmstadt prepared the RPD. Written statements, suggestions, and doubts raised during the whole proceeding were taken into account in the preparation of the final RPD (RPDA 2002b: 25–29). On June 10, 2002, the RPD was released after eight months of further negotiations. The conclusion was that the spatial compatibility is given, albeit with conditions regarding a more improved reduction of environmental impacts (RPDA 2002b: 1, 256).

After the release of the RPD, especially the environmental associations still demanded the non-extension. Nevertheless, also alternatives such as the establishment of an airport system between Frankfurt Airport and Frankfurt-Hahn Airport to increase the onsite capacities were refused by the environmental associations (BUND 2002b; RW 2002). Based on their opinion, they further criticized the fact that several detailed and relevant questions were not discussed. They argued that the RP Darmstadt just referred these questions to the following RSE (BUND 2002a). The local citizens' groups shared this opinion (FNP 2002c). The neighboring municipalities reacted partly with heavy critique (FAZ 2002a). Moreover, the major parties were still torn (FAZ 2002e; see footnote for further details).[133] Even though the RPD included conditions to improve the environmental impact of the extension, the Fraport AG was still satisfied with the RPD and optimistic for the following RSE (DVZ 2002; Fraport 2002). Other air transport institutions also welcomed the decision (FNP 2002a). With the approval, the change of the then valid regional planning also had to be initiated by the *Regionalversammlung*

---

132 E.g. the *Flughafenkonzept der Bundesregierung* (BMVBW 2000), environmental reports of the Federal Government of Germany (Bundestag 1998, Bundestag 2000), and the EU (EC 1999).

133 FNP (2002a); FR (2002a); FR (2002c).

*Südhessen*, a committee of 99 members from different local authorities in FRM (FAZ 2002b).

#### *4.3.5.2 Interim findings: RPP and RPD*

In the following, the RPP and RPD of 2000–2002 are analyzed by the theoretical derived concept of ERG. The latent content of the five aspects of ERG is reconstructed in the regulatory governance structure and decision-making process of the RPP and RPD, 2000–2002. This analysis is followed by establishing the missing link between the structural perspective of political science and the procedural character of policy-making, which Mayntz already proclaimed in the 1980s (Mayntz 1983), by discussing the institutional arrangement of that period.

*Discussion of shared perceptions of joint challenges*

With the nomination of the RP Darmstadt as the responsible hearing authority, the RPP became the institutional proceeding for spatial planning including recursive and participatory policy-making. As institution already in place and together with the official hearings as well as the link between the formal proceedings of the RPP and the more informal process of mediation of the RDF, the RP Darmstadt initiated open multi-actor discussions across hierarchical and traditional borders. This link and the lawful integration of affected state counties and municipalities, the involved public agencies and the environmental associations, the support of recursive policy-making, and learning by experience from lower-level were promoted. Integrating the sum of all relevant actors by requesting statements and displaying the planning documents in public, the decision-making process was accompanied by learning from experience and tacit knowledge of the lower levels. In particular, the provision of the planning documents and the hearings fostered the exchange and pooling of information during the official RPP. With the review of the planning documents, shared perceptions of joint challenges were laid bare. These challenges, as well as the best available techniques to reduce the spatial impact, were discussed throughout the whole RPP.

*Definition of goals and measurement systems*

High-level goals were initially set by the request of Fraport AG to extend the airport. Common objectives, based on Fraport AG's planning documents and expert opinions, especially from environmental associations, were set and further negotiated. These negotiations led to the formulation of reference documents and recommendations based on expert opinions and further tacit knowledge. The proposals, reference documents, and recommendations were discussed with the respective addressees, especially during the hearings between April 8–25, 2002. Goals and the measurement systems established, were challenged against extensive information to establish a practical and workable implementation plan and feasible regulations, resulting in the RPD published on June 10, 2002. The final planning documents and the RPD represent the guidance documents to clarify the approach of the extension.

*Implementation*

With the hearings from April 7–11, 2003, and between January 15, and March 5, 2004, a binding overall framework was established to further harmonize directives and regulations. As the responsible hearing authority, the RP Darmstadt delegated the task of implementation – the adaption of the planning documents during the negotiations – back to Fraport AG. With that, the RP Darmstadt entrusted lower-level units and delegate responsibilities and power to lower-level authorities or experts. Based on the discussions and gathered information, Fraport AG prepared a revised edition of the planning documents, integrating the findings of hearings into the project planning. The obligatory hearings between April 8–25, 2002 created the capacities for the most objective and imperial process to review and reassess the planning documents and counter opinions. Fostering cooperation while implementation can only be identified as a matter of fact, rather than actively promoted.

Taking into account the contributed tacit knowledge, new developments, and innovations discussed during the RPP, Fraport AG adjusted existing rules based on tacit knowledge and subsequent experience. Fraport AG adopted and modeled the goals, the measurement systems, and rules in its second edition of the planning documents for the final RPD. Nevertheless, regarding the technical nature of the planning documents and the legally binding quality of the RPD, both are less characterized by soft guidance to fulfill a binding overall framework. However, the Fraport

AG had the most freedom to implement the demands into the overall approach and final execution of the airport extension.

The same has to be mentioned for the adaption of the RSP2000 and the LEPH2000. Both regional spatial planning acts had to be adapted more based on the legal decision than on soft guidance.

*Report and review*

The RPD as such presents an institutionalized assessment, including mandatory peer reviews of the plans. Besides the RP Darmstadt – as the responsible hearing authority – especially members of the RDF peer-reviewed the planning documents in detail. As institutions already in place, the RP Darmstadt and the RDF – partly common, partly separate – monitored and evaluated the decision-making process. Based on cooperation and legal demand, a constitutional and collective regulatory dialog was initiated, mainly revealed in the obligatory hearings from April 8–25, 2002. The link between the official proceedings of decision-making and mediation created an environment of continuous adjustment and learning during the RPP. By sending the planning documents to the affected state counties and municipalities for review and the public display of the planning documents between October 22, and November 2, 2001, as well as the integration of environmental associations, local citizens' groups, and chambers in the hearings, the processing of information up- and downwards hierarchy levels and the consultation of additional opinions becomes most evident. During the decision-making process, expert opinions and comments on the implementation, the defined goals, and the measurement system were exchanged. Notably, Fraport AG – as the project proponent – was required extensively to justify planned and taken activities, again the hearings are the most obvious indicator.

*Review of the measurement system, the decision-making process, and the stakeholders*

The initial planning documents with the intended goals and measurement system were assessed, questioned, and cross-checked. During the decision-making process of the RPP, especially during the hearings in April 2002 and several lawsuits, also policy objectives, procedures of political programming, legislation, and the decision-making process of the planned extension of Frankfurt Airport were continually reviewed. This revealed areas where adjustments were needed. That becomes most visible in the legal actions taken regarding the then-valid regional spatial planning acts – RSP2000 and LEPH2000.

The hearings are also evidence that further stakeholders were formally integrated into the overall regulatory structure of governance and the decision-making process of the RPP and RPD. With the revised edition of the planning documents, taking new developments, and expert opinions into account, Fraport AG presented a suitable amendment that was presented to the responsible authority, the RP Darmstadt. The following RPD, based on the planning documents, contained legally binding and non-binding rules for further regional planning and the extension of Frankfurt Airport.

**Institutional arrangement of the RPP Frankfurt Airport**

In addition to the analysis of the decision-making process, the former paragraph presented a detailed picture of the institutional arrangement of the RPP Frankfurt Airport. Integrating several private and non-private institutions as well as the general public, the RPP Frankfurt Airport was a multi-actor decision-making proceeding across hierarchical and traditional borders. This led to heterarchical relations within the regulatory structure of governance. These relations linked institutions from all governance levels and private individuals. For this reason, it was possible to discuss shared perceptions of joint challenges and common issues regarding the extension of Frankfurt Airport. This highlights the multilevel governance character beyond the levels, including the sum of relevant actors.

However, if the non-private metropolitan governance bodies were involved actively by requesting a statement on the planning documents, their overall participation was weak during the RPP. No active engagement is identifiable. There was no indication that even the *Planungsverband* – as the then-officially-recognized metropolitan governance body of FRM – actively engaged in the hearing. Enforced by law and based on the voluntary cooperation of stakeholders, a metropolitan perspective was at least partly integrated. Not being coordinated, the geographic-functional aspect of FRM stepped behind the single interests of the involved institutions.

Still, for the private metropolitan logistics actor, it has to be mentioned, that they were not established during the RPP.

**Confirmation of the research-guiding hypothesis for the RPP and RPD, 2000–2002**

This interim conclusion shows that the theoretical derived concept of ERG is a suitable analytical framework for analyzing the regulatory structure of governance, the decision-making processes, and the institutional arrangements in the case of the RPP and RPD, 2000–2002. With that, the research-guiding hypothesis can be confirmed for the RPP and RPD, 2000–2002. It also shows that the concept of ERG counterfeits the challenge of a missing link between the structural perspective of political science and the procedural character of policy-making, which Mayntz already proclaimed in the 1980s (Mayntz 1983), by including the analysis of the institutional arrangement of that period.

## 4.3.6 Road scheme evaluation and approval of plans, 2002–2004

With the letters as of December 6, 2002, and January 23, 2003, Fraport AG announced its plans to build the A380 Shipyard.[134] The RP Darmstadt – as the responsible hearing authority – ascertained that the actions have to be proven by an RSE. On January 20, 2003, Fraport AG dispatched the application plan to the RP Darmstadt (Fraport 2003a).[135]

### *4.3.6.1 RSE A380 Shipyard*

For the RSE the Fraport AG had to execute an environmental impact assessment.[136] The main interest of this assessment was the impact of the planned clearing of woodland and further spatial use. In this environmental impact assessment, the general public was involved by a

---

134 According to §45 II LuftVZO.

135 The Fraport AG requested permission to build the A380 Shipyard with all necessary facilities, the construction of a new taxiway and in this context the realignment of Tor 31 and the street K 152/K 823 as well as the construction of a rainwater basin and a new service road. The Fraport AG requested the construction of a new parking structure as well (HMWVL 2004: 52).

136 According to §3 e I 2 UVPG, together with §3 c I and III UVPG.

hearing from April 7–11, 2003.[137] The goal of the hearing was to integrate further opinions and tacit knowledge into the overall process for a more objective assessment (RPDA 2003e: 38, 160). The hearing was criticized by local citizens' groups from the very beginning and was accompanied by protests (BBI 2004; FAZ 2004c).

Besides the impact on woodland, a further major topic was noise emission. Different opinions and assessments of noise emission were elaborated. The results were discussed jointly during the hearing of the environmental impact assessment (RPDA 2003c: 76). If necessary, information and data were reconfirmed by further research (RPDA 2003c: 119). Another major topic was pollution. During the whole hearing, the functionality and accuracy of the goals as well as the presented measurement systems of Fraport AG to implement the project were discussed (RPDA 2003a; see footnote for further details).[138]

Again, especially the environmental associations argued for a stronger focus on the environmental impact (BUND 2003a; BUND 2003b). The local citizens' groups argued in the same manner (BVF 2003; WUT 2003). Municipalities and local citizens' groups further criticized the dealings with the ban on night flights and discussed a maximum of aircraft movement (FR 2003a; FR 2003b; WT 2003). Moreover, the local citizens' groups claimed for more engagement of the municipalities against the extension. In this context, Fraport AG was accused of blockading the joint discussion (FR 2003c; WT 2003).

Furthermore, the actual approach of the hearing was criticized. It was argued by the objectors that the hearing was only used to complete the planning documents and not to discuss and prepared a detailed environmental impact analysis (FNP 2003b). The completion of the overall planning process was still scrutinized at the end of the hearing. By contrast, it was also argued that the identification of the gaps in the planning documents was exactly the goal of the hearing (FNP 2003a). One of the main challenges according to all stakeholders was the overflight of the chemical company Ticona, which had to be reassessed after the hearing

137 See protocols RPDA (2003e), RPDA (2003a), RPDA (2003b), RPDA (2003c), and RPDA (2003d).

138 RPDA 2003b: 61-64; RPDA 2003c; RPDA 2003d: 110f, 143).

(FAZ 2003). Fraport AG declared, that the results of the hearing would be taken into consideration and that the extension plan would be adjusted according to the findings of the hearing (Fraport 2003b).

After the hearing and the further public discussion, the plan was completed by Fraport AG. On May 27, 2003, the RP Darmstadt officially initiated the RSE. Already one day later, the planning documents were sent to the directly affected cities of Frankfurt am Main, Kelsterbach, Mörfelden-Walldorf, Raunheim, and Rüsselsheim. With the letter as of June 10, 2003, the documents were also dispatched to several further private and non-private institutions for peer review. The affected municipalities announced and displayed the planning documents to the general public between June 23, and July 22, 2003.[139] During this time and two weeks after, everybody who was affected by the planned construction had the chance to appeal against the plans (HMWVL 2004: 52–54). In this context, the municipalities had different opinions about their involvement. On the one hand, some complained about their noninvolvement. On the other hand, an excessive involvement was mentioned (*Beteiligungswahn*) (FNP 2004h).

Several heads of divisions or departments of RP Darmstadt, whose scope of duties were affected, were also involved. Moreover, the RDF and the Hessian State Council were informed by the RP Darmstadt. The RP Darmstadt actively requested 69 out of all directly involved institutions, to give statements on the documents provided by Fraport AG (HMWVL 2004: 60–63; Annex A.10).

The majority of involved institutions were non-private administrative. However, further institutions – above all associations – were involved in almost even terms of absolute numbers. With the *Planungsverband*, one metropolitan governance body was actively involved in the RSE A380 Shipyard. The private metropolitan logistics actors were not yet established. Affected private economic institutions were also requested for a statement (Fig. 4.18).[140] Besides the statement of the actively involved

139 According to §10 II (3) LuftVG, together with §73 V *Hessisches Verwaltungsverfahrensgesetz* (HVwVfG).

140 See Annex A.10 for more detail.

| # | Institutional form | | Institution | Delegates |
|---|---|---|---|---|
| | **Non-private** | | | |
| | | **Administrative** | ------------------------------------------------------- | **31** |
| 1 | | | Federal Republic of Germany | 5 |
| 2 | | | Federal State of Hesse | 14 |
| 3 | | | RPDA | 1 |
| 4 | | | Sub-federal state level | 11 |
| | | **Metropolitan** | ------------------------------------------------------- | **1** |
| 5 | | | Metropolitan Governance Bodies | 1 |
| | **Private** | | | |
| | | **Metropolitan** | ------------------------------------------------------- | **0** |
| | | | - | |
| | | **Economic** | ------------------------------------------------------- | **9** |
| 6 | | | Air transport | 1 |
| 7 | | | Others | 8 |
| | **Further** | | ------------------------------------------------------- | **28** |
| 8 | | Association | | 24 |
| 9 | | Chamber | | 4 |

*Figure 4.18: Institutional aggregation of stakeholders of the RSE A380 Frankfurt Airport, 1st evaluation, own depiction, based on HMWVL (2004: 54–57).*

institutions, private individuals also made objections based on the public disclosure of the planning documents.

Including the RP Darmstadt itself, 44 out of the 69 institutions submitted a written statement. Overall, 40,839 objections were made by the general public.

The statements of the private and non-private institutions as well as the objections made by private individuals were jointly discussed during the main hearing between January 15, and March 5, 2004. The documents of the planned extension and the executed assessments for the planning documents were reviewed intensively (HMWVL 2004: 58–63).[141]

141 See protocols RPDA (2004a); RPDA (2004b); RPDA (2004c); RPDA (2004d); RPDA (2004e); RPDA (2004f); RPDA (2004g); RPDA (2004h); RPDA (2004i); RPDA (2004j); RPDA (2004k); RPDA (2004l); RPDA (2004m); RPDA (2004n); RPDA (2004o); RPDA (2004p); RPDA (2004q); RPDA (2004r); RPDA (2004s); RPDA (2004t); RPDA (2004u); RPDA (2004v);

During the 30 trial days, 167 petitions were proposed (HMWVL 2004: 64). Many verbal statements of objectors were debated between the involved institutions and the individual private objectors. At the beginning, mainly the methods of the actual proceeding of the hearing were discussed (FAZ 2004d; see footnote for further detail).[142] The non-disclosure of the hearing agenda upfront was criticized as ridiculous and illegal (FNP 2004a; FNP 2004i; FR 2004a). The adjustment and the cancellation of the hearing was requested throughout the whole hearing (RPDA 2004d; RPDA 2004i: 7, 33, 52; RPDA 2004m: 41, 44). In addition, the cancellation due to prejudice was postulated by the environmental associations and local citizens' groups (BIMW 2004; BUND 2004a; FNP 2004e). All the petitions were discussed and later properly refused by the responsible authorities (FAZ 2004a; RPDA 2004a(a); RPDA 2004a(b)).

The RSE was further marked by the assessment and joint discussion of different challenges, resulting from the goals of regional planning at the local, regional, national, and European levels. The goals and the adjustment of the affected frameworks like the RPS2000 (RPDA 1999) and the LEPH2000 (HMWVL 2000) were broadly debated (RPDA 2004i: 17, 27; see footnote for further detail).[143] At the same time, a benchmark to other legally binding goals of the LuftVG was made (RPDA 2004t: 23f). In connection with that, lower-level planning frameworks of affected municipalities were also brought into the discussion. These frameworks also had to be adjusted (RPDA 2004i: 38–46). This discussions and adjustments were characterized by a defect in previous conversations, missing information pooling, and inadequate coordination processes between Fraport AG and the objectors (RPDA 2004l: 5f), Additionally, overlapping interests of national (RPDA 2004w: 19f), European (RPDA 2004s: 54f) and global (UN) goals were debated during the hearing (RPDA 2004aa: 37). Furthermore, the methodology concerning how necessary

RPDA (2004w); RPDA (2004x); RPDA (2004y); RPDA (2004aa); RPDA (2004a[c]); RPDA (2004a[d]); RPDA (2004a[e]); RPDA (2004a[f]).

142 FNP (2004h); RPDA (2004d); RPDA (2004f); RPDA (2004i); RPDA (2004l); RPDA (2004m); RPDA (2004o); RPDA (2004a(b)).

143 RPDA (2004m: 41, 56); RPDA (2004o: 23f, 32–35, 38–50); RPDA (2004q: 6); RPDA (2004r: 11f, 23).

knowledge for the planning documents was generated and provided, was criticized (RPDA 2004c: 3, 17; RPDA 2004p: 1–7, 16, 34f; RPDA 2004w: 10). As a matter of fact, the discussion of overall methodological questions was recognized under an extra topic of the RSE. For two days, only critiques regarding methodological questions were discussed (RPDA 2004a: 28–62; RPDA 2004b: 1–48).

The hearing also provided the opportunity to discuss the general necessity of the project at the site of Frankfurt Airport. Alternatives for the construction of the A380 Shipyard at Frankfurt Airport and the benchmark with other sites in Germany as well as Europe were largely discussed (RPDA 2004j; see footnote for further details).[144]

Likewise, the RSE jointly reviewed the consideration of the findings of the RPP in the extension plan, provided by Fraport AG (RPDA 2004o: 59–62; RPDA 2004r: 1–24). Not only was the RPP reassessed, but also the approach and results of RDF were discussed (RPDA 2004y: 33). Moreover, best practices were brought into the discussions (RPDA 2004a(c): 15).

Directly after the hearing, Fraport AG announced that it would integrate the new insights and the information gathered during the hearing into its further planning (Fraport 2004c).

Parted from the formal proceedings, the hearing, and the RSE were also heavily discussed in the general public. It was accompanied by the protests of affected citizens, supported by local citizens' groups and environmental associations (FR 2004f). Being part of the mediation process, the church was also involved in public discussions (FR 2004e). Regarding environmental issues, especially environmental associations, as well as municipalities and the *BÜNDNIS 90/DIE GRÜNEN* made further objections during and after the hearing (BUND 2004c). Additionally, statutory violations were remarked regarding the hearing as well as due to the plans (BUND 2004d; FNP 2004f; FNP 2004g). The *Planungsverband* did not actively engage in the hearing and took only a minor role in the public discussion (FR 2004b).

After the trial days, the Fraport AG revised the initial plan and took new insights of the hearing into account. This led to the decision not to build

144 RPDA (2004r: 32, 52); RPDA (2004u); RPDA (2004w); RPDA (2004x).

| # | Institutional form | | Institution | Delegates |
|---|---|---|---|---|
| | **Non-private** | | | |
| | | **Administrative** | ------------------------------------------------ | 17 |
| 1 | | | Federal Republic of Germany | 2 |
| 2 | | | Federal State of Hesse | 7 |
| 3 | | | RPDA | 1 |
| 4 | | | Sub-federal state level | 10 |
| | | **Metropolitan** | ------------------------------------------------ | 1 |
| 5 | | | Metropolitan Governance Bodies | 1 |
| | **Private** | | | |
| | | **Metropolitan** | ------------------------------------------------ | 0 |
| | | | - | |
| | | **Economic** | ------------------------------------------------ | 4 |
| 6 | | | Air transport | 1 |
| 7 | | | Others | 3 |
| | **Further** | | ------------------------------------------------ | 5 |
| 8 | | Association | | 5 |

*Figure 4.19: Institutional aggregation of stakeholders of the RSE A380 Frankfurt Airport, 2nd evaluation, own depiction, based on HMWVL 2004: 69–71.*

the planned car park, which reduced the impact on the protected forest areas. However, the environmental associations, the political parties, and the general public requested the repeated disclosure of the adjusted plans (GRÜNE 2004). On July 17, 2004, the Fraport AG dispatched the adjusted extension plan to the responsible authority again (Fraport 2004a). The adjusted plan contained – according to Fraport AG – ideas and knowledge, created during the previous hearing (Fraport 2004e).

These documents were sent again to the relevant institutions for peer review (HMWVL 2004: 65–69).

Disrespecting that for the second evaluation the singled departments of the RDPA were named, fewer institutions were included. However, non-private administrative institutions again represented the majority of institutions requested for a statement. The *Planungsverband* – as the metropolitan governance body – was actively involved, albeit no other non-private metropolitan governance body. Nonetheless, the private metropolitan logistics actors were not yet established. Compared with the previous hearing, less economic institutions were included. With the

*Bundesvermögensamt Frankfurt*, one institution not involved in the first peer review was integrated in the second.[145]

In the following, one further private institution was also involved by submitting the plans and further subject-specific documents.[146] The environmental associations were informed with an additional interim report on the flora-fauna compatibility study by the RP Darmstadt. The integration of further institutions and a second hearing was decided as unnecessary (HMWVL 2004: 73).

After the submission of the revised planning documents, the HMWVL was nominated as the responsible planning authority for the AoP (HMWVL 2004: 74).[147] It was also decided that a further environmental impacted assessment, including the opinion of the general public, had to be carried out.[148] Besides the official proceedings, Fraport AG and the environmental associations met to discuss and solve further challenges regarding the environment and pollution (Fraport 2004d). Nevertheless, the extension plan was discussed controversial and several municipalities still disapproved the extension due to the environmental impact and the expected pollution (FNP 2004b; FNP 2004c). Linked to several conditions, the HMWVL enacted the AoP on November 26, 2004 (HMWVL 2004; HMWVL 2004a).

Environmental associations announced legal action against the AoP immediately after the release (BUND 2004b; FAZ 2004b; FNP 2004d). Several municipalities took the case to court (FNP 2005b; FNP 2005m). This legal action was taken to all court instances, the HessVGH and BVerwG. Besides the environmental associations and the affected municipalities, *BÜNDNIS 90/DIE GRÜNEN* also argued in favor of legal action (BUND 2005a; BUND 2005b; FAZ 2005h; GRÜNE 2005). After the decision of the HessVGH, as of June 28, 2005, the AoP of

145 See Annex A.11 for more detail.

146 With the letters as of August 26, August 31, and September 10, 2004, the documents were dispatched to the legal representative of the BUND e.V. (HMWVL 2004: 72).

147 According to §10 II (1) s3 LuftVG, together with §1 II s1 *Verordnung zur Bestimmung von luftverkehrsrechtlichen Zuständigkeiten* (LuftV-ZustVO).

148 According §§8 I LuftVG, 3e I (2) UVPG, together with §3 c I UVPG. The inclusion of the general public had to be carried out according to §9 I UVPG occurred according to §10 II LuftVG, together with §73 III-VIII HVwVfG.

the A380 Shipyard was finally approved by the BVerwG on September 7, 2005 (BVerwG 4 B 49.05; HessVGH 12 A 8/05); see footnote for further detail).[149]. The demanded step toward the ECJ by environmental associations was rejected by the BVerwG (FAZ 2005b; FNP 2005h). However, the EC itself argued in favor of the extension of Frankfurt Airport (FR 2005g).

#### 4.3.6.2 *Interim findings: RSE and AoP, 2002–2004*

In the following, the RSE and AoP 2002–2004 are analyzed by the theoretical derived concept of ERG. The latent content of the five aspects of ERG is reconstructed in the regulatory governance structure and decision-making process of the RSE A380 Shipyard. This analysis is followed by establishing the missing link between the structural perspective of political science and the procedural character of policy-making, which Mayntz already proclaimed in the 1980s (Mayntz 1983), by discussing the institutional arrangement of that period.

*Discussion of shared perceptions of joint challenges*

The RSE A380 Shipyard was the institutionalized proceeding for policy-making to implement the planned project and design the political programming of spatial planning. With the dispatch of the planning documents, the preliminary hearing on the environmental impact assessment from April 7–11, 2003, and the actual RSE from January 15, and March 5, 2004, open, transparent multi-actor discussions across hierarchical and traditional borders were initiated. With the hearings, the RP Darmstadt – as the institution already in place – supported a recursive policy and decision-making processes as well as learning by experience from lower-level. The hearings integrated the sum of all relevant actors for a recursive policy-making in the decision-making process. The goal was the identification of topics, the best available techniques, the joint advancement, and to generate a common understanding of challenges regarding the airport extension and the RSE A380 Shipyard. Especially the hearings, but also with the surrounding discussion of the official proceedings, e.g. the meetings of Fraport AG and the environmental associations,

149 BVerwG (2005); FAZ (2005a); FAZ (2005d); FNP (2005i); FR (2005a).

created a heterarchical environment and were accompanied by learning from experience and tacit knowledge of the lower levels. Above all, the second hearing during the RSE linked stakeholders from all governance levels. The discussions led to continuous adjustment and learning during the proceeding. On the one hand, higher-level goals were discussed and adjusted, most remarkably the regional planning frameworks of RSP2000 and LEHP2000. On the other hand, the publication of the initial planning documents of the RSE A380 Shipyard led to the discussion of shared perceptions of joint challenges. Furthermore, the previous RPP as well as the approach and results of the RDF were debated. This knowledge found entry into the revised edition of the planning documents. Provision and the display of the planning documents, supported by the hearings are the central indications of the exchange and pooling of information.

However local citizens' groups and environmental associations were more closely involved, the delegation of mandate and a proactive advancing of mandate holders to increase ownership is hardly identified.

*Definition of goals and measurement systems*

The planning documents of the RSE A380 Shipyard and especially the related discussions in the preliminary hearing of the environmental impact assessment from April 7–11, 2003, and the actual hearing of the RSE from January 15, and March 5, 2004, indicate the identification and setting of high-level goals and common objectives. During the hearings, technical, but also procedural measurement systems were programmed. Both found their way in the decision-making process. Based on the most objective data, goals and the measurement systems were elaborated during the hearings. These hearings also served as the platform to discuss proposals with addressees. The already-proposed goals and measurement systems presented in the planning documents were extensively discussed together with the addressees and affected stakeholders. Especially the hearings of the RSE A380 Shipyard and the revision of the planning documents illustrate the establishment of a practical and workable implementation plan and feasible regulations for the airport extension. The planning documents and the AoP – finally approved on September 7, 2005 – are the main reference documents with recommendations and sets of legally (non-)binding rules. With the AoP and the planning documents, the RP

Darmstadt and Fraport AG presented guidance documents to clarify the approach of the airport extension.

*Implementation*

With the hearings from April 7–11, 2003, and between January 15, and March 5, 2004, a binding overall framework was established to further harmonize directives and regulations. The obligatory hearings created the capacities to have the most objective and imperial process to review and assess the planning documents and counter opinions. On the one hand, based on the hearing and statements given, the Fraport AG edited its planning documents and implemented the findings of the hearing accordingly. On the other hand, regional planning was adapted, most evident in the adjustment of the regional planning frameworks of RSP2000 and LEHP2000. The adjustment of existing rules was based on tacit knowledge and subsequent experience that became explicit during the decision-making process. As the responsible hearing authority, the RP Darmstadt delegated the task of implementation – the adaption of the planning documents during the negotiations – back to Fraport AG. The responsible regional planning authorities were entrusted with the implementation. With that, the RP Darmstadt entrusted lower-level units and delegate responsibilities and power to lower-level authorities or experts. Based on both, soft guidance and order, the actors and addressees had the most possible freedom and autonomy for implementation.

Active fostering of cooperation while implementation is hardly identified during the RSE A380 Shipyard. It can only be identified as a matter of fact, rather than actively promoted.

*Report and review*

The RSE A380 Shipyard, with the preliminary hearing on the environmental impact assessment from April 7–11, 2003, and the actual RSE from January 15, and March 5, 2004, presents an institutionalized assessment, including mandatory peer reviews of the plans. Besides the RP Darmstadt – as the responsible hearing authority – the involved institutions, such as non-private administrative and metropolitan as well as private economic and associations, peer-reviewed the planning documents and the decision-making process in detail. As institutions already in place, the RP Darmstadt and the RDF – partly common, partly separate – monitored and evaluated the proceeding. Based on cooperation and legal demand, a constitutional and collective regulatory dialog was initiated, mainly revealed in the obligatory hearings from April 7–11, 2003, and between January 15, and March 5, 2004. On the one hand, the Fraport AG and the RP Darmstadt had to justify planned and taken activities as well as performance, especially during both hearings. On the other hand, objectors were also demanded to continually justify their approaches. With that process, additional opinions and relevant scientific issues have been integrated into the decision-making process. Also, information was processed up- and downwards the levels, which was especially supported by the environmental associations and the *Bündnis90/Die Grünen* during the RSE A380 Shipyard.

By sending the planning documents to the affected state counties, municipalities, and further private and non-private institutions for review and the public display of the planning documents between June 23, and July 22, 2003, the processing of information up- and downwards hierarchy levels and the consultation of additional opinions becomes most evident. During the whole decision-making process, expert opinions, comments, and objections on the airport extension, the defined goals, and the measurement system were exchanged, especially regarding the challenges of pollution and environmental impact. Also, the decisions of the former RPP and the approach and results of the RDF were evaluated and peer-reviewed again.

*Review of the measurement system, the decision-making process, and the stakeholder*

The initial planning documents, sent by the Fraport AG to the RP Darmstadt on January 20, 2003, the intended goals and measurement system were assessed, questioned, and cross-checked. Also, the adjusted planning documents as of July 17, 2004, were assessed. During the decision-making process of the RSE A380 Shipyard, especially with the preliminary hearing on the environmental impact assessment from April 7–11, 2003, the actual RSE from January 15, and March 5, 2004, and several lawsuits technical, but also policy objectives, procedures of political programming, legislation, and the decision-making process of the planned extension of Frankfurt Airport were continually reviewed. The review came to a head by taking the planned extension in front of the BVerwG and the demand to take the case also in front of the ECJ.

This process revealed areas where adjustments were needed. That becomes most visible in the legal actions taken regarding the then-valid regional spatial planning acts (RSP2000 and LEPH2000), the adaption of lower-level land use planning, and the adaption of the planning documents, mainly conspicuous in the waiver of the planned car park to reduce the impact on the protected forest areas. However a third hearing was denied, the two official hearings give evidence to the integration of further stakeholders, e.g. by requesting further statements. The additional stakeholders have been formally addressed in the second hearing. With that, these institutions were officially part of the overall regulatory structure of governance and the decision-making process of the RSE A380 Shipyard and the AoP.

With the revised edition of the planning documents, taking new developments, and expert opinions into account, Fraport AG presented a suitable amendment. It was presented to the responsible authority – the RP Darmstadt. The following AoP, based on the planning documents, formulated by the HMWVL as the responsible planning authority, contained legally binding and non-binding rules for the further regional planning and the extension of Frankfurt Airport.

### Institutional arrangement of the RSE A380 Shipyard

Besides the sketch of the decision-making process, the previous paragraph also drew a clear picture of the institutional arrangement of the RSE A380 Shipyard. As presented, the RSE represents a process to establish heterarchical relations between private and non-private institutions as well as private individuals. They were linked across hierarchical and traditional borders. Hence, the RSE A380 Shipyard was characterized by an open multi-actor discussion.

Especially the hearings created the capacity to discuss the proposal of the planning documents with the addressees. In addition, the heterarchical character of the overall proceeding was fostered by the meetings of Fraport AG and the environmental associations to discuss common challenges and further special issues of the extension. These efforts involved institutions of all governance levels.

However, no non-private metropolitan governance body was associated with the RSE, neither by a statement nor by an active engagement during the trial days. Nonetheless, the private metropolitan logistics actors were not yet established.

The noninvolvement is remarkable. By analyzing the hearing, it becomes obvious that directly affected municipalities have common metropolitan denominators. They united and acted jointly; however, represented by lawyers and not by a metropolitan governance body.

**Confirmation of the research-guiding hypothesis for the RSE and AoP, 2002–2004**

This interim conclusion shows that the theoretical derived concept of ERG is a suitable analytical framework for analyzing the regulatory structure of governance, the decision-making processes, and the institutional arrangements in the case of the RSE and AoP, 2002–2004. With that, the research-guiding hypothesis can be confirmed for the RSE and AoP, 2002–2004. It again shows that the concept of ERG counterfeits the challenge of a missing link between the structural perspective of political science and the procedural character of policy-making, which Mayntz already proclaimed in the 1980s (Mayntz 1983), by including the analysis of the institutional arrangement of that period.

### 4.3.7 Road scheme evaluation and approval of plans, 2004–2007

With the letter as of February 8, 2000, Fraport AG requested the extension of the existing runway system of Frankfurt Airport.[150] As the planned extension represented a substantial modification, there was the need for an RSE.[151]

---

150 According to §45 II LuftVZO. Fraport AG requested permission to construct the Runway North-West including all necessary facilities, a fast taxiway to Runway North-West as well as two further taxiways. Additionally, the Fraport AG requested permission to build a fire station in the area of Runway North-West, several apron facilities, further fast taxiways in the existing runway system, the installation of outer marker beacons, the construction of Terminal 3, as well as the rebuilding of existing constructions and development of further new facilities (HMWVL 2007: 184–195).

151 According to §8 LuftVG.

#### *4.3.7.1 RSE Runway North-West/Terminal 3*

Fraport AG presented the planning documents of the planned extension with the letter as of September 9, 2003. After the RP Darmstadt requested additional information, the first draft of the planning documents comprised 60 files (Fraport 2004b). As the responsible hearing authority, the RP Darmstadt officially initiated the RSE after this first adjustment.[152]

From December 14–17, 2004, upfront the display of the planning documents, 206 institutions were requested directly for a statement on the planned extension (Annex A.12). The RP Darmstadt requested the involved municipalities to display the planning documents from January 17, to February 16, 2005, to comply with the obligation of public participation.[153] In this time and two weeks after, everybody who was affected by the planned extension had the chance to appeal against the plans. For the RSE Runway North-West/Terminal 3, non-residents of fifteen municipalities were also informed about the extension plan.[154] The planning documents were dispatched to further private and non-private institutions, with the request of a statement.

The majority of the requested institutions were non-private administrative. With the *Regierungspräsidium Karlsruhe* (RP Karlsruhe), the jurisdiction was not only restricted to the Federal State of Hesse. Additionally, two metropolitan governance bodies were requested to state the planning documents. With the *Planungsverband*, the officially recognized metropolitan governance body for FRM was included. Even though LRM was in its planning phase, the private metropolitan logistics actors were still not fully established. Most of the involved private economic institutions were non-air transport companies. Only one company came from that sector. The further institutions were represented by associations, chambers, and the church (Fig. 4.20).[155]

Together with the official statements, the open discussion led to 127,198 objections by 123,577 objectors. These objections were systematically

---

152 According to §10 II LuftVG, together with §73 II HVwVfG.
153 According to §10 II (3) LuftVG, together with §73 V HVwVfG.
154 According to §11 II LuftVG, together with §73 II HVwVfG and §35 *Hessisches Naturschutzgesetz* (HENatG).
155 See Annex A.12 for more detail.

| # | Institutional form | | Institution | Delegates |
|---|---|---|---|---|
| | Non-private | | | |
| | | Administrative | ------------------------------------------------ | 136 |
| 1 | | | Federal Republic of Germany | 8 |
| 2 | | | Federal State of Hesse | 21 |
| 3 | | | RPDA | 34 |
| | | | RP Karlsruhe | 1 |
| | | | RP Kassel | 1 |
| 4 | | | Sub-federal state level | 71 |
| | | Metropolitan | ------------------------------------------------ | 2 |
| 5 | | | Metropolitan Governance Bodies | 2 |
| | Private | | | |
| | | Metropolitan | ------------------------------------------------ | 0 |
| | | | - | |
| | | Economic | ------------------------------------------------ | 41 |
| 6 | | | Air transport | 1 |
| 7 | | | Others | 40 |
| | Further | | ------------------------------------------------ | 27 |
| 8 | | Association | | 19 |
| | | Chamber | | 7 |
| | | Church | | 1 |

*Figure 4.20: Institutional aggregation of stakeholders of the RSE Runway North-West/Terminal 3 Frankfurt Airport, 1st evaluation, own depiction, based on HMWVL 2007: 212–221.*

processed for the following hearing. Of the objectors, 116 were granted anonymity (HMWVL 2007: 209–211).

After the objection period, all objections made in a timely manner, the planning documents, and executed assessments were discussed jointly at the hearing from September 12, 2005,[156] to March 27, 2006, on 101 trail days (HMWVL 2007: 223f).[157] This included further trial days, which

156 With the start of the hearing, the constructions and clearing of woodland for the A380 Shipyard started simultaneously (FR 2005n; FR 2005o).

157 See protocols RPDA (2005a), RPDA (2005b), RPDA (2005c), RPDA (2005d), RPDA (2005e), RPDA (2005f), RPDA (2005g), RPDA (2005h), RPDA (2005i), RPDA (2005j), RPDA (2005k), RPDA (2005l), RPDA (2005m), RPDA (2005n), RPDA (2005o), RPDA (2005p), RPDA (2005q),

became necessary due to of new insights discovered (RPDA 2006a(y); RPDA 2006a(z); RPDA 2006ba). The trial days were announced by the RP Darmstadt and the date was made public in around 50 daily newspapers and further media in FRM (RPDA 2005be). Already ahead of the hearing, petitions of several municipalities to stop the proceeding were announced (FR 2005e).

Nevertheless, the hearing was started accordingly. It followed an agenda made public upfront (RPDA 2015bg: 11–17). This was the first time that an agenda of a hearing in the extension process of Frankfurt Airport had been made public *ex ante*. Its publication made it easier for private individuals, not exempt by their employer, to participate. It was now assessable when the own topic and objection were jointly discussed. During the 101 trial days, besides the 65 files of planning documents (Fraport 2004b), an additional 104 files[158] were displayed and jointly discussed. Also, a telephone hotline was established, and an official

RPDA (2005r), RPDA (2005s), RPDA (2005t), RPDA (2005u), RPDA (2005v), RPDA (2005w), RPDA (2005x), RPDA (2005y), RPDA (2005aa), RPDA (2005ab), RPDA (2005ac), RPDA (2005ad), RPDA (2005ae), RPDA (2005af), RPDA (2005ab), RPDA (2005ah), RPDA (2005ai), RPDA (2005aj), RPDA (2005ak), RPDA (2005al), RPDA (2005am), RPDA (2005an), RPDA (2005ao), RPDA (2005ap), RPDA (2005aq), RPDA (2005ar), RPDA (2005as), RPDA (2005at), RPDA (2005au), RPDA (2005av), RPDA (2005aw), RPDA (2005ax), RPDA (2005ay), RPDA (2005az), RPDA (2005ba), RPDA (2005bb), RPDA (2005bc), RPDA (2005bd), RPDA (2006a), RPDA (2006b), RPDA (2006c), RPDA (2006d), RPDA (2006e), RPDA (2006f), RPDA (2006g), RPDA (2006h), RPDA (2006i), RPDA (2006j), RPDA (2006k), RPDA (2006l), RPDA (2006m), RPDA (2006n), RPDA (2006o), RPDA (2006p), RPDA (2006q), RPDA (2006r), RPDA (2006s), RPDA (2006t), RPDA (2006u), RPDA (2006v), RPDA (2006w), RPDA (2006x), RPDA (2006aa), RPDA (2006ab), RPDA (2006ac), RPDA (2006ad), RPDA (2006ae), RPDA (2006af), RPDA (2006ag) RPDA (2006ah), RPDA (2006ai), RPDA (2006aj), RPDA (2006ak), RPDA (2006al), RPDA (2006am), RPDA (2005an), RPDA (2006ao), RPDA (2006ap), RPDA (2006aq), RPDA (2006ar), RPDA (2006as), RPDA (2006at), RPDA (2006au), RPDA (2006av), RPDA (2006aw).

158 Including 85 files of statements and objections, 11 files of basic files, four files of scoping papers of the environmental impact assessment, and four files of the flora-fauna-compatibility study (HMWVL 2007: 235). However, the access to the statements and objections is restricted by the RP Darmstadt, the

website informed about the status quo of the hearing (HMWVL 2007: 234–236). Both, the hotline and the webpage were promoted throughout the hearing by the media (FNP 2005f). An information point was set up, where further information that became relevant during the hearing was made available to the participants of the hearing (RPDA 2005ah: 97; RPDA 2005ai: 88). Nevertheless, the distribution and processing of information before and during the hearing was criticized. This issue was also taken in front of the court (RPDA 2006i; RPDA 2006n: 10–15, 57). Several petitions to completely stop the hearing and the RSE were made. Possible statutory violations were remarked, especially with respect to the processing of relevant (environmental) information up- and downwards (RPDA 2005a; see footnote for further detail).[159]

All petitions proclaimed in the hearing were processed to the responsible authority and – if necessary – legal action was initiated (HMWVL 2007: 236; RPDA 2005bf).

As announced by the RP Darmstadt, the participation should be easier, than in the former RSE A380 Shipyard hearing (FR 2005m).[160] The objectors criticized the agenda set upfront. They requested an open discussion, whereby each objector could speak freely about his/her objections whenever desired (RPDA 2005h: 34). In the further hearing – as one of the major facet – challenges of paramount regional, national, European, and international goals were discussed from the very beginning (RPDA 2005d: 9, 12, 38; see footnote for further details).[161] Furthermore, the challenges regarding spatial planning were discussed in detail on extra trial days, focusing only on the RPS2000 and the LEPH2000 (RPDA

---

documents were made available by a lawyer for the presented research by signing a confidential note.

159 RPDA 2005b: 49; RPDA (2005d); RPDA (2005f); RPDA (2005x): 23, 79.

160 The media, partly perceived it from another perspective, see, e.g. taz (2005).

161 RPDA 2005f: 33f, 44; RPDA 2005i; RPDA 2005l: 135; RPDA 2005n: 116; RPDA 2005r: 126-130; RPDA 2005x: 55, 102; RPDA 2005y: 51f; RPDA 2005af: 8-10, 13f, 17, 35f, 45f; RPDA 2005ab: 14, 33, 60f, 97-99, 114f, 78f, 129; RPDA 2005al; RPDA 2005am; RPDA 2005ba: 3-7; RPDA 2005bb: 91f; RPDA 2006b: 44f; RPDA 2006h: 48-51; RPDA 2006j: 58; RPDA 2006l; RPDA 2006n: 77; RPDA 2006o: 90-92; RPDA 2006v; RPDA 2006aj; RPDA 2006ak.

2006e; RPDA 2006f; RPDA 2006g). Additionally, the administrative stipulated goals of the Fraport AG were discussed in this context (RPDA 2005f: 86; RPDA 2005i).[162] The description and presentation of the goals presented in the RSE Runway North-West/Terminal 3 planning documents were contested and jointly discussed as well (RPDA 2005y: 9). It was further criticized, that the methodological options of the official proceedings were not utilized in full (RPDA 2005d: 75). However, not only the non-utilization but also how methods and measurement systems were utilized was critically discussed – e.g. regarding noise emission (RPDA 2005d: 13). In this manner, planned goals and measurement systems of executed studies and assessments of different relevant fields, which were the basis for the RSE, were peer-reviewed and broadly debated. Their adjustment was requested and discussed likewise (RPDA 2005c: 123; see footnote for further details).[163]

The topic of measurement was not only strongly discussed in the official proceeding. Additionally, in the media the executed assessments were widely criticized, especially in terms of noise emission (FAZ 2005c; see footnote for further details).[164] The concrete situation and opportunities of

162 The operating company of an airport has to provide the needed infrastructure to operate an airport (RPDA 2005f: 86). This understanding can be derived from Art. 87e IV GG. The Federal State of Germany has a governmental responsibility for infrastructure provision, including airports (Hösch 2009: 73).

163 RPDA 2005i; RPDA 2005k: 16f, 104f, 113; RPDA 2005l: 89, 137f; RPDA 2005m: 67f, 74–76, 107–109, 127–129; RPDA 2005n: 112–115, 123; RPDA 2005o: 124; RPDA 2005p: 14, 51f, 61, 66–86; RPDA 2005q; RPDA 2005s; RPDA 2005t: 65, 148, 155; RPDA 2005v: 11–20, 23f, 30–37; RPDA 2005x: 5, 63, 95–97; RPDA 2005aa; RPDA 2005ab; RPDA 2005ac; RPDA 2005ad; RPDA 2005af: 5–7; RPDA 2005ai; RPDA 2005aj; RPDA 2005ak; RPDA 2005am; RPDA 2005ao; RPDA 2005ap; RPDA 2005aq; RPDA 2005ar: RPDA 2005as; RPDA 2005at; RPDA 2005au: 29–31; RPDA 2005aw; RPDA 2005ax; RPDA 2005ay; RPDA 2005az: 21–23; RPDA 2005ba: 38f; RPDA 2005bb: 38f, 68–71; RPDA 2006h; RPDA 2006j; RPDA 2006l; RPDA 2006o: 89f; RPDA 2006v; RPDA 2006x; RPDA 2006aa; RPDA 2006ad; RPDA 2006ag; RPDA 2006ah; RPDA 2006ai: 15–32, 36, 48, 72–77; RPDA 2006aj; RPDA 2006al; RPDA 2005an: 77, 82; RPDA 2006ao; RPDA 2006ap; RPDA 2006as; RPDA 2006av: 73f.

164 FNP (2005e); FR (2005f); FR (2005h); MS (2005).

noise reduction at Frankfurt Airport were also reviewed and benchmarked with other cases (RPDA 2005e: 100; RPDA 2005y: 47; RPDA 2005av).

After the hearing, the RP Darmstadt – as the hearing authority – released a full statement on the trial days. Additionally, it made the basic results of the hearing available to the broader public, not directly involved (RPDA 2006ax).[165] With the end of the hearing, the HMWVL became the responsible planning authority (HMWVL 2007: 237).

Before and during the hearing, the local citizens' groups have started to engage more again. In addition to the assessments presented by Fraport AG, the local citizens' groups assigned assessments and studies likewise (FNP 2005c; FR 2005i; FR 2005o). Objections in terms of stopping the hearing and the RSE, appealing on concessions made in the RSE of 1971, were refused (FNP 2005l; FR 2005l). Besides, the supporters started to engage and act in favor of the extension (VhU 2005).

However, the expected resistance and the public participation in the meetings remained behind the exceptions (FAZ 2005i; see footnote for further details).[166] Some voices argued, that this is due to the chosen locality, where the hearing was held (FAZ 2005f; FNP 2005d; FR 2005d). Not only the local citizens' groups but also the political parties asked the general public to participate in the hearing with increased intensity (FNP 2005f).

The hearing was characterized by the presence and argumentation of lawyers, especially in the beginning (AZ 2005; see footnote for further details).[167] This became especially applicable with the request of environmental impact information, which was taken to court (WK 2005a). The HessVGH decided in this regard, that the RP Darmstadt had to guarantee inspection of the files compiled by Fraport AG (HessVGH 2006).[168] This was also the time when the hearing was most strongly discussed, both in public and by politicians (CDU 2006; see footnote for further details).[169]

---

165 See RPDA (2015).
166 FNP 2005g; FR 2005c; FR 2005k.
167 FAZ 2005g; FAZ 2005i; FNP 2005a; FNP 2005k; WK 2005b.
168 Decision 12 Q 2828/05.
169 FAZ 2006; FDP 2006; FNP 2006; FR 2006; GRÜNE 2006; SPD 2006.

Concerning the schedule, it was criticized that the affected citizens could not attend the trial days, first regarding the location and second in terms of their working hours (FAZ 2005e; FR 2005b). The schedule was adjusted during the hearing. This made it easier for private objectors to express their objections (FNP 2005j).

Fraport AG already announced during the hearing that it would adjust the planning documents after the hearing (FR 2005j; Fraport 2004f). Based on the gathered information in the hearing, the statements, and objections, Fraport AG adjusted the planning documents accordingly. After the requested changes – especially regarding maneuvering surface, structural engineering, road infrastructure, supply, and disposal, as well as environmental issues and further minor subjects – the planning documents covered 65 files (Fraport 2007a). From March 23, to April 23, 2007, the adjusted planning documents were displayed again in 65 affected municipalities. It was again possible to appeal against the plans until two weeks after the public disclosure. Non-residents of 15 municipalities were additionally informed about the extension plan (HMWVL 2007: 237–244).[170] With the letter as of March 1, 2007, again several private and non-private institutions were directly requested for a statement.

As in the first request, most of the requests were sent to non-private administrative institutions. Again, the *Planungsverband* – as the official metropolitan governance body – was involved. Furthermore, a third regional governance body was integrated this time. Again, only one company – as a private economic institution from the sector of air transport – was requested to give a statement. The number of further institutions has risen. What is most conspicuous is the active integration of legal representatives (Fig. 4.21).[171]

Of the 261 institutions requested to give a statement 116 responded accordingly.

Besides the low return rate, one aspect is most noticeable. No legal representative stated at this point. However, this time the *Planungsverband* actively engaged with a statement. LRM was not finally established. The

170 According to §11 II LuftVG, together with §73 II HVwVfG and §35 HENatG.

171 See Annex A.13 for more detail.

| # | Institutional form | | Institution | Delegates |
|---|---|---|---|---|
| | **Non-private** | | | |
| | | **Administrative** | ---------------------------------------- | **135** |
| 1 | | | Federal Republic of Germany | 9 |
| 2 | | | Federal State of Hesse | 25 |
| 3 | | | Federal State of Rhineland-Palatinate | 3 |
| 4 | | | RPDA | 8 |
| 5 | | | Regierungspräsidium Karlsruhe | 1 |
| 6 | | | Regierungspräsidium Kassel | 1 |
| 7 | | | Sub-federal state level | 88 |
| | | **Metropolitan** | ---------------------------------------- | **3** |
| 8 | | | Metropolitan Governance Bodies | 3 |
| | **Private** | | | |
| | | **Metropolitan** | ---------------------------------------- | **0** |
| | | | - | |
| | | **Economic** | ---------------------------------------- | **58** |
| 9 | | | Air transport | 1 |
| 10 | | | Others | 57 |
| | **Further** | | ---------------------------------------- | **65** |
| 11 | | Association | | 39 |
| 12 | | Chamber | | 9 |
| 13 | | Church | | 2 |
| 14 | | Legal Representatives | | 15 |

*Figure 4.21: Institutional aggregation of stakeholders of the RSE Runway North-West/Terminal 3 Frankfurt Airport, 2nd evaluation, own depiction, based on HMWVL (2007): 245–257.*

other non-private logistics actors were also not yet initiated. Again non-public administrative institutions were the most commonly-represented group (Fig. 4.22).[172]

Overall, 23,984 different objections of 27,818 objectors were made regarding the revised planning documents (HMWVL 2007: 245–258). The objections of the general public were again systematically processed, and the information was given to Fraport AG (HMWVL 2007: 260).

After the adjustment and the dispatch of the planning documents, a second full RSE was decided as being overly extensive according to

172 See Annex A.14 for more detail.

| # | Institutional form | | Institution | Delegates |
|---|---|---|---|---|
| | Non-private | | | |
| | | Administrative | ---------------------------------------- | 81 |
| 1 | | | Federal Republic of Germany | 7 |
| 2 | | | Federal State of Hesse | 12 |
| 3 | | | RPDA | 7 |
| 4 | | | Sub-federal state level | 55 |
| | | Metropolitan | ---------------------------------------- | 2 |
| 5 | | | Metropolitan Governance Bodies | 2 |
| | Private | | | |
| | | Metropolitan | ---------------------------------------- | 0 |
| | | | - | |
| | | Economic | ---------------------------------------- | 19 |
| 6 | | | Air transport | 1 |
| 7 | | | Others | 18 |
| | Further | | ---------------------------------------- | 14 |
| 8 | | Association | | 8 |
| 9 | | Chamber | | 5 |
| 10 | | Church | | 1 |

*Figure 4.22: Institutional aggregation of responding institutions of the RSE Runway North-West/Terminal 3 Frankfurt Airport, 2nd evaluation, own depiction, based on HMWVL (2007): 260–265.*

the requested changes. However, in cases where threatening of existence occurred by the planned extension, single hearings were arranged with the affected stakeholders.[173] The further public discussions focused on environmental questions in particular (HMWVL 2007: 266–269).

Besides this approach, private and non-private institutions, the general public, and the planning authority requested further clarification regarding several issues. In its 47 clarification requests between 2005 to 2007, the RP Darmstadt requested further explanations on topics like air traffic foresight, environmental impact, as well as risk and security (Annex A.15).[174] In addition, non-private institutions, the *Hessisches*

173 According to §10 II No 6 LuftVG, together with §73 VI HVwVfG.

174 The Fraport AG dispatched the first clarification paper on December 16, 2005. The last clarification paper was transmitted on November 30, 2007.

*Landesanstamt für Umwelt und Geologie*, the *Hessisches Ministerium für Umwelt, ländlicher Raum und Verbraucherschutz* (HMULV) and the *Hessisches Landesanstamt für Straßen- und Verkehrswesen* requested additional clarification. Further discussions with the DFS, regarding air traffic, were held and twelve comprehensive assessments were requested additionally (HMWVL 2007: 289–293).

With the letter as of November 9, 2007, the RP Darmstadt dispatched all necessary documents to the BMVBS for its assessment.[175] After the BMVBS informed the RP Darmstadt with its letter as of December 12, 2007, that no public interest of the Federal State Germany is affected by the extension and also the HMULV confirmed the environmental compatibly, the HMWVL released the AoP (HMWVL 2007) on December 18, 2007 (HMWVL 2007: 293f).

After the approval, supporters of the extension appreciated the decision in favor of the extension (BARIG 2007; see footnote for further details).[176] Opponents criticized the AoP and announced that they would take the decision to court (BUND 2007; GRÜNE 2007; FR 2007).

Nevertheless, not only opponents wanted to take legal action. Moreover supporters such as DLH, DLH Cargo, TUI, and *Deutsche Post*, announced that they would take legal action to counterfeit conditions of night flights and emission (FR 2008b; FR 2008d). Furthermore, several municipalities announced legal action, including Frankfurt am Main, which was forced by its city council to prepare a lawsuit (FAZ 2008; see footnote for further detail).[177] However, the lawsuit was decided and rejected by the HessVGH on August 21, 2009 (HessVGH 2009).[178] Additionally, on April 4, 2012,

175 According to §31 II 4 LuftVG.

176 CDU 2007; FDP 2007; Fraport 2007b; VhU 2007.

177 FNP 2008a; FNP 2008b; FR 2008a; FR 2008c.

178 The general decision refereed to the following lawsuits: 11 C 227/08.T (Stadt Offenbach am Main); 11 C 305/08.T (Tanklager Raunheim GmbH); 11 C 312/08.T (Stadt Mörfelden-Walldorf); 11 C 318/08.T (BUND); 11 C 321/08.T (Stadt Neu-Isenburg); 11 C 329/08.T (Stadt Raunheim); 11 C 336/08.T (Stadt Rüsselsheim); 11 C 349/08.T (Lufthansa AG und Lufthansa Cargo AG); 11 C 359/08.T (Gründstückseigentümer aus Frankfurt-Sachsenhausen); 11 C 499/08.T (Klinikum Offenbach GmbH); 11 C 509/08.T (Gründstückseigentümer aus Kelsterbach) (HessVGH 2009).

the BVerwG confirmed the decision of the HessVGH and decided upon the last open question regarding the ban of night flights (BVerwG 2012).[179]

### *4.3.7.2 Interim findings: RSE and AoP, 2004–2007*

In the following, the RSE and AoP 2004–2007 are analyzed by the theoretical derived concept of ERG. The latent content of the five aspects of ERG is reconstructed in the regulatory governance structure and decision-making process of the RSE Runway North-West/Terminal 3. This analysis is followed by establishing the missing link between the structural perspective of political science and the procedural character of policy-making, which Mayntz already proclaimed in the 1980s (Mayntz 1983), by discussing the institutional arrangement of that period.

*Discussion of shared perceptions of joint challenges*

The RSE Runway North-West/Terminal 3 was the institutionalized proceeding for policy-making to implement the planned project and design the political programming of spatial planning. With the dispatch of the planning documents on September 9, 2003, by Fraport AG, the request for statements on the already updated planning documents in December 2004, the public display between January 17, and February 16, 2005, the hearing between September 12, 2005, and March 27, 2006, the additional trail days, which become necessary during the decision-making process, as well as the 47 clarification requests between 2005 and 2007, open, transparent multi-actor discussions across hierarchical and traditional borders were initiated. With the hearings, the RP Darmstadt – as the institution already in place – supported a recursive policy- and decision-making processes as well as learning by experience from lower-level. The hearings between September 12, 2005, and March 27, 2006, integrated the sum of all relevant actors for a recursive policy-making in the decision-making process. The goal was the identification of topics, best available techniques, the joint advancement, and to generate a common understanding of challenges regarding the airport extension and the RSE Runway North-West/Terminal 3. Especially the hearings, but also the further activities like the request for statements, the information

179 Decision BVerwG 4 C 8.09.

of further non-residents or the additional hearings of extensively affected stakeholders, created a heterarchical environment and were accompanied by learning from experience and tacit knowledge of the lower levels. Above all, the 101 trail days of the official proceeding and decision-making process linked stakeholders from all governance levels. The discussions, but also several lawsuits led to continuous adjustment and learning during the proceeding. On the one hand, higher-level goals were discussed and adjusted. On the other hand, the discussion of the initial planning documents of the RSE Runway North-West/Terminal 3, already beforehand the official proceedings between the RP Darmstadt and the Fraport AG, but especially during the hearings between September 12, 2005, and March 27, 2006, led to the discussion of shared perceptions of joint challenges. By that, tacit knowledge found entry into the revised edition of the planning documents. Provision and the display of the planning documents, supported by the hearings are the central indications of the exchange and pooling of information. Furthermore, the official decision-making process and the pooling of information was supported by a telephone hotline, an official website, and an information booth during the hearing.

However, local citizens' groups and environmental associations were more closely involved and political parties asked the general public to participate in the hearing with higher intensity, the delegation of mandate and a proactive advancing of mandate holder to increase ownership is hardly identified. Rather, the discussion of shared perceptions of joint challenges – especially in the beginning of the hearings – was more characterized by a legal dispute and discussion during the hearing.

*Definition of goals and measurement systems*

The planning documents of the RSE Runway North-West/Terminal 3 and especially the related discussions beforehand the official proceeding of the RSE, the request for statements on the already updated planning documents in December 2004, the hearings between September 12, 2005, and March 27, 2006, the additional trail days, and the 47 clarification requests between 2005 and 2007, indicate the identification and setting of high-level goals and common objectives. During the hearing, technical, but also procedural measurement systems were analyzed and programmed.

Both found their way in the decision-making process. Based on the most objective data, goals, and the measurement systems were elaborated during the hearings. These hearings also served as the platform to discuss proposals with addressees. The already-proposed goals and measurement systems presented in the planning documents were extensively discussed together with the addressees and affected stakeholders. Especially the prior discussions between the RP Darmstadt and the Fraport AG, the 101 trail days of the RSE Runway North-West/Terminal 3 hearings, the additional hearings, and the revision of the planning documents indicate the establishment of a practical and workable implementation plan and feasible regulations for the airport extension. The planning documents and the AoP, finally approved on September 7, 2005, are the main reference documents with recommendations and sets of legally (non-)binding rules. On December 18, 2007, with the AoP and the revised edition of the planning documents, the RP Darmstadt and Fraport AG presented guidance documents to clarify the approach of the airport extension.

*Implementation*

With 101 trail days between September 12, 2005, and March 27, 2006, a binding overall framework was established to further harmonize directives and regulations. Together with the telephone hotline, an official website, and an information booth, the obligatory hearings created the capacities to have the most objective and imperial process to review and assess the planning documents and counter opinions. Based on the hearing and statements given, the Fraport AG edited its planning documents and implemented the findings of the hearing accordingly – especially regarding technical topics and environmental challenges identified during the decision-making process. The adjustment of existing rules was based on tacit knowledge and subsequent experience that became explicit during the proceeding. As the responsible hearing authority, the RP Darmstadt delegated the task of implementation – the adaption of the planning documents during the negotiations – back to Fraport AG. With that, the RP Darmstadt entrusted lower-level units and delegate responsibilities and power to lower-level authorities or experts. Based on soft guidance and order, the Fraport AG had the most possible freedom and autonomy for implementation.

Also, the active fostering of cooperation was identifiable during the RSE Runway North-West/Terminal 3. Besides official efforts, such as the telephone hotline, the official website, and the information booth, local citizens' groups and political parties asked the general public to participate in the hearing with increased intensity.

*Report and review*

The RSE Runway North-West/Terminal 3, especially the requests for statement and the trail days between September 12, 2005, and March 27, 2006, presents an institutionalized assessment, including mandatory peer reviews of the plans. Besides the RP Darmstadt – as the responsible hearing authority – the involved institutions, such as non-private administrative and metropolitan as well as private economic and associations peer-reviewed the planning documents and the decision-making process in detail. As institutions already in place and responsible for the decision-making process, the RP Darmstadt monitored and evaluated the proceeding. Besides the RP Darmstadt, local citizens' groups and the general public observed the decision-making process. Based on cooperation and legal demand, a constitutional and collective regulatory dialog was initiated, mainly revealed in the obligatory hearings from September 12, 2005, and March 27, 2006. On the one hand, the Fraport AG and the RP Darmstadt had to justify planned and taken activities as well as performance, especially during the hearings. On the other hand, objectors were also demanded to continually justify their approaches. With that process, additional opinions and relevant scientific issues have been integrated into the decision-making process. Also, information was processed up- and downwards the levels. Utilizing the possibilities provided by ICT, the formal proceeding included a telephone hotline, an official website, and an information booth. By sending the planning documents to the affected state counties, municipalities and further private and non-private institutions for review and the public display of the planning documents between January 17, and February 16, 2005, and actively requesting statements, the processing of information up- and downwards hierarchy levels and the consultation of additional opinions becomes most evident.

During the whole decision-making process, expert opinions, comments, and objections on the airport extension, the defined goals, and the measurement system were exchanged, especially regarding the challenges of pollution and environmental impact. By taking several issues to court, also the legal review was conspicuous in the RSE Runway North-West/Terminal 3.

*Review of the measurement system, the decision-making process, and the stakeholders*

The initial planning documents, sent by the Fraport AG to the RP Darmstadt on September 9, 2003, the intended goals and measurement system were assessed, questioned, and cross-checked. During the decision-making process of the RSE Runway North-West/Terminal 3, especially with the hearings between September 12, 2005, and March 27, 2006, and several lawsuits technical, but also policy objectives, procedures of political programming, legislation, and the decision-making process of the planned extension of Frankfurt Airport were continually reviewed. The review came to a head by taking the planned extension in front of the BVerwG, which finally confirmed the decision of the HessVGH and decided upon the last open question regarding the ban of night flights on April 4, 2012. This process revealed areas where adjustments were needed. However a third hearing was decided as unnecessary, the official decision-making process gives evidence to the integration of further stakeholders. This is indicated by requesting further statements and the direct involvement of stakeholders who were threatening of existence occurred during the decision-making process with additional hearings. With that, all these institutions and private individuals were officially part of the overall regulatory structure of governance and the decision-making process of the RSE Runway North-West/Terminal 3 and the AoP.

The revised edition of the planning documents, taking new developments and expert opinions into account, Fraport AG presented a suitable amendment. It was presented to the responsible authority – the RP Darmstadt. The HMWVL – as the responsible planning authority – formulated the AoP as of December 18, 2007. It contained legally binding and non-binding rules for further regional planning and the extension of Frankfurt Airport.

**Institutional arrangement of the RSE Runway North-West/Terminal 3**

Besides the clear depiction of the decision-making process, the previous analysis also determined the institutional arrangement of the regulatory structure of governance during the RSE Runway North-West/Terminal 3. Establishing heterarchical capacities between private and non-private institutions as well as private individuals to discuss shared perceptions of common challenges, the RSE Runway North-West/Terminal 3 represented the institutionalized process to plan and execute the further extension. The RSE and the related hearings created an open multi-actor discussion beyond hierarchical and traditional borders. It included all governance levels and the sum of relevant actors.

After not being involved in the former RSE, one non-private governance body was engaged in the RSE Runway North-West/Terminal 3. Besides its official statement, the Planungsverband was also actively involved in the hearing, albeit not extensively. Most of the private metropolitan logistics actors were still not yet established. Only the LRM was initiated at the very end of the official proceeding, although it was not involved.

More far-reaching than before, a joint approach of directly affected municipalities is identifiable. Still arguing for their own sake, on the one hand, they were extensively represented jointly by lawyers on the other hand. Additionally, private individuals were associated with the joint approach of the non-private administrative institutions, being represented by lawyers.

**Confirmation of the research-guiding hypothesis for the RSE and AoP, 2004–2007**

This interim conclusion shows that the theoretical derived concept of ERG is a suitable analytical framework for analyzing the regulatory structure of governance, the decision-making processes, and the institutional arrangements in the case of the RSE and AoP, 2004–2007. With that, the research-guiding hypothesis can be confirmed for the RSE and AoP, 2004–2007. It shows that the concept of ERG counterfeits the challenge of a missing link between the structural perspective of political science and the procedural character of policy-making, which Mayntz already proclaimed in the 1980s (Mayntz 1983), by including the analysis of the institutional arrangement of that period.

### 4.3.8 Adjustment of plans, 2007 and beyond

After the AoP of December 18, 2007, and the decision of HMWVL (2007), the first constructions were carried out and the Runway North-West was brought into service in 2011. In the following years, several adjustments were made. Some had to be decided by a supplementary RSE and AoP. These minor proceedings do not necessarily include a broader involvement of third parties. Nevertheless, in all cases, the plans of adjustment were discussed with the probably-affected institutions.

This paragraph examines four supplementary RSEs (HMWVL 2013a; HMWVL 2013b, HMWVL 2013c; HMWEVL 2014) in their historical appearance to reconstruct and analyze the adjustments according the theoretical derived concept of ERG.

#### *4.3.8.1 Adjustment 66p 01.03.04/27, April 30, 2013*

With the confirmation of the decision of the HessVGH on April 4, 2012, the BVerwG decided also that the AoP regarding regulations concerning noise reduction measures and soundproofing for commercial real estate properties, as defined in A XI 5.1.3 of the AoP, was rescinded (HMWVL 2007: 143f). According to the decision, the planning had to be adjusted (BVerwG 2012).[180] Therefore, the HMWVL initiated an adjustment of the AoP.

After the necessary preliminary investigation, the HMWVL informed Fraport AG with the letter as of January 15, 2013, that the adjustment of plans is carried out. Due to that, Fraport AG had to gather and pool the relevant information regarding the findings of the preliminary investigation concerning regulations concerning noise reduction measures and soundproofing for commercial real estate properties. The authorization to correct the findings and errors is accepted by German given law.[181] As the

180 Decision BVerwG 4 C 8.09.
181 BVerwG 4 B 17.05.

responsible authority, the HMWVL provided a draft of the adjustments respecting the judgment of the BVerwG to Fraport AG. Fraport AG and the affected private economic institutions were requested for statements. All institutions took the possibility to reply. Afterward, nine major topics of the draft to adjust the plans were jointly discussed mainly between the Fraport AG and HMWVL. Requests for further assessments and expert opinions were refused. A further environmental impacted assessment was decided as not necessary.

For the requested correction of defects identified by the BVerwG, an additional supplementary RSE was executed.[182] The participation of further parties or the general public was not necessary, because the identity of the AoP – meaning the general planning structure – was not changed and the correction of defects had to be implemented regardless. Therefore, the further involvement of the affected institutions above the statement was also decided as not necessary (HMWVL 2013a). Moreover, all affected institutions would become beneficiaries (HessL 2013).

Accordingly, no non-private metropolitan governance body and no private metropolitan logistics actor was involved.

As requested by the BVerwG, the HMWVL established several safety objectives for noise protection, that are not related to any further negative environmental impact. Therefore, the supplementary AoP was decided on April 30, 2013, and the HMWVL corrected the mistakes, identified in the AoP. The Fraport AG was requested to fulfill the correction (HMWVL 2013a).

### *4.3.8.2 Adjustment 66p 01.03.04/29, May 10, 2013*

A further area of adjustment was the improvement of safety issues regarding wake turbulence. According to A XI 2.3 of the AoP, Fraport AG – as the operating company of Frankfurt Airport – is responsible for the remedying of damage caused probably by wake turbulence (HMWVL 2007: 130). Due to the increased reported damages probably caused by wake turbulence[183], the HMWVL decided to adjust the AoP regarding A

182 According to §10 VIII LuftVG, together with §10 I and §1 II LuftV-ZustVO.
183 FNP (2013b); FR (2013c), FR (2013d).

XI 2.3. With the letter of April 24, 2013, the HMWVL initiated an official supplementary RSE.

After the assessment, the HMWVL as the responsible authority provided a draft of the adjustments to Fraport AG and requested a statement. Fraport AG gave two statements on the topic of damages caused by wake turbulence and the adjustment of the safety measures of the AoP. These statements were discussed jointly between the stakeholders. Afterward, the measurement system regarding crossing altitude was adjusted accordingly.

Given the fact, that the indicated adjustment was insignificant[184] and the identity of the AoP was not changed, the involvement of further institutions and the general public was decided as not necessary. No non-private metropolitan governance bodies and no private metropolitan logistics actors were involved in the discussions of the Adjustment 66p 01.03.04/29, May 10, 2013. Furthermore, the adjustment of the AoP had only positive effects on the safety objectives in terms of wake turbulence, which made the involvement of further institutions unnecessary.

The adjustment was decided with the supplementary AoP as of May 10, 2013 (HMWVL 2013b). After the release, the decision was made public in all affected municipalities (HMWVL 2013d). In addition, Fraport AG itself informed about the decision and the rights of the citizens (Fraport 2013).

Compared with the previous adjustment, the improvement regarding wake turbulence gained more attention from local citizens' groups and the media. Additionally, the challenges were discussed more extensively besides the official proceedings (BBI 2013; FNP 2013a; HMWVL 2013d). Local citizens' groups also announced that they would bring the adjustment to court (FR 2013a; FR 2013b; MS 2013). However, the HessVGH refused the urgent motion (HessVGH 2013).[185]

---

184 According to §76 II HvwVfG.

185 Decisions 9 B 1362/13.T and 9 B 1363/13.T.

#### 4.3.8.3 *Adjustment 66p 01.03.04/26, September 6, 2013*

In the letter as of July 30, 2013, Fraport AG itself requested an adjustment of the AoP, especially regarding Terminal 3, the passenger transfer system, the baggage handling system, the traffic and transport infrastructure, and the drainage system.

Due to the request, the HMWVL invited several private and non-private institutions to give statements on the planned adjustment.[186] Non-private metropolitan governance bodies and private metropolitan logistics actors were not actively involved.

After the joint discussions, it was asserted that the adjustment does not change the identity of the original AoP and it was characterized as insignificant.[187] The arguments of the opponents of the extension that Terminal 3 would change in its usability and capacity were disproven. In addition, the change of the spatial usage was refuted, because no neighbor would be affected and the changes would happen on the current site of Frankfurt Airport.

With the proof of the pre-assessment, that no further environmental impact assessment had to be executed, the supplementary AoP was decided on September 6, 2013 (HMWVL 2013c).

With the release, but not directly linked to the supplementary AoP, Thorsten Schäfer-Gümbel, the then-top candidate of Hessian SPD, requested a reassessment of the planned Terminal 3 extension (Schäfer-Gümbel ny).

Furthermore, especially public discussions regarding the planned changes did not appear.

#### 4.3.8.4 *Adjustment 66p 01.03.04/31, May 26, 2014*

This adjustment is closely linked to the adjustment 66p 01.03.04/29 presented above. After the AoP was adjusted regarding wake turbulence and new measures were established, further cases of damaged roofs appeared outside the defined measurement area. For this reason and with

186 According to §63 II BNatSchG.
187 According to §76 III HVwVfG.

the letter as of May 9, 2014, the HMWEVL[188] – as responsible authority – recommended enlarging the measurement area to protect roofs against wake turbulence. As further roofs were damaged by wake turbulence[189], Fraport AG announced on May 19, 2014, that the measures of the AoP would be adjusted and the area enlarged (HMWEVL 2014: 1–4). This was also discussed and requested by the HMWEVL (FAZ 2014b; FR 2014b). After the assessment, the HMWEVL provided a draft of the adjusted measures to Fraport AG and requested a statement.

Given that the indicated adjustment was insignificant[190] and the identity of the original AoP was not changed, the involvement of further institutions and the general public was not necessary. Furthermore, enlarging the defined measurement area, the adjustment of the AoP had only positive effects on the safety objectives regarding wake turbulence.

Accordingly, the adjustment was decided on May 26, 2014 (HMWEVL 2014: 5f). Non-private metropolitan governance institutions and private metropolitan logistics actors were not involved.[191]

#### *4.3.8.5 Interim findings: adjustment of plans, 2007 and beyond*

In the following, the adjustment of plans from 2007 and beyond is analyzed by the theoretical derived concept of ERG. The latent content of the five aspects of ERG is reconstructed in the regulatory governance structure and decision-making process of the adjustment of plans, 2007 and beyond. This analysis is followed by establishing the missing link between the structural perspective of political science and the procedural character of policy-making, which Mayntz already proclaimed in the 1980s (Mayntz 1983), by discussing the institutional arrangement of that period.

*Discussion of shared perceptions of joint challenges*

The supplementary RSEs and the adjustments of plans represent more an act of order to adjust high-level goals and measurement systems

188 Formally HMWVL.
189 FAZ (2014a) and FR (2014a).
190 According to §76 II HvwVfG.
191 Wake turbulence are still the case of the HessVGH (HessVGH 2015).

without an extensive participation process. However, further economic actors, in case of the adjustment 66p 01.03.04/27, April 30, 2013, and affected private individual as the general public in the cases of adjustment 66p 01.03.04/29, May 10, 2013, and 66p 01.03.04/26, September 6, 2013, were only involved if decided necessary, not based on open, transparent multi-actor discussions across hierarchical and traditional borders. The regulatory structure of governance and the decision-making process was mainly hierarchical – usually starting with the decision of the BVerG, through the request to correct the mistakes by the HMWVL and its successor to the execution of the adjustment by Fraport AG. Even affected stakeholders were requested for statements to create a common understanding of the challenges, the identification of topics and best available techniques was more based on the decisions of the HessVGH and BVerG on August 21, 2009, and April 4, 2012. Still, limited dialogues were partly initiated and the requests of statements allowed integrating tacit knowledge and to learn from lower-level experience. However, the establishment of heterarchical relations is not identifiable during the adjustments of plans. Due to the hierarchical character of order, the delegation of mandate in certain topics and proactive advancing of mandate holder to increase ownership can only be identified as a matter of fact regarding technical specifications – e.g. concerning measurement systems concerning the damage caused by wake turbulence. Information was mainly provided by HMWVL to Fraport AG for further pooling and specification.

*Definition of goals and measurement systems*

High-level goals and common objectives discussed during the adjustments were set during the RSE Runway North-West/Terminal 3. The adjustments concerned the specification of these high-level goals. With the request to correct the mistakes by the HMWVL, the focus was more on the programming of measurement systems – e.g. enlarging the area to protect roofs against wake turbulence. Again, the decision-making process was more characterized by order than the discussion of proposals with addressees. Negotiations were mostly characterized by a bi-lateral setting of the responsible authority and Fraport AG.

With the published adjustment plans, reference documents were disclosed, containing legally binding rules clarify the approach.

*Implementation*

The decision-making process was operationalized by the given law. It was hardly characterized by soft guidance to fulfill a binding overall framework to further harmonize directives and regulations. It represents more the needed capacities for implementation based on order, not the design of an objective and imperial organizations fostering cooperation while implementation. With that, its implementation was rather regulated by order than characterized by entrustment of lower-level units and voluntary delegation responsibilities and power to lower-level. The HMWVL and Fraport AG did not have the most possible freedom and autonomy for the implementation of goals. Not only the decision-making process but also the implementation of the defined goals and measures was more defined by the fulfilling of specific demands. Especially the first adjustment regarding wake turbulence (66p 01.03.04/29, May 10, 2013) was characterized by order. With that, the published adjustments were more hard rather than soft guidance to fulfill a binding overall framework and further harmonization of directives and regulations.

*Report and review, review of the measurement system, the decision-making process, and the stakeholder*

As a matter of fact and the character of the supplementary AoPs as duty fulfillment, neither goals nor metric systems had to be reviewed after the first discussions. Reporting and reviewing were restricted to the successful implementation of the adjustments. Neither a constitutional and collective regulatory dialog was initiated, nor an environment of continuous adjustment and learning was created. Due to that, no periodical assessments, audits, or peer reviews were institutionalized during the decision-making process of the adjustments. Taken activities and performance were not to justify. It was actively decided against those elements, mainly based on the argument that the adjustments are minor, entirely positive, and that the identity of the AoP of December 18, 2007, is not subject to change. However, tacit knowledge was requested and integrated by requesting statements of affected stakeholders, information was mainly processed downwards – usually starting with the decision of the BVerG and the request to correct the mistakes by the HMWVL. Only the adjustment 66p 01.03.04/26, September 6, 2013, presents a decision-making process where the information was initially processed upwards by the request of Fraport AG for adjustment.

**Institutional arrangement of ERG in the supplementary RSEs**

In addition to the decision-making process, the presented analysis also presents a clear picture of the institutional arrangements. The analysis shows that the supplementary RSEs involved the actual relevant addressees. What was most apparent is that all institutions or private individuals were all involved more directly. In both cases, neither non-private administrative governance bodies, nor private metropolitan logistics actors were involved.

Nevertheless, the analysis exhibits that the presented analytical framework is suitable to analyze the institutional arrangement of the geographic-functional region of FRM.

**Confirmation of the research-guiding hypothesis for the adjustment of plans, 2007 and beyond**

This interim conclusion shows that the theoretical derived concept of ERG is a suitable analytical framework for analyzing the regulatory structure of governance, the decision-making processes, and the institutional arrangements in the case of the adjustment of plans, 2007 and beyond. With that, the research-guiding hypothesis can be confirmed for the adjustment of plans, 2007 and beyond. In addition, it proves that the theoretical derived concept of ERG is also suitable to analyze hierarchical regulatory structures of governance. Also, it shows that the concept of ERG counterfeits the challenge of a missing link between the structural perspective of political science and the procedural character of policy-making, which Mayntz already proclaimed in the 1980s (Mayntz 1983), by including the analysis of the institutional arrangement of that period.

# 5 Discussion

*A simple idea is elaborated [...] people can coordinate with each other without anyone's coordinating them [...] and without rules that fully prescribe their relations to each other.*
(Lindblom 1965: 3)

## 5.1 ERG: an analytical framework in policy analysis

The first result of the presented research is the elaboration of the theoretical derived concept of ERG. ERG is a suitable analytical framework to analyze regulatory structures of governance, decision-making processes, and institutional arrangements. Based on its five aspects – "discussion of shared perceptions of joint challenges," "definition of goals and the measurement system," "implementation," "report and peer review," and "review of the measurement system, the decision-making processes, and the stakeholder" – ERG presents a theoretical-laden but feasible consecutive framework for policy analysis. These five aspects are defined by latent content mainly based on the findings of Sabel and Zeitlin (2010a), but also of further research in the field of experimentalist governance such as De Búrca, Keohane, and Sabel (2014) and Sabel and Zeitlin (2012b). As presented, the theoretical derived concept of ERG was detailed and successfully applied.

The second result of the analysis shows that the elaborated and theoretical derived concept of ERG is a suitable framework to analyze the regulatory structure of governance, decision-making processes, and institutional arrangements in the geographic-functional region of FRM. Applying the framework on the case of the extension of Airport Frankfurt answered the research-guiding question, how the regulatory structure of governance of the airport extension in the geographic-functional region of FRM is organized. Further, the research-guiding hypothesis that the theoretical derived concept of ERG is a suitable analytical framework for analyzing the regulatory structure of governance, decision-making processes, and institutional arrangements in geographic-functional regions is confirmed by the presented research.

With that, the presented research addresses the described research problem (Sec. 1.1) and the identified white spots (Sec. 1.2). At the same time it builds on previous research and enhances the insights regarding governance in geographic-functional regions in general and the metropolitan region of FRM in particular, also presented in Sects. 1.1 and 1.2. With that, the presented research contributes to contemporary political science as presented in Sect. 1.3.

As presented in the interim findings of all periods of the extension – Airport extension, 1960s–1980s, Mediation process, since 1997, RPP and RPD, 2000–2002, RSE and AoP, 2002–2004, RSE and AoP, 2004–2007, and Adjustment of plans, 2007 and beyond – it is possible to analyze the regulatory structure of governance and decision-making processes by the manifest and latent content of the theoretical derived concept of ERG. On the one hand, it was possible to identify the latent content defining the manifest content in the respective periods, such as "open, transparent multi-actor discussions across hierarchical and traditional borders" for the aspect of "discussion of shared perceptions of joint challenges" or the "joint monitoring and evaluation, peer reviews as well as reporting" for the aspect of "report and peer review," e.g. in the mediation process and the RSE A380 Shipyard. On the other hand, it was possible to delimit cases in which the latent content was hardly or not traceable, e.g. open, transparent multi-actor discussions within a heterarchical environment during the airport extension, 1960s–1980s.

As shown, the elaborated and theoretical derived concept of ERG allows to include the institutional arrangement of the geographic-functional in the analysis by examining the involved institutions and stakeholders formally brought into the overall process, e.g. by officially requesting further stakeholders for statements during the RSE A380 Shipyard and RSE Runway North-West/Terminal 3. The latent content is again possible to identify or it was possible to delimit cases in which further stakeholders were not integrated or after the official decision-making process, e.g. the RSE of Runway 18-West.

As an analytical framework in policy analysis, the elaborated concept of ERG overcomes the identified challenges of the previous approaches of regional and metropolitan governance. The theoretical derived concept of ERG structures the decision-making process to its five aspects accordingly. This opens the regulatory structure of governance for policy analysis and makes the decision-making processes traceable. Against the previous approaches of regional and metropolitan governance, ERG depicts a pragmatic approach to analyze the regulatory structure of governance. Applying the analytical framework, governance cannot be described as simply failed in metropolitan FRM. Following Dewey, it is rather based on pragmatism. This identified pragmatism has the advantage that the theoretical derived concept of ERG is not in the need of the definition of the region *ex ante*, different from the approaches of regional and metropolitan governance. Based on the consecutive character, ERG makes a definition of the region and the respective actors up front obsolete. Applying the theoretical derived analytical framework of ERG allows the integration and exclusion of actors by necessity without questioning the regulatory structure of governance and the decision-making process in general. The presented analysis made this visible. It shows that applying the theoretical ideas of ERG allows the adaptation of the institutional arrangement. With that, the region – the research object – is adaptable according to functional aspects and needs, without questioning the regulatory structure of governance. Utilizing the theoretical derived concept of ERG, the adjustment is rather a need regarding decision and policy-making than an obstacle for analysis. For the case of the airport extension in FRM, especially in the RPP and the later RSEs after the 1960s and 1980s prove that point. By enhancing the group of addressees or exchanging stakeholders this possibility was utilized during the hearings without changing the governance structure. Non-responding stakeholders were not further officially included in the regulatory structure of governance and the decision process in case of not replying to requests. In these cases, further inclusion was only progressed if their integration was enforced by law. As a result, the region can be defined pragmatically subject to the relevant stakeholders without jeopardizing the regulatory framework of governance. It has not to be defined according to traditional inflexible political borders. This is especially visible in situations where stakeholders organize themselves around a given challenge, e.g. the mediation process or the joint representation by lawyers in the hearings. This makes regulatory structures of governance and decision-making processes more efficient because the non-affected stakeholders do not have to be involved only because they are "accidentally" embraced by political borders.

Furthermore, the theoretical derived analytical framework of ERG allows breaking down the decision-making process in the presented consecutive steps. On the one hand regarding historical stages and timely correlation – Airport extension, 1960s–1980s; Mediation process, since 1997; RPP and RPD, 2000–2002; RSE and AoP, 2002–2004; RSE and AoP, 2004–2007; Adjustment of plans, 2007 and beyond – as well as in a timely correlation. On the other hand, each historical stage can be analyzed regarding the consecutive stages, the five aspects of ERG – "discussion of shared perceptions of joint challenges", "definition of goals and the measurement system", "implementation," "report and peer review," and "review of the measurement system, the decision-making processes, and the stakeholder." With that, regulatory structures of governance and decision-making processes can be analyzed separately. ERG allows narrowing down the regulatory structure of governance and the decision-making process by time and its five aspects for analysis.

## 5.2 Metropolitan governance bodies in FRM

As stated the presented research had a focus on the involvement of metropolitan governance bodies during the airport extension in FRM. Here, the involvement of non-private metropolitan governance bodies and private metropolitan logistics actors is noteworthy. Against the argument of governance failure in metropolitan regions, especially in FRM (Hoyler, Freytag, and Mager 2008a; Lackowska 2011a: 111), the metropolitan perspective is identifiable in the regulatory structure of governance, decision-making processes, and the institutional arrangements. However, it is torn. On the one hand, non-private metropolitan governance bodies are actively integrated into the decision-making processes of the airport extension by the responsible authorities, e.g. requesting statements on the planning documents or the direct involvement in the mediation process. On the other hand, with their current setting, the non-private metropolitan governance bodies are unable or unwilling to provide the capacity to engage in the extension topic more extensively. First, the analysis shows that the non-private metropolitan governance bodies were unable to unify their members having a common approach, even not with a minimal common metropolitan denominator. Second, compared to the engagement of other institutions and private objectors, their involvement in the discussions and the decision-making process was weak. Finally, this position was not strengthened, although they were recognized as the major metropolitan governance body of the region.

For the private metropolitan logistics actors, this picture is even stronger. The research shows that the private metropolitan logistics actors in FRM were not involved in the decision-making process and the institutional arrangements. On the one side, not a single private metropolitan logistics actor was actively involved in any mediation process or official proceeding by the responsible authorities. On the other side, they were unable or unwilling to proactively provide adequate capacities for a joint approach.

Under these circumstances, the general approach of the non-private metropolitan governance bodies and the private metropolitan logistics actors is questioned. In the case of inability, the need for a change in the institutional setting – which is according to Ebner (2008) one of the major challenges for institutions (Ebner 2008: 5–8) – comes to the fore. The non-private metropolitan governance bodies and the private metropolitan logistics actors have to change their approach or business model if they want to engage in the metropolitan decision-making process. In the case of unwillingness to engage, both will ultimately prove insignificant in the decision-making processes and for the metropolitan decision-making. This finding is aligned with Walter-Rogg (2013) and Lackowska (2011a). On the one hand, in line with the results of the presented research, Walter-Rogg (2013) identified a rudimentary involvement of the metropolitan actors. No metropolitan institution "has the power to decide or implement policies relevant to the whole region" (Walter-Rogg 2013: 139f). On the other hand, unavailable capacities or unwillingness to coordinate joint actions raises "doubts about characterizing them as engaging in governance" (Lackowska 2011a: 111).

Further metropolitan institutions, such as economic development departments and chambers – which were identified as relevant by Lackowska (2011a) and Zimmermann and Heinelt (2002) – did not play an important role in the decision-making processes of the airport extension. The analysis of the institutional arrangements shows that most of them were not even involved. Summarizing this, even though the metropolitan perspective is integrated into the decision-making process, these findings also speak against the metropolitan reform tradition arguing for the need of a metropolitan governance body.

In the case of FRM, the litmus test of their relevance in the decision-making process will be testified in the following years by their involvement in the discussions around the construction of Terminal 3 and the further extension of Frankfurt Airport.[1] To take the position of adequate metropolitan governance institutions, non-private metropolitan governance bodies and the private metropolitan logistics actors have to face the challenges of the "Changing Governance of Local Economies" (Crouch et al. 2004) and they have to actively participate in the decision-making processes. Otherwise, the previously mentioned "doubts about characterizing them as engaging in governance" (Lackowska 2011a: 111), will remain. Metropolitan governance bodies and private metropolitan logistics actors will continue to be numerous loose and uncoordinated networks of private and non-private institutions unable to actively participate in regulatory structures of governance and decision-making processes.

The notion that such an approach is needed and possible, is shown by the fact that the non-private administrative institutions agglomerated around common topics without a traditional, hierarchical organized approach. This is exposed by their joint representation by lawyers in the RSE Runway North-West/Terminal 3 of Frankfurt Airport. Without being organized upfront, non-private administrative institutions – above all municipalities – organized around the jointly identified and defined objectives (goals). Another key point is that by linking further private individuals, traditional borders between the private and public

1 Its construction will start in 2015 (FAZ 2015; FR 2015).

sphere were overcome. As a result, the decision-making process is not only located between the hierarchical governance levels of multilevel governance; moreover, it is transformed into a mélange of hierarchical and vertical public and private interests, whereby decision-making is negotiated between both spheres.

This is also identifiable in the mediation process. Here, especially the established working groups created this mélange. The working group of the Anti-Noise-Pact at Frankfurt Airport is an impressive example. On the one hand, private and non-private institutions jointly identified and defined common goals and objectives. On the other hand, they also established measurement systems and agreed to them in the Anti-Noise-Pact.

## 5.3 Governance learning in FRM

Against the common approach, combining comparative public policy and different modes of governance, the presented research shows that there are several aspects of postulating governance learning. Based on fallibilism the regulatory structure of governance and the decision-making process is identified as an evolutionary approach. From a general perspective, this can already be seen in the change from the top-down approach of steering politics – which was immanent between the 1960s and 1980s – to governance, including a bottom-up perspective, that became more relevant in the further approach and the establishment of the parallel mediation process. With a multi-dimensional network perspective, this differentiation was taken into account. In a multilevel system between administrative and private actors, as well as civil society decisions are made (Benz 2003a; Geißel 2012b).

Following the theoretical derived concept of ERG, governance is a reciprocal discussion of repeated objectives and the revision of objectives and measurement systems (De Búrca, Keohane, and Sabel 2014; Sabel and Zeitlin 2010a; Sabel and Zeitlin 2012b). It structures fluid practices of governance within any possible variation of institutions (Overdevest and Zeitlin 2014). Accordingly, policy-making is an evolutionary, recursive process in which uncertain institutional innovations and a multi-actor environment is created.

The following discussion will first focus on the mediation process and second present the findings regarding the official proceedings of the RPP and RSEs.

### 5.3.1 Governance learning during mediation

First, regarding the mediation process, the major indication for governance learning is the initiation itself. Based on the experiences of the 1960 and 1980s the administration was eager to improve the extension process. On the one hand, the administration was keen to avoid the riots related to the extension process of the Runway 18-West in the 1970s and 1980s. On the other hand, the efficiency of the extension process between the 1960s and 1980s related to the Runway 18-West was questioned. In relation to this, in 1997 the *Gesprächskreis Flughafen* was initiated, even before the official procedures started (4.3.4.1).

Accordingly, the major change is the direct and active involvement of civil society in the decision-making processes. By establishing the expert groups, the mediation took into account the notion that not all topics can be clarified at the highest level of the hierarchy. Tasks and the achievement of goals were delegated. As a result, the ownership to take responsibility at the lower levels was enhanced, coordination and voluntarism were facilitated.

Focusing on the decision-making processes, the orientation on supposed best practices can be seen as an aspect of governance learning. The inclusion of national and international best practices – e.g. BER and Seattle-Tacoma – supported the diffusion of (implicit) knowledge, practices, and role models. Thus, the decision-making processes follow the facets of compliance, emulation, and herding.

In addition, governance learning can be presupposed by the more direct addressing of actors of civil society. After actors of the civil society – mainly environmental associations – had redeemed the cooperation with the *Mediationsgruppe* from 1998 to 2000, those actors were addressed more directly by the administration. This again facilitated coordination and voluntarism. Equally important is information sharing and transparency. On the one hand, information was shared more frequently between the levels. It was agreed to exchange quarterly status reports between the mediators, Fraport, and the RP Darmstadt. On the other hand, especially by the mediators and further actors of the civil society, information was provided more comprehensively, transparently, and sustainably for the civil society. This goes in hand with the then-growing availability of modern ICT, which strengthened the effect over time. Today social media channels like Facebook, wikis, or Twitter as well as e-Governance (Meier 2012; Piagessi, Sund, and Castelnovo 2011; Scholl 2010) – meaning the provision of governance services by utilizing modern ICT – enforce the provision of information. Consequently, dense communication and reflexivity were taken into account.

Second, for the institutional arrangement of the mediation process, the change from a personage focus, concentrating on individuals to a multi-actor environment seems to be a strong indication of learning. Improvements regarding the quantity and quality are identifiable. The beginning of the mediation process was characterized by personage rather than institutions. This changed with the establishment of the *Mediationsgruppe* in 1998, the RDF in 2000, and finally with the FFR in 2008. During this time, a multi-actor environment was established. Starting with a very restricted and closed circle of 16 institutional and private individual members, the contemporary FFR includes 75 institutional members, including metropolitan perspectives. In the beginning, the mediation process – as the extension process between 1960s and 1980s – was mainly characterized by individuals personally associated with the mediation process. Moreover, the following proceedings – both, the official proceedings as well as the mediation processes – were marked by the political sphere. On the one hand, the then-Hessian Minister-president Roland Koch played a major role in the extension. On the other hand, the period of the *Mediationsgruppe* was also marked by the mediators Kurt Oeser – who gained a major role in the mediation process – Frank Niethammer and Klaus Hänsch.

However, this subjectivity changed during the process. Today, the FFR is more characterized by an institutional perspective, including politics, private economic and non-economic institutions as well as society. The central position of the individual stepped behind the institutions' characteristics. That improved the objectivity of the discussion.

With attention to the implementation of further expert groups, this effect was reinforced. These expert groups created heterarchical links between the stakeholders and initiated a recursive process of bottom-up learning, including implicit and explicit knowledge. As a result, the lower levels supported the definition of challenges, goals, and measures for the extension process, as well as the achievement of results. The multi-actor environment of today's FFR embraces administrative and private actors (A.8).[2]

Besides the exception of the temporary engagement in a sub-expert group in the RDF, the metropolitan institutions were not engaged in the mediation process. Only with the last period of the FFR was, the *Regionalverband* – as the metropolitan governance institution – officially included. With that, the quality was further improved by including the metropolitan perspective.

Furthermore, the integration of representatives of the major political parties – independent of the current state government – improved the regulatory structure of governance, the decision-making process, and the institutional arrangement of the mediation process. In the last period – the FFR – the fractions of CDU, SPD, *Bündnis90/Die Grünen* and *Die Linke* of the *Hessische Landtag* also formally entered the mediation process. It is important to realize that, elected-members of the political parties were officially involved in the mediation process for the very first time. Hence, the institutional arrangement was continuously adapted, including the facets of hierarchy, the "shadow of hierarchy," or the "shadow of law." Recognizing that not all identified topics and challenges were discussed on the upper level and the creation of the task forces – including further institutions outside the mediation process – also improved the overall process. This integration of lower-level units and delegation of responsibilities, improved the institutional arrangement and the approach to overcome a possible lock-in, based on the knowledge that is taken for granted.

As presented, governance learning can be identified in the mediation process. For the mediation process in FRM, problem definition, the

2 See A.4 for the *Gesprächskreis Flughafen*, A.5, A.6 for the *Mediationsgruppe*, and A.7 for the RDF.

achievement of results, goals and the discussion of implicit as well as explicit knowledge can be identified. Furthermore, the facets of hierarchy, the "shadow of hierarchy," or the "shadow of law," facilitating coordination, and voluntarism are deducible. Finally, the facets of compliance, emulation, and herding as well as dense communication and reflexivity are derivable from the presented analysis.

### 5.3.2 Governance learning during the official proceedings

For the RPP and the RSEs, this continuous learning is not unambiguously identifiable. Here, governance learning is much more multi-faceted.

The extension during the 1960s and 1980s present the starting point for analyzing governance learning in the presented research. Examining the official proceedings between the 1960s and 1980s and the construction of the Runway 18-West it becomes evident, that the official proceedings were characterized by a few institutions. As the presented research shows, the initial decision-making process is only localized in the sphere of non-private administrative institutions. No private institutions and no metropolitan, neither non-private nor private, were involved. In addition, it is noticeable, that the period was marked by a few individuals, such as Kurt Oeser or the then-Hessian Minister-president Holger Börner.

This led to the circumstance that the general public felt unheard. Besides the protests and riots, that got high attention, many cases to integrate further stakeholders were brought in front of the court. Up today, the construction of Runway 18-West is a symbol of a party politics that ruthlessly ignored the interests of citizens (RP 2004). At that time, the extension process gripped the people in FRM in almost every situation. Not only in front of the court and party politics the construction of Runway 18-West discussed. Also in schools, the topic and the issues of (non-)involvement were reviewed in social studies lessons (SPIEGEL 2008). In front of that, a local-regional conflict of interest in economic and environmental issues turned into a general political conflict about the understanding of democracy. The original conflict faded into the background. Discussions were partly led by arguing about the fundamental understanding of the constitution, the relationship between state organs – parliament, government, judiciary, and police forces, citizens' movements, democracy, non-institutionally regulated resistance as well as violence and law (Jahn 1984: 309).

Based on the felt non-inclusion, the general public solidarized with the protestors.

In front of the public discussions, the protests, and political pressure the official proceedings were enhanced with the request of further statements and the facilitation of public hearings. However, they both followed only downstream the official decision-making process.

Where discussions during the extension process and the exchange of arguments took a back seat and violence came to the fore, up until today people partly speak of civil war-like conditions rather than demonstrations. Polices forces themselves argue today, due to the experience made during the demonstrations, they learned to better handle similar situations, e.g. handling demonstrators against tuition fees (SPIEGEL 2008). Then-protestors argue in the same direction. "We have gained experience and learned" (Hechler quoted by SPIEGEL (2008)).

In party politics the biggest riots in German history over an airport expansion contribute to the fact that the *Bündnis90/Die Grünen* and alternative conceptions entered the Hessian State Parliament in 1982 for the first time (SWR 2013.)

In front of that, one major learning was the fostering of official hearings during the decision-making process to take tacit knowledge and the general public into account. The execution of the RPP and the mandatory inclusion if environmental associations as well as the hearings upfront the following RSEs – RSE A380 Shipyard and RSE Runway North-West/Terminal 3 – with the final approval of the decisions taken in the AoP present a major step regarding governance learning. Not only taking tacit knowledge and the general public into account, this made it possible to divide the decision-making process in two consecutive steps – the analysis and decision focusing on the environmental impact and to realize spatial planning (RPP and RPD) and the evaluation and decision on the authorization of the actual project plan (RSE and AoP).

In addition, regarding the official proceedings and with a throwback to the extension between the 1960s and 1980s and the construction of the Runway 18-West, the interlocking with the mediation process and integration of its results one of the major visible features of governance learning. The integration of the mediation and its results in the official proceedings supported the non-escalation as it happened in connection with the construction of the Runway 18-West in the 1980s.

Further, the orientation towards national and international supposed best practices is another aspect of governance learning. Other extension processes in Berlin and Munich, but also international practices from Seattle-Tacoma International Airport, the initiated Part Study 150, the voluntary planning to identify current and future noise impacts, as presented above were reviewed during the respective hearings.[3] They were benchmarked and peer-reviewed to adopt improvements and supposed best practices.

Additionally, the adoption of an upfront released agenda is one of the most prominent examples. In the case of Frankfurt Airport and the RSE A380 Shipyard, the agenda of the hearing was not published upfront. This was criticized and the decision-making process was benchmarked to other regions in which the agenda was made public upfront. For this reason, the agenda of the RSE Runway North-West/Terminal 3 was published *ex ante*. With the publication of an agenda released upfront, the participatory process of the RSE Runway North-West/Terminal 3 was structured and improved. Private individuals, not represented by a lawyer who attended all trial days, had the possibility to join the hearing when it was relevant for their objections.

Another learning can be seen in the increase and the timing of the public hearings. During the extension between the 1960s and 1980s the hearings initially took place downstream the official decision-making process. In the following, the general public was already included during the official decision-making process. Further, the capacities of the official proceedings were extended with each period. Starting with overall 18 days in the first RPP, the second already comprised 35 days of negotiation. With the RSEs, the negotiation was extended again. The last RSE was extended during the already-started proceeding. Overall, it comprised 106 days of trial.

Finally, the RSE Runway North-West/Terminal 3 was heavily characterized by the negotiations of legal representatives. This marked the related

3 E.g. RPDA (2004d), RPDA (2004i), RPDA (2005l), or RPDA (2005n)).

hearing with a very formal and objectified methodology. Nevertheless, from a learning perspective the raising juridification also hampered the improvements. Both the RPP and the RSE proceedings represent administrative tools not only to integrate institutions with huge capacities to follow an extensive formal proceeding; moreover, the proceedings were also designed to integrate civil society, literally John Doe. However, due to the formalization and the juridification John Doe was unable to participate in full. Additionally, this exclusion holds true for the quantitative dimensions of the proceeding. Due to the scope of the proceeding, the ordinary citizen – not represented by a lawyer or other experts – was excluded from the RSE Runway North-West/Terminal 3 *de facto*. Apart from the duration, this is also apparent for the relevant information. While the hearing of the RPP took fifteen trial days and the RPD had 284 pages, the number of trial days doubled for the RSE A380 Shipyard to 30 and the AoP had 396 pages. These numbers exceeded in the RSE Runway North-West/Terminal 3. The hearing lasted for 101 trial days and the AoP had 2,515 pages.[4] The additionally-presented statement of the RP Darmstadt included another 1,799 pages. In addition, most of the requested statements of the municipalities had about 150 to 200 pages. However, some of the 116 replies had 2,000 to 3,000 pages. One statement of a single municipality had 8,369 pages. Already the final planning documents presented to the public for review included 65 files. This volume of data hampered or made it impossible for a single private individual to exercise the right of participation – beside the physical attendance.

Analyzing the quantity and quality of involved institutions as indicators, governance learning varies. Except from a major increase from the extension between the 1960s and 1980s to the RPP from 2000 to 2002, there was no constant rise of involved institutions.

As already stated above, the official proceedings of the extension during the 1960s and 1980s and the construction of the Runway 18-West were partly characterized by a few individuals. This perspective changed with the following proceedings of the RPP and RPD, 2000–2002, RSE and

4 As comparison: the AoP of the extension in 1971 had nine pages.

AoP, 2002–2004 and 2004–2007 as well as the adjustments of plans, 2007 and beyond. The integration of further stakeholders and taking out institutions not replying to the request for a statement can be considered as a further facet of governance learning. On the one hand, upfront neglected institutions were taken into account, by integrating them afterward. On the other hand, the noninvolvement of institutions not replying to the request of a statement supported a leaner decision-making process and institutional arrangement. Finally, governance learning is identifiable regarding the ICT and the information exchange through the governance levels. During the trial days of the second RSE, a hotline was installed so that those people who were affected and not having internet access yet were also able to obtain information from their home. In addition, at least one major private website or wiki and later even Facebook-groups of a local citizens' group pooled all relevant data in almost real time, at least from a certain perspective.[5] With the increasing importance of social media, the discussions were enlarged to Facebook and other web-based discussion forums.[6] Additionally, Fraport as the operating company of Frankfurt Airport informed about their extension plans on their websites. The RP Darmstadt – as the responsible authority – shared at least the RPD and the AoP or the latest steps in the extension process. This result also confirms the hypothesis of Bröchler and Blumenthal (2006) that new dimensions to program political objectives and networks are relevant.

Finally, the raising juridification and the *de facto* exclusion of John Doe as a singled ordinary citizen is apparent. For example, the latest hearing on the extension of Frankfurt Airport was strongly characterized by the discussion of the legal representatives. According to the technical jargon as well as the huge amount of data generated, the ordinary citizen is *de facto* excluded from the decision-making process.

Summarizing the facet of governance learning for the official proceedings, the analysis is torn. On the one hand, the decision-making process was improved, especially by taking supposed best practices into account.

---

5 Frankfurt: *Zukunft Rhein-Main* (www.zukunft-rhein-main.de). The DFLD (www.dfld.de) focuses on the case of Frankfurt, but also has a national perspective.

6 E.g. www.facebook.com/pages/Keine-dritte-Startbahn/117688718249663.

On the other hand, they were downgraded, especially by the fact of the *de facto* exclusion of John Doe.

This variation is also recognizable by analyzing the involvement of non-private metropolitan governance bodies. The active involvement in the official proceedings improved the awareness of the metropolitan perspective. However, taking the non-integration of the private metropolitan logistics actors into account, the advancement has to be relativized.

Besides the improvements and learning regarding the quality and quantity of the institutional arrangements, the already-remarked juridification is relevant in the context of governance learning. While there is no common approach promoted by the non-private metropolitan governance bodies or private metropolitan logistics actors, the unifying of the stakeholders of the extension process by legal representatives is conspicuous. This joint approach – especially of directly affected municipalities – has risen during the whole extension process. Neither the non-private metropolitan governance bodies, nor the private metropolitan logistics actors were able to unify institutions and carry out this task. They were unable to create the capacities to handle this major metropolitan challenge in the geographic-functional region of metropolitan FRM. This indicates two conditions. On the one hand, there is a need for a joint metropolitan approach in FRM. On the other hand, the current non-private metropolitan governance bodies and private metropolitan logistics actors are not set up sufficiently to respond to this demand.

### 5.3.3 Governance learning: conclusion

Taking all of these aspects into consideration, the result and the answer of the question, of whether governance learning is identifiable during the extension process of Frankfurt Airport can only be positive. However, on the one hand, the processes improved. On the other hand, there was also deterioration, especially with respect to the participation of John Doe.

Although the presented approach can be criticized (Fabrizio Gilardi and Radaelli 2012: 161–164), it has to be taken as a starting point to analyze the learning aspect in geographic-functional regions. The presented analysis has shown that within the same successive and evolutionary decision-making process of the mediation in FRM the facets of governance learning of comparative public policy and different modes of governance can be analyzed. Taking further decision-making processes into account, such as the one of *Bahnprojekt Stuttgart-Ulm* – better known as Stuttgart 21 – governance learning also has to be questioned. Not analyzing the regulatory governance framework and decision-making process of Stuttgart 21 in detail, it has to be noted, that governance learning is oscillating. The mediation process in FRM was initiated even before the official proceedings of the extension started, while the arbitration in Stuttgart was only initiated after the excesses in the *Schlossgarten*.[7]

Finally, two major questions remain open. The first question is, how the decision-making process of further large-scale projects – such as the high-voltage transmission line from the north to the south Germany or the *Kulturcampus* in Frankfurt – utilizes generated knowledge. This question will be answered over time.

The second major question is who is learning in a decision-making process? Following Geißel (2012b) and the definition made in the introduction (Ch. 1), that decision-making is negotiated between individuals coming from administrative and nonadministrative institutions as well as citizens (Geißel 2012b: 165), a hypothesis for future research could be that: governance learning degenerates if involved individuals coming from administrative and nonadministrative institutions as well as citizens are no longer part of the regulatory structure of governance and the decision-making processes.

## 5.4 Further findings and deductions – ERG

The presented research shows that the theoretical derived analytical framework of ERG helps to overcome several challenges of the analysis of a traditional governance approach. Nine major aspects are outlined in the following.

First, as presented, for traditional governance analysis, it is difficult to identify the relevant stakeholders of a certain topic, namely, who are – by which characteristic – the relevant institutions or mandate holders in the

7 September 30, 2010, the *Schlossgarten* was evicted by police. Several people were injured, some were injured heavily. The day is also called Black Thursday.

decision-making process. On the one hand, in regional governance they are commonly chosen bottom-up by the prematurely-involved institutions themselves. This is usually the case when actors agglomerate around a common project, such as the extension of an airport. On the other hand, institutions agglomerate if collective action is promoted or enforced by the administration top-down, including during the extension of the analyzed airports. This finding makes the cooperation between the bottom-up and the top-down perspective not only favorable. Again, the mélange of hierarchical and vertical public and private interests has to be governed. Sometimes this cooperative approach can become a must. In situations where a need for democratic legitimacy exists, but private veto players are involved, public and private needs have to be negotiated and agreed. In these situations, traditional governance analysis as well as policy-making reach their limits.

As it is evident from the presented research, the theoretical derived analytical framework of ERG overcomes this challenge. First, the analytical framework of ERG involves private and non-private institutions as well as private individuals where necessary. Second, pragmatism and fallibilism which lie within ERG consider further stakeholders after the decision-making process has been initiated. Finally, based on pragmatism and fallibilism, stakeholders that become irrelevant for the further decision-making process can be rejected. As the research presented, this approach solves the challenges of traditional governance analysis regarding the setting of institutional arrangement. Theoretical derived concept of ERG combines the bottom-up perspective and the top-down approach and stakeholders are involved where necessary. Following Dewey (2012), this involvement is the case when being affected by plans and measures create a joint interest "that is, those affected by the consequences are perforce concerned in conduct of all those who along with themselves share in bringing about the results" (Dewey 2012: 58).

Summarizing this, the theoretical derived concept of ERG is a suitable analytical framework for governance analysis in situations where actors agglomerate around a common topic throughout hierarchical and traditional borders, e.g. global city regions, industrial districts, innovation systems, milieus, and clusters.

Second, regarding the metropolitan perspective an additional facet of agglomeration is interesting which is revealed by analysis of metropolitan regions with the presented analytical framework of ERG. However, the metropolitan perspective is relevant, major metropolitan institutions are less relevant for the decision-making in geographic-functional regions. The analysis shows that in situations where actors identify essential challenges and are aware of the situation, they will engage in joint, reciprocal actions. Accordingly, the finding of Sabel and Zeitlin (2012b) – who came to this result at the European level – is also confirmed on the geographic-functional level, like the metropolitan region. On the one hand, this confirms for the geographic-functional regions, that the theoretical derived concept of ERG approach appears particularly well-suited for governance analysis to domains, in which a sovereign with authority to set common goals is not yet established. On the other hand, it is also suitable to analyze governance in situations where the diversity of local and regional conditions and institutions make an adoption and enforcement of *ex ante* fixed rules less- or unfeasible (Overdevest and Zeitlin 2014: 26; Sabel and Zeitlin 2012b: 177).

The presented analysis shows this by the following results especially. First, affected institutions and private individuals agglomerated around a given challenge (airport extension). Second, initially enforced by political engagements, the affected stakeholders gathered in the mediation process to discuss the common challenges, guidelines, goals and measurement systems. Finally, this approach is identifiable within the hearings of the RPP and RSEs. Here, the agglomeration was stipulated by law. Especially in the case of FRM it is recognizable, that affected stakeholders additionally agglomerated under the umbrella of legal representatives to act jointly. On the one side, this fact is striking, because these stakeholders had access to the non-private metropolitan governance bodies and private metropolitan logistics actors, with the possibility to join under their patronage. On the other side, even if these capacities were established by the administration for the purpose of discussing common regional topics, these capacities were not utilized to discuss the metropolitan challenges regarding the airport extension. Against this background, the non-private metropolitan governance bodies as well as the private metropolitan logistics actors have missed their self-set goals to be the platform to discuss joint regional (logistics) challenges.

Summarizing that facet, the presented research points out that the affected stakeholders are willing to agglomerate discussing joint challenges. Reasons why they do not agglomerate under the patronage of non-private metropolitan governance bodies or private metropolitan logistics actors have to be identified somewhere else, the presented research sees indications in the configuration of the business models and in their proactive engagement.

Third, the analysis highlights that the decision-making processes touched all governance levels and the phenomenon of glo*c*alization. Especially the decision-making process of the RSE A380 Shipyard has shown that local phenomena were discussed beyond the governance levels. The local construction of the A380 Shipyard is needed to secure the international air transport system. As a result, global challenges are brought to the regional and local level in FRM. Effects of global issues are discussed and solved on these lower levels. *Vice versa*, the identified challenges and the benchmarking with other European locations brought the occurring challenges at the regional and local level back to the European and international level.

To summarize that facet, the presented analysis outlines that the theoretical concept of ERG can handle the challenges occurring by embracing the governance levels. Additionally, it exposes that ERG tackles the challenges regarding glo*c*al phenomena of governance analysis.

Fourth, the presented analysis confirms several findings of Blumenthal (2005) introduced in the theoretical discussion of governance: first, governance in dissociation to government in geographic-functional regions concentrates more on nonhierarchical systems of problem-solving, whereby societal self-control becomes important; second, the distinction between subject and object of steering blurs; third, the view changes from an actor to an institutional focus, and the institutional arrangement comes into the fore; and finally, governance does not always require a problem to solve. Governance also takes the divergent sets of rules between different institutions of the institutional arrangement into account.

Fifth, as stated in the introduction of this research, the general governance discussion sees change as one of the main challenges in governance and requests that future governance analysis takes changing framework conditions into account. Especially governance as a process is confronted with changing framework conditions caused by time. These changes can be exogenous or endogenous (Ahrend, Gamper, and Schumann 2014: 5; see footnote for further details).[8]

As presented, the theoretical concept of ERG can recognize these changes and is its consecutive character is capable of handling them in the analysis. On the one hand, ERG focuses on the integration of further information, tacit knowledge, and experience from the lower levels. Accordingly, the theoretical deduced approach of ERG can take new developments and innovations occurring during the decision-making process into account. On the other hand, by being able to integrate further stakeholders if necessary, ERG reflects the changing institutional arrangement of the geographic-function region. Integrating the embeddedness perspective, the latter aspect of change is even more highlighted. As a result, the change of institutions – which is according to Ebner (2008) one of the major challenges in recent institutional theory – is also taken into account.

Sixth, this goes in hand with the demand of continuous modernization of the administration formulated in general governance research (Héritier 2002; Mayntz 1996). First, by integrating lower-level experience, justifying taken activities against most objective information, and taking new developments and innovations into account, the theoretical derived concept of ERG promotes this continuous modernization topic-wise. Second, by respecting an open, transparent multi-actor discussion and the further integration of stakeholders, ERG promotes this modernization and differentiation, also institutional-wise. The decision-making process makes it possible to link the administrative system to further sectoral and regional policy networks. Given that theoretical deduced approach of ERG includes the geographic-functional perspective and the embeddedness approach, the decision-making process also includes stakeholders outside the initial institutional arrangement. This makes an consecutive approach – as the theoretical concept of ERG presents – highly relevant in

8 Crouch and Trigilia (2003); Koch (2013); Salet, Thornley, and Kreukels (2003).

cases where sectoral and regional networks promote vested interests with a quality similar to a public good, not yet integrated into the institutional arrangement. It provides a scientific approach where such links are missing. This approach allows systematizing content according to a sound methodology. To summarize this facet, the theoretical derived concept of ERG provides a governance framework respecting the formulated requests by general governance research to modernize administration.

Seventh, theoretical deduced approach of ERG partly opposes aspects regarding the challenges formulated by Mayntz (2003) in the broader governance discussion. First, the challenge of blockade still persists. However, ERG promotes the possibility that the decision-making processes eliminates information taken for granted that hampered and blocked the decision-making process by integrating lower-level information and tacit knowledge. Second, suboptimal compromises also partially persist. It is the nature of compromises being not the maximum result for the single party. However, integrating knowledge form all governance levels objectifies information to minimize suboptimal results of all stakeholders. Third, by having the possibility to integrate further relevant stakeholders, ERG opposes the challenge of agreements neglecting third parties not involved in the decision-making process. Nonetheless, deliberated exclusion is not prevented, although integrating the sum of all relevant stakeholders presumes the opportunity of participation. Otherwise, it is not to speak of governance regardless. Finally, the analysis shows that it is difficult to promote and enforce agreements made. However, this challenge is opposed by joint discussions of shared perceptions. ERG objectifies information based on the basic principle of the better argument.

Eighth, the presented research confirms that the analytical framework of ERG combines the structural and procedural aspects of governance. As presented for the mediation, the RPP, and the RSEs, the theoretical deduced concept of ERG allows analyzing decision-making processes and the surrounding institutional arrangements in which the decision-making processes are embedded. It counterfeits the challenge of a missing link between the procedural character of policy-making and the structural perspective of political science, which Mayntz already proclaimed in the 1980s (Mayntz 1983).

Finally, the research shows that the concept of ERG is a deliberative democratic approach. The aspects of ERG focus on the participation of all relevant stakeholders and sum of all co-existing forms of collective regulation of social issues. However, when it comes to legislation and the decision-making process steps out of the shadow of hierarchy, the final decision is left to the elected representatives and the official proceedings. Accordingly, the argument of Sabel and Zeitlin, that the experimentalist (regional) governance approach is most suitable for environments where no central, final decider is available and clear jurisdiction is missing (Sabel and Zeitlin 2010b: 6; Sabel and Zeitlin 2012b: 170), is questioned. Even though the theoretical deduced approach of ERG does not require a central, final decider *ex ante*, this aspect becomes problematic when it comes to the question of decision. Each case showed that a central, final decider was involved or requested during the decision-making process. Two aspects have to be named especially. On the one side, each final decision – the RPD and the AoP – was finally released by a single authority. In not one case, other participants signed the final decision. On the other side, each stage of the extension was taken in front of the court for a final decision. Even the discussion of the official proceedings, the jointly discussed challenges, and the decisions based on those discussions were taken to all the court instances. In respect of this finding, several assertions can be made. First, in cases where no central, final decider is needed upfront it becomes necessary after a certain point. Second, when there is the possibility to make use of a central, final decider, the final decision is requested for the decision. Notwithstanding with Lackowska (2011a) and Zimmermann and Heinelt (2002) arguing that the central, final decider is or should be a metropolitan governance body, the research shows that it was always the court making the final decision. As already outlined in this conclusion, the analysis exhibits that the metropolitan governance bodies do not play a significant role in the decision-making process.

Now, one might ask whether the theoretical deduced approach of ERG is applicable if a central, final decider is available at all. The answer is: yes, it is. Referring to its theoretical basis of pragmatism and fallibilism, ERG does not know a final decision. Ends and means are mutually transformable. Serving democracy ends and means have to be reconsidered as conditions and surrounding arrangements change (Dewey 2012; Sabel 2012).

Nonetheless, this is not the focus of the presented analysis and further research is needed in this field.

This and further future fields in the field of ERG are outlined in the following paragraphs.

## 5.5 Future research areas: ERG in political science

The discussion of the initial concept of experimentalist governance presented by Sabel and Zeitlin (2012b) is mainly characterized by a law perspective. The presented research linked the theoretical concept to political science. By focusing on the decision-making processes and the institutional arrangements in the geographic-functional region of FRM, the analysis contributes to close the identified gaps of policy analysis in political science. However, the presented contribution can only be seen as the tip of the iceberg. More research has to be carried out to improve and establish ERG as an analytical framework in policy analysis. This future research on ERG in political science is promising. In the following paragraphs further points of contact are presented. First, two research paths are especially interesting for future research: ERG and the democratic innovation as well as ERG and the entrepreneurial state. Both ideas are outlined and linked to the analytical framework of ERG.

Second, some additional promising research fields are named. Finally, the practical implications of ERG in political programming are outlined.

### ERG and innovations in democracy

Geißel (2013) ascertains that "[d]espite the world-wide triumph of democracy, the quest for an optimal "politike" has not yet reached the "end of history" (Fukuyama). It turned out that representative democracies do not necessarily satisfy the citizens" (Geißel 2013: 8). Man's life did not resemble that of Fukuyama's dog or Nietzsche's last man (Fukuyama 2006: 311f). Moreover, in relation to today's "Globalia" and the rising importance of geographic-functional regions, the focus is more on "Human Nature and the Reconstruction of Social Order" (Fukuyama 2000). In other words, "[…] representative democracy, as we know it today, is changing" (Geißel 2012a: 209). These changes and the combination of participatory elements above voting become more relevant for the future (Geißel 2012a; Geißel et al. 2014a; Geißel et al. 2014b).

This broad background is the context for Geißel to discuss democratic and participatory innovations.[9] Nevertheless, it is difficult to define innovation. Geißel (2013) "refer[s] to participatory innovations as new procedures consciously and purposefully introduced with the aim of mending current democratic malaise and improving the quality of democracy" (Geißel 2013: 10). Based on a comprehensive literature survey she identifies cooperative governance, deliberative procedures, and direct democratic procedures as the three paramount participatory innovations (Geißel 2013: 10-13; (Geißel 2012b: 164–166; 170–177).

The theoretical derived concept of ERG fits in this field, whereby this fit is outlined in the following. However, in the future there is the need to analyze ERG in depth in the light of innovative democracy.

In the first place Geißel argues, that according to G. Smith (2005) or Talpin (2012) cooperative governance is characterized by the direct influence of citizens on the political decision-making. The lower levels are empowered and the power is shared between the private and the non-private institutions as well as the private individuals. This empowerment goes beyond the inclusion of trade unions and employers' representatives (Geißel 2013: 10f). Decision-making is negotiated between private and non-private institutions as well as citizens (Geißel 2012b: 165). These aspects are identifiable in the theoretical deduced ERG approach. On the one hand, it includes an open, transparent multi-actor environment, covering private and non-private institutions as well as private individuals. On the other hand, as the presented analysis shows, institutional arrangements go well beyond trade unions and employers' representatives and the concept of ERG includes the entrustment and the delegation of power to lower-level authorities and other experts.

In the second place, according to Geißel, deliberative procedures focus on the discursiveness of will-formation. Against the European

9 For further readings see Geißel (2012a), Geißel (2012b), Geißel (2011a), Geißel (2011b), Geißel (2008), Geißel and Joas (2013), and Geißel and Newton (2012).

perspective, including strict rules of arguing between free and equal citizens, the American perspective only demands discussions (Habermas 2001). Following the latter, deliberative proceedings are consultative "and provide elaborated advice which is submitted to decision-making bodies" (Geißel 2013: 12). Moreover, the theoretical derived concept of ERG includes discussions to formulate reference documents with legally (non-)binding rules and suitable proposals for the higher authority. For the presented analysis the Anti-Noise-Pact represents such an elaborated advice. Additionally, the examined cases represent promising examples to research deliberative participation because they cover the whole range of consultative-discursive procedures, from a minimum of information exchange to well-prepared and continuously-improved deliberative discussions (Geißel 2012b: 165).

Finally, Geißel characterizes direct democratic procedures by casting votes, which are constitutionally required or initiated bottom-up. Their utilization ranks from the agenda setting over the attempts to influence political decisions to the purpose of securing the official mandate from the citizens (Geißel 2012b: 164; Geißel 2013: 12f). Although the theoretical derived concept of ERG does not specifically include these procedures, the presented analysis shows that the analytical framework is still open for casting votes.

This short overview briefly links the concept of ERG with studies on democratic innovations and participation. However, this is by far not sufficient. Future research has to analyze the concept of ERG from a democratic innovation perspective in depth.

Based on her previous work (Geißel 2009), her subsequent synoptic findings (Geißel 2012a) and as presented in the anthology of Geißel and Newton (2012), Geißel further introduces a rich analytical approach with five consecutive pillars to evaluate participatory innovations. The framework – similar to Michels (2011) – embraces "inclusion and meaningful participation," "legitimacy and political support," 'effectiveness' and 'enlightened citizens and democratic education' (Geißel 2013: 13–22).

With a short analysis, the productivity of in-depth research of ERG in the light of Geißel's framework is outlined in the following.

This first pillar is presented by open, transparent multi-actor discussions. Based on the wider understanding of governance and including a multilevel governance perspective, ERG entrusts the lower levels and includes further stakeholders if necessary. For the "inclusion and meaningful participation," there is the question of whether ERG approach is able to activate the inactive. This activation is – according to Geißel, based on the findings of Newton (2012) and Talpin (2012) – the best way to strengthen inclusion and meaningful participation (Geißel 2012a: 210).

The second pillar of "legitimacy and political support" is promoted by the open, transparent multi-actor discussions of the theoretical deduced approach ERG. On the one hand, discussing shared perceptions beyond the hierarchical levels and the traditional borders, as well as the delegation of mandates, provides a maximum of political support and legitimacy. On the other hand, political support is fostered by creating capacities to discuss shared perceptions and joint challenges. Nevertheless, regarding legitimacy the "shadow of hierarchy" should not to be ignored. Additionally, regarding the political support, the real consideration and implementation of new findings into political programming has to be discussed in the light of innovative participation. By neglecting these findings this innovative participation might lead to negative effects (Geißel 2012a: 213).

The third pillar is represented by "deliberation." Discussions, the joint identification of common high-level goals and objectified, accessible information – all of which are latent content of ERG – enable citizens to reflect their demands and preferences in a deliberative way.

The fourth pillar of Geißel's framework is "effectiveness," which covers the achievement of common high-level goals. First, the theoretical concept of ERG covers this pillar by setting the high-level goals as well as its measurement systems. Second, peer-reviewing and justification of taken actives examine the performance in reaching the high-level goals. Finally, it includes the possibility to adjust the goals and measurement systems to secure effectiveness.

The fifth pillar that Geißel identified in her research – "enlightened citizens and democratic education" – dates back to the US Declaration of Independence. Research on democracy has shown that stable democracies can only consolidate if citizens have learned democratic skills. Here, following the theoretical derived concept ERG improves these democratic skills. In the manner of "taking part in a choral society or a bird-watching club [...]" (Putnam 1993: 90), open and transparent multi-actor discussions in institutionalized capacities for political programming – as part of the latent content of ERG – foster participation (Beetham 2012), civic education (Geißel 2012b), public spiritedness (Fishkin 2012) and civic competences (Talpin 2012), which Geißel (2012a) identifies as major assets of democratic and participatory innovations (Geißel 2012a: 211). However, it is questionable in this example, whether citizens have to have at least some democratic skills if they want to participate in the discussions.

The overview shows that the theoretical derived concept of ERG includes the yardsticks and findings of democratic innovations presented by Geißel (2012a). An in-depth analysis based on the framework of Geißel (2013) is promising for research in political science in the future. On the one hand, the future research will foster the basic knowledge on ERG. On the other hand, the question, of whether ERG can be regarded as an innovative democratic participation approach is answered.

Additionally, the presented research has opened further questions regarding democratic participation. In the first place and in light of Grießel's postdoctoral thesis (Geißel 2007), the researched case of FRM, the aspect of legitimacy and the phenomenon of the "shadow of hierarchy" sets a further question: has the critical citizen improved the overall decision-making process of the extension or does John Doe endanger the implementation of supra-regional infrastructure projects? A detailed analysis of this question is also a promising field of engagement regarding the legitimacy of current spatial planning in supra-regional infrastructure projects in Germany. The gathered data of the presented analysis is usable to answer this question.

In the second place, regarding direct democracy, a further case of an airport extension, namely Munich Airport and the referendum as of June 17, 2012, is interesting. Here, the citizens refused the extension of Munich Airport (Merkur 2012a; Merkur 2012b). Due to the supportive attitude regarding the extension of the major shareholders of *Flughafen München*

*GmbH*, the result of the referendum was neglected.[10]. Referring to the hypothesis of Setälä (2006) – taken up again by Geißel – namely, that there has never been a case rejecting the result of a formal referendum in Western European democracies (Geißel 2012b: 165; Setälä 2006: 713) further questions are: have direct democratic procedures failed in the aftermath of June 17, 2012? How efficient are direct democratic decisions if the company shares matter for decisions?

In the final place, referring to the phenomenon of raising juridification of civic participation, the predicted and perceived changes in the participation and the democracy should be discussed (Geißel et al. 2014a; Geißel et al. 2014b). Does the identified juridification not cut across the request of more participation, transparency and the government statement of Willy Bandt: "We want to dare more democracy [...] [so that] each citzen has the possibility to be involved in reforms of the state and the society" (Brandt 1969)?[11] Is this juridification not a return to representation (by lawyers in the administrative system) away from a multi-faceted participation process?

This brief discussion of innovations in democracy exposed several starting points in the research field of innovative democracy and participation for research in the future.

### ERG and the entrepreneurial state

Following Ebner (2009b), the Schumpeterian entrepreneurial state is "defined as the exercise of entrepreneurial functions by organs of the state" (Ebner 2009b: 382). According to Schumpeter (2003), the entrepreneurial function "is to reform or revolutionize the pattern of production by exploiting an invention or, more generally, an untried technological possibility for producing a new commodity or producing an old one in a new way, by opening up a new source of supply of materials or a new outlet for products, by reorganizing an industry and so on" (Schumpeter 2003: 385). Further, following Ebner (2009b),

10 The Free State of Bavaria holds 51 percent, the Federal Government of Germany 26 percent, and the city of Munich holds 23 percent of the FMGs' shares (FMG 2014: 102).

11 Translated the by author.

the entrepreneurial state not only exercise the entrepreneurial functions itself; moreover, it also provides the capacities to carry out these tasks in an environment of novelty and uncertainty. "The actual capacity of the entrepreneurial state addresses the potential for exercising policies that promote innovation as a socio-economic process on an economy-wide scale. In Schumpeterian terms, it defines the performance of the entrepreneurial state in the carrying out the entrepreneurial function of introducing novelty in an economic system – either by direct interventions in the economic process or by its conditioning through institutional and physical infrastructures" (Ebner 2009b: 385).

Regarding the presented research the question is, whether the theoretical deduced approach of ERG supports the entrepreneurial state. For future research two possibilities seem to be feasible. One the one hand, the cases of East Asia, innovation, and the entrepreneurial state – extensively researched and presented by Ebner (Ebner 2007; Ebner 2009a; Ebner 2013)[12] – can be analyzed from the theoretical derived ERG perspective. In this manner the ASEAN or the SAARC appear as good examples of geographic-functional regions. On the other hand the case analyzed in the presented research can be analyzed by the concept of the entrepreneurial state. At least the extension of Frankfurt Airport and the extensive data gathered during the research represents a possible case to analyze. Both paths seem to be fruitful for future research for either the concept of the entrepreneurial state or ERG approach.

For the former Ebner presents a comprehensive theory-laden framework based on six preliminary propositions that can be linked to the latent content of ERG (Ebner 2009a: 387f; Ebner 2013: 10f). These linkages are outlined in the following.

The first proposition focuses on the formation of entrepreneurial capacities of the entrepreneurial state. The presented theoretical concept of ERG is linked to this aspect by the initiation of open and transparent multi-actor environments and the establishment of implementation capacities.

12 See Ebner (2002), Ebner (2004), Ebner (2006b), Ebner (2006a), Ebner 2014, Ebner (forthcoming) 2015[b], Ebner (forthcoming) 2015a, and Ebner (forthcoming) 2016 for further reading.

The question is whether it is able to create an environment and decision-making process to foster and promote innovation. Policy efforts to foster technological leadership in an uncertain environment are the second proposition. This aspect of the entrepreneurial state is closely connected to the setting of high-level goals and the establishment of practical and workable implementation plans and feasible regulations as promoted by the theoretical deduced approach of ERG. These policy efforts foster self-sustaining knowledge-based interactions to promote competitiveness on the one hand and institutional change of non-private institutions on the other. This third proposition is also traceable in ERG. Here, heterarchical links are established to support information exchange up- and downwards the governance levels. Close to this, the fourth preliminary proposition emphasizing the knowledge transfer beyond the levels and between state and society. The theoretical derived concept of ERG refers to the provision, exchange, and pooling of information at many stages and it also requests the processing of information up- and downwards the governance levels. Taking geographic-functional aspects into account, the fifth proposition reflects the diversity of structural, institutional, and spatial patterns of the entrepreneurial state referring to global production networks and local agglomerated innovation activities. In relation to the increasing globalization and focusing on the geographic-functional region as the governance environment, the theoretical deduced approach of ERG is directly linked to this preposition. The final preliminary proposition is also marked by glocalizational aspects. Building capacities for global and local firms simultaneously and addressing their entrepreneurial orientations, the entrepreneurial state links international competitiveness with local strategies. The theoretical derived concept of ERG covers this aspect by establishing an open and transparent multi-actors environment across hierarchical and traditional borders.

Accordingly, the theoretical derived analytical framework of ERG is suitable to examine the entrepreneurial state in a progressing glob-alication taking global and local phenomena as well as multilevel aspects of geographic-functionality into account. Both seem relevant to the entrepreneurial state. For governance, the theoretical derived concept of ERG is a fruitful approach to establish and organize the entrepreneurial state, respectively the entrepreneurial geographic-functional region.

Despite being a superficial scratch at the tip of the framework presented by Ebner, it promises a worthwhile broader research based on the gathered data of the presented research. In the manner of Ebner ((forthcoming) 2015[b]), one of the most interesting elements would be to reconstruct the innovation processes during extension processes of the presented cases. How were identified technical innovations promoted by the entrepreneurial state throughout the extension process? Has the administration promoted these innovations by a suitable decision-making process? On the one hand, the data gathered allows analyzing – at least for the major case of Frankfurt Airport – how the state created or failed to create capacities for entrepreneurial orientations during the overall extension process. On the other hand, there are indications that the data allows retracing the "career" of innovative ideas born during the extension process, how the ideas were recognized by the official proceedings and whether or how they were ultimately implemented.

**Further fields of interest**

The previous paragraphs have shown that there are links to other research fields, especially in political and social science. The following brief list presents further ideas where further research on ERG appears as fruitful in the future or the gathered data is a solid basis for future research.

- Based on the presented research, it appears that the self-image or the corporate purpose of non-private metropolitan governance bodies and private metropolitan logistics actors being a relevant metropolitan institution differ from social reality. Although the case of the airport extension is a major topic, especially for an airport region such as FRM, it is just one (even important) case to analyze the involvement, performance, and relevance these institutions. In the light of the ascertained performance of regional agglomerations (M. E. Porter 2014; M. E. Porter 2014) and the increasing importance of geographic-functional regions the relevance of non-private metropolitan governance bodies and private metropolitan logistics actors has to be proven by further fields of engagement. The question is whether non-private metropolitan governance bodies and private metropolitan logistics actors are relevant in the decision-making processes in geographic-functional regions, how an optimal metropolitan governance institution has to be set up?

- Throughout the extension process the global relevance of the airports was always highlighted. The connection between the global relevance and the local consequences was already named in previous research. However, due to the gathered data, including material, partly not available for research before, the presented analysis of glocalization seems to be a fruitful research field in the future. For example the perception of global and local issues of the different governance levels by analyzing the argumentation in the hearings is promising to obtain further insights.
- In the light of civic participation, the identified raising juridification is very interesting to analyze. The German law stipulates the participation of the broader public in the planning proceedings of the RPP and the RSE, whereas the participation only excludes public bodies. Now, the question is how the increasing juridification is compatible with the indicated broad participation the ordinary citizen. Does this juridification lead back to simple steering without involving John Doe as a critical citizen?
- The current discussions on the Transatlantic Trade and Investment Partnership (TTIP) also appear to be very interesting to analyze by the theoretical concept of ERG. How is the TTIP negotiated in a bi-polar environment without a central, final decider? Moreover, how will the TTIP be implemented in the single states of Europe?
- The UN-level seems also to be very interesting to analyze by experimentalist (regional) governance. One country, one vote represents exactly the core idea of decisions with no central, final decider.
- In the field of international cooperation, the Problem Driven Iterative Adaptation (PDIA) approach (Andrews 2014; Andrews, Pritchett, and Woolcock 2012) is comparable with ERG. Focusing on decision-making processes in development cooperation PDIA also represents an analytical framework for governance in less developed countries. Further research here is fruitful for the future to enhance both concepts.

## 5.6 Conclusion

Finally, by analyzing the regulatory structure of governance, decision-making processes, and the institutional arrangements based on the theoretical derived concept of ERG for the very first time, the presented research question and subordinated research questions have been answered by extensive content analysis. The analysis examined unique data, which was partly unavailable for research before. Overall, 12,737 files with a total of approximately 150,000 pages were screened during the pre-analysis of all data gathered. For the actual analysis documents with more than 31,000 pages as the primary resource of information were examined by the theoretical derived concept of ERG. One the one hand, the official documents of the hearing and planning authorities of the airport extension process – the RPD, the AoPs, and the transcripts of the respective hearings – were analyzed in detail. One the other hand, the official documentation of the mediation – final reports and the deliverables of the task forces – were comprehensively examined.

Taking Sabel and Zeitlin's ideas as the basis and integrating the geographic-functional aspect, the theoretical concept of ERG is deduced from the concept of experimentalist governance. It provides a research framework to analyze regulatory structures of governance in geographic-functional regions – such as FRM. The presented analysis shows, that the elaborated approach of ERG is a suitable analytical approach in policy analysis to examine the regulatory structure of governance, decision-making processes, and the institutional arrangements in geographic-functional regions.

The regulatory structure of governance and the decision-making processes are not restricted to the administrative institutions of a classical administrative system – e.g. the state. Additionally, with the integration of the embeddedness perspective, it closes the gap between process-oriented policy analysis and classical political research, focusing on institutions and the structure of institutional arrangements. By placing a stronger focus on geographic-functionality, the theoretical deduced concept of ERG copes with the demands formulated by new regionalism, "call[ing] for a differentiate repertoire of political organizations including a variety of loosely institutionalized arrangements such as [...] interlocal coordination,

| | |
|---|---|
| Discussion of Shared Perceptions of Joint Challenges | • Open, transparent multi-actor discussions across hierarchical and traditional borders (formal and informal)<br>• Identification of topics, best available techniques, joint advancement, and common understanding of challenges<br>• Delegation of mandate in certain topics; proactive advancing of mandate holder to increase ownership<br>• Establishment of heterarichical relations and integration of stakeholders<br>• Institutions already in place support recursive policy-making and learning by experience from lower level<br>• Provision, exchange and pooling of information |
| Definition of Goals and the Measurement System | • Setting high-level goals and common objectives<br>• Programming of measurement system<br>• Establishment of practical and workable implementation plan and feasible regulations<br>• Discussion proposals with addressees<br>• Formulation of reference documents, recommendations, and elaboration of legally (non-)binding rules<br>• Provision and discussion of guidance documents to clarify approaches |
| Implementation | • Operationalization of implementation to adopt or model the goals, the measurement system, and rules<br>• Entrust lower level units and delegate responsibilities and power to lower level authorities or experts<br>• Actors and addressees have most possible freedom and autonomy for implementation of goals<br>• Adjustment of existing rules based on tacit knowledge and subsequent experience<br>• Establishment of implementation capacities; design objective and imperial organizations, private/non-private<br>• Soft guidance to fulfill a binding overall framework and further harmonization of directives and regulations<br>• Fostering cooperation while implementation |
| Report and Peer Review | • Joint monitoring and evaluation, peer reviews as well as reporting<br>• Constitutional and collective regulatory dialogs; environment of continuous adjustment and learning<br>• Justification of taken activities and performance<br>• Institutionalization of periodical assessments, audits, and peer reviews<br>• Release and discussion of opinions and comments on implementation process, goals, and measurement system<br>• Process information up- and downwards; consultation of additional opinions and any relevant scientific issue |
| Review of the Measurement System, the Decision-making Processes, and the Stakeholder | • Reassessment, questioning and cross-check of intended goals and measurement system<br>• Recursive revision of policy objectives, procedures of political programming, legislation, and best practices<br>• Identification of areas that need adjustments, if necessary initiate legal action<br>• Present suitable proposals for amendments to (higher) authorities<br>• Take new developments and innovations into account<br>• If necessary bring in further stakeholders formally into the overall process |

*Figure 5.1: Five aspects of experimentalist regional governance.*

public-private partnerships, and policy-specific arrangements" (Koch 2013: 400).

Therefore, the framework represents a theoretical deduced and empirical verified analytical concept to analyze regulatory structures of governance and evolutionary decision-making processes in geographical-functional regions. In contrast to regional and metropolitan governance, the theoretical deduced concept of ERG has a clear analytical framework and does not lack vague definitions. The framework makes it possible to analyze regulatory structures of governance, its decision-making processes and institutional arrangements in geographic-functional regions according to manifest and latent content of five aspects.

Based on the consecutive character, ERG makes a definition of the region and the respective actors up front obsolete. Applying the theoretical derived analytical framework of ERG allows the integration and exclusion of actors by necessity without questioning the regulatory structure of governance and the decision-making process in general. The presented analysis made this visible. It shows that applying the theoretical ideas of ERG allows the adaptation of the institutional arrangement. With that, the region – the research object – is adaptable according to functional aspects and needs, without questioning the regulatory structure of governance. Utilizing the theoretical derived concept of ERG, the adjustment is rather a need regarding decision and policy-making than an obstacle for analysis. For the case of the airport extension in FRM, especially in the RPP and the later RSEs after the 1960s and 1980s prove that point. By enhancing the group of addressees or exchanging stakeholders this possibility was utilized during the hearings without changing the governance structure. Non-responding stakeholders were not further officially included in the regulatory structure of governance and the decision process in case of not replying to requests. In these cases, further inclusion was only progressed if their integration was enforced by law. As a result, the region can be defined pragmatically subject to the relevant stakeholders without jeopardizing the regulatory framework of governance. It has not to be defined according to traditional inflexible political borders. This is especially visible in situations where stakeholders organize themselves around a given challenge, e.g. the mediation process or the joint representation by lawyers in the hearings. This makes regulatory structures of governance and decision-making processes more efficient because the non-affected stakeholders do not have to be involved only because they are "accidentally" embraced by political borders.

Furthermore, the theoretical derived analytical framework of ERG allows breaking down the decision-making process in the presented consecutive steps. On the one hand regarding historical stages and timely correlation – Airport extension, 1960s–1980s; Mediation process, since 1997; RPP and RPD, 2000–2002; RSE and AoP, 2002–2004; RSE and AoP, 2004–2007; Adjustment of plans, 2007 and beyond – as well as in a timely correlation. On the other hand, each historical stage can be analyzed regarding the consecutive stages – the five aspects of ERG.

Concerning other approaches, the pragmatic character of ERG has the advantage of not being in the need of the final definition of the region upfront. The region itself is defined by the institutional arrangement of the regulatory governance structure. Based on pragmatism, stakeholders are linked to the regulatory governance structure if they are affected by the decision-making process and its results and not qua a "random jurisdiction." By integrating the characteristics of embeddedness, the theoretical concept of ERG enables policy analysis to take the surrounding institutional arrangement of the geographic-functional region into account. As a result, the theoretical derived concept of ERG allows identifying relevant institutions in the decision-making process independent of an upfront defined region.

With that, the theoretical deduced approach of ERG makes the institutional arrangement of "Metropolis" analyzable and connects the structural aspect of political research with the process-oriented policy analysis. As stated, the presented research had a focus on the involvement of metropolitan governance bodies during the airport extension in FRM. Here, the involvement of non-private metropolitan governance bodies and private metropolitan logistics actors is noteworthy. Against the argument of governance failure in metropolitan FRM, the metropolitan perspective is identifiable in the regulatory structure of governance, decision-making processes, and the institutional arrangements. However, it is torn. On the one hand, non-private metropolitan governance bodies are actively integrated into the decision-making processes of the airport extension by the responsible authorities, e.g. requesting statements on the planning documents or the direct involvement in the mediation process. On the other hand, with their current setting, the non-private metropolitan governance bodies are unable or unwilling to provide the capacity to engage in the extension topic more extensively. First, the analysis shows that the non-private metropolitan governance bodies were unable to unify their members having a common approach, even not with a minimal common metropolitan denominator. Second, compared to the engagement of other institutions and private objectors, their involvement in the discussions and the decision-making process was weak. Finally, this position was not strengthened, although they were recognized as the major metropolitan governance body of the region.

For the private metropolitan logistics actors, it has to be noted that they were only established at the end of the observation period. Still, the picture of non-involvement is even stronger. The research shows that the private metropolitan logistics actors in FRM were not involved in the decision-making process and the institutional arrangements. On the one side, not a single private metropolitan logistics actor was actively involved in any mediation process or official proceeding by the responsible authorities. On the other side, they were unable or unwilling to proactively provide adequate capacities for a joint approach.

Regarding governance learning the major insight is, that it is identifiable. However, the question of who is learning has to be investigated in-depth in the future.

From a general perspective, it can already be identified in the change from the top-down approach of steering politics to governance, including a bottom-up perspective, that became more relevant in the further approach and the establishment of the parallel mediation process. The question of whether governance learning is identifiable during the extension process of Frankfurt Airport can only be positive. However, on the one hand, the processes improved. On the other hand, there was also deterioration, especially for the participation of John Doe.

Although the presented approach can be criticized, it has to be taken as a starting point to analyze the learning aspect in geographic-functional regions. The presented analysis has shown that within the same successive and evolutionary decision-making process of the mediation in FRM the facets of governance learning of comparative public policy and different modes of governance can be analyzed.

Besides this, proving that the theoretical concept of ERG is valuable to analyze the regulatory framework of governance in the context of the airport extension initially links social science in and researches in logistics, especially air transport. The gathered data is fruitful to enhance these currently-weak ties of both research fields in the future.

Finally, the presented results of the research have highlighted more promising research fields were further analysis and insights are needed in the future.

# A Appendix

## A.1 Keywords of content analysis

### General

Benchmark; Dialog; Diskussion; Einführung; Entscheidung; gemeinsam; gemeinschaftlich; Information; Implementierung; Kooperation; Methodik; Methode; Monitoring; Partizipation; Peer Review; Review; Überprüfung; zusammen

### Airport Frankfurt

Air Cargo Community Frankfurt e.V.; CargoCity Frankfurt Task Force; House of Logistics and Mobility; HOLM; Logisitk RheinMain; Logistik Rhein Main; LRM; Planungsverband; Regionalverband; Umlandverband

## A.2 Further regional institutions FRM

| Institution | Members/Participants | Main field of engagement | Year |
|---|---|---|---|
| Frankfurt Main Finance e.V. | Private/non-private | • (International) promotion of the financial district of Frankfurt | since 2008 |
| Frankfurt Rhein-Main 2012 GmbH | Private/non-private | • Promotion of FRM for the Olympic Games 2012 | 2001 |
| Frankfurter Verkehrs- und Tarifverbund (FVV) | | • Regional public transport | 1976 – 1995 |
| Gemeinnützige Kulturfonds Frankfurt RheinMain GmbH | Non-private | • Promotion of FRM as a cultural region | since 2006 |
| KulturRegion Frankfurt RheinMain gGmbH (Kulturinitiative Rhein-Main (kirm), since 1998 | Private/non-private | • Promotion of culture in FRM | since 2005 |
| Material Valley e.V. | Private/non-private | • Promotion of FRM as high-tech location for materials science and materials technology. | since 2002 |
| Medienmittwoch-das Bild der Region | Private | • Promotion of the media in FRM | since 2001 |
| Naturalia Frankfurt/Rhein Main | Private/non-private | • Protection of the natural resources of the region | 2002 |
| Rat der Regionen | Non-private | • Communal conferences<br>• Release am annual report of communal collaboration<br>• Promote a joint regional appearance<br>• Integrate further rural districts | since 2011 |
| Regionalpark RheinMain | Private/non-private | • Secure remaining open areas in FRM for the relaxation and leisure of the people of the region | since 1995 |
| Rhein-Main Verkehrsverbund (RMV) | Non-private | • Regional public transport<br>• Secure remaining open areas in FRM for the relaxation and leisure of the people of the region | since 1995 |
| Rhein-Main-Abfall-GmbH | Non-private | • Regional waste management | since 1998 |

*Figure A.1: Further regional institutions FRM, based on Lackowska (2011a) and Zimmermann and Heinelt (2002), own depiction, missing information was added by statutes, annual reports, or institutions' website.*

| Institution | Members/Participants | Main field of engagement | Year |
|---|---|---|---|
| Automatisierungsregion Rhein Main Neckar | Private | • Regional network promotion in the field of automation | since 2007 |
| Automotive Cluster RheinMainNeckar | Private/non-private | • Regional promotion and coordination of automotive actors | since 2003 |
| Bionik-Netzwerk Hessen | Private/non-private | • Regional promotion of cooperation in the field of bionics | since 2013 |
| Cluster für Individualisierte ImmunIntervention (CI3) e.V. | Private/non-private | • Promotion of Research and application of individualized immune interventions | since 2010 |
| Competence Center Aerospace Kassel-Calden (CCA) | Private | • Cooperation, marketing, and innovation in the fields of aviation and airport | since 2010 |
| Competence Center for Applied Security Technology - CAST e.V. | Private/non-private | • Regional promotion of IT-security | since 2004 |
| ConsultingRegion FrankfurtRheinMain | Non-private | • Promotion of the consulting sector and niche supplier | |
| deENet - Kompetenznetzwerk dezentrale Energietechnologien e.V. | Private/non-private | • Joint development of integrated system solutions for energy supply | since 2003 |
| DieMaschinenbauPartner | Private | • Promotion of the field of mechanical engineering, technology transfer and joint project development | since 2008 |
| E-Finance Lab | Private | • Joint development of methods for redesigning processes and value chains developed in the industry | since 2002 |
| Energiegenossenschaft Odenwald Cluster | Private | • Regional promotion of energy topics and the energy sector | since 2009 |
| Engineering-High-Tech-Cluster (EHC) Fulda e.V. | Private/non-private | • Promotion of the engineering and production of plant systems and applications in high technologies | since 2008 |
| Frankfurt Biotech Alliance e.V. (FBA) | Private | • Promotion of the research field of biotechnology | since 2013 |
| gamearea-FRM e.V. | Private | • Promotion and marketing of the digital entertainment industry | since 2009 |
| HESSENMETALL | Private | • Regional promotion of Hessian metal and electrical industry | since 1947 |
| HeRo - Kompetenzzentrum HessenRohstoffe e.V. ( | Private/non-private | • Promotion raw material sector and improvement of technology and technology transfer<br>• Fostering communal cooperation | since 2004 |
| Initiative gesundheitswirtschaft rhein-main e.V. | Private | • Regional and Economic promotion in the field of health | since 2006 |

*Figure A.2: Further regional institutions FRM, based on HMWEVL (2015a), part I, own depiction, additional to former, missing information was added by statutes, annual reports, or institutions' website.*

| Institution | Members/Participants | Main field of engagement | Year |
|---|---|---|---|
| IT for Work e.V. | Private/non-private | • Promotion of regional cooperation of the IT-sector | since 2012 |
| IT-Netzwerk e.V. | Private | • Promotion or regional cooperation and innovation in the IT-sector | since 2004 |
| ITS Hessen (Intelligente Transport- und Verkehrssysteme Hessen) e.V | Private | • Promotion of innovative intelligent transport systems | since 2010 |
| KompetenzNetz UmweltTechnologie - KNUT | Private | • Fostering regional cooperation in the field of environmental technology | since 2007 |
| MetallverarbeitungsCluster Waldeck-Frankenberg | Private | • Promotion of regional cooperation in the field of metal industry | since 2005 |
| mst-Netzwerk Rhein-Main e.V. – Kompetenznetzwerk Mikrosystemtechnik | Private | • Promotion of micro system technology in the FRM | since 2004 |
| Optence e.V | Private | • Networking in and knowledge transfer in photonics | since 2001 |
| Regionalmanagement Nordhessen GmbH (Gesundheitsregion Nordhessen; GrimmHeimat NordHessen; MoWiN.net e.V. - Netzwerk der Mobilitätswirtschaft Nordhessen) | Private/non-private | • Regional promotion of cooperation and the economic structural change<br>• Economic promotion of the region of Northern Hesse and its sectors | since 2002 |
| Rhein-Main Adaptronik e.V. | Private | • Promotion of cooperation in the field of adaptive systems | since 2007 |
| Rhein-Main-Cluster Chemie und Pharma | Private | • Promotion of cooperation and innovations in the fields of chemistry and pharma | since 2008 |
| Technologie & Innovation Medizinregion Mittelhessen (timm) | Private/non-private | • Promotion of cooperation and knowledge transfer in the medical economyy | since 2008 |
| Verpackungscluster Mittelhessen | Private/non-private | • Promotion competiveness of its members in the field of packaging | since 2009 |
| Wasserstoff- und Brennstoffzellen-Initiative Hessen e.V. (H2BZ) | Private/non-private | • Promotion of collaboration and market and technology marketing in the field of hydrogen and fuel cell technology | since 2002 |
| Wetzlar Network e.V. - Industrienetzwerk Optik Elektronik Mechanik | Private/Non-private | • Promotion of cooperation in the fields of optics, electronics, and mechanics | since 2010 |
| Zeitsprung IT-Forum Fulda e.V | Private | • Promotion of education research, and science in the IT sector | since 1998 |

*Figure A.3: Further regional institutions FRM, based on HMWEVL (2015a), part II, own depiction, additional to former, missing information was added by statutes, annual reports, or institutions' website.*

| Actor | Actor |
|---|---|
| AG Mobil - Netzwerk der Zulieferindustrie | House of IT (HIT) |
| Arbeitsgemeinschaften des RKW Hessen | House of Pharma (in Planung) |
| Arbeitskreis der hessischen IT-Initiativen und Cluster | Hessisches Telemedia Technologie Kompetenz-Center e.V. |
| BIEG Hessen | InGeoForum |
| BioFrankfurt | Initiative Gesundheitswirtschaft Rheinland-Pfalz |
| BioRegion Rhein-Neckar-Dreieck e.V. | Initiative Wissensregion FrankfurtRheinMain |
| CEF - Cluster of Excellence Frankfurt "Macromolecular Complexes" | Innovationsplattform Magnetische Mikrosysteme INNOMAG e.V. |
| Center for Advanced Security Research Darmstadt (CASED) | It4work |
| Chemie- & Pharma-Cluster Industriepark Höchst/Rhein-Main | IT-Forum Mainz |
| CLUK Cluster der Kreativwirtschaft in Hessen | IT-Klub Mainz |
| Cluster Erneuerbare Energien Odenwald (CLEO) | IT-Netz Bayerischer Untermain |
| Cluster Integrierte Bioindustrie (CIB) Frankfurt | KLARA-Net |
| consulting@eschborn.net e.V. | KompetenzNetz Elektromobilität Bayerischer Untermain |
| ConsultingNetz Bad Homburg e.V. | Kompetenznetz Mechatronik & Automation Bayerischer Untermain Großwallstadt |
| Digital Hub FrankfurtRheinMain e.V. | Kompetenzzentrum Materialcharakterisierung |
| EC-M Beratungszentrum Elektronischer Geschäftsverkehr Mittelhessen | Kooperationsverbund Fahrzeugsicherheit Bayerischer Untermain |
| Excellence Cluster Cardio-Pulmonary System (ECCPS) | Logistik-Netz Bayerischer Untermain Aschaffenburg |
| FIV - Förderverein für integrierte Verkehrssysteme e.V. | Marketing-Club Mainz-Wiesbaden e.V. |
| Frankfurt Bio Tech Alliance e.V. | MaterialForschungsverbund Rhein-Main (MatFoRM-TUD) |
| Frankfurt Institute for Advanced Studies (FIAS) | Medienkompetenznetz RheinMain Mainz |
| Fraunhofer-Allianz Ambient Assisted Living | Medizintechnik Rhein-Neckar-Dreieck |
| Gesundheits-Netzwerk Deutschland Mitte-Südwest | Mikrosystemtechnik-Netzwerk Rhein-Main e.V. |
| Gesundheitswirtschaft Rhein-Main e.V. | NanoNetzwerkHessen |
| H-A-T Hessische Arbeitsgemeinschaft der Technologie-, Innovations- und Gründerzentren e.V. | Nationales Genomforschungsnetz |
| Haus der Kreativwirtschaft (in Planung) | Optence e.V. – Kompetenznetz Optische Technologien |
| Hessen Design e.V. | Prozesslernfabrik CIP |
| Hessen-Biotech | Rhine-Main Neuroscience Network ($rmn^2$) |
| Hessen-IT | Transferzentrum Adaptronik |
| Hessen-Nanotech | Umwelttechnik und EnergieWasserstoff- und Brennstoffzellen Initiative Hessen e.V. (H2BZ) |
| Hessen-Umwelttech Wiesbaden | Water Engineering Network e.V. |
| Hessische Film- und Medienakademie Offenbach am Main | Zeitsprung IT-Forum Fulda e.V. |
| House of Clean Energy | Zentrum für Arzneimittelforschung, Entwicklung und Sicherheit (ZAFES) |
| House of Finance (HOF) | Zentrum für Graphische Datenverarbeitung e.V. (ZGDV) |

*Figure A.4: Further regional institutions FRM, based on Regionalverband (2015), own depiction, additional to former, missing information was added by statutes, annual reports, or institutions' website.*

## A.3 Involved institutions of the RSE Runway 18-West

| # | Institutional form | | Institution | Institutional Unit |
|---|---|---|---|---|
| | Non-private | | | |
| | | Administrative | | |
| 1 | | | Federal Republic of Germany | Bundesanstalt für Flugsicherung |
| 2 | | | | Bundesvermögensstelle Frankfurt (Main) |
| 3 | | | | Bundesbahndirektion Frankfurt (Main) |
| 4 | | | | Oberpostdirektion Frankfurt (Main) |
| 5 | | | | Wehrbereichsverwaltung IV |
| 6 | | | Federal State of Hesse | Amt für Bodendenkmalpflege |
| 7 | | | | Hessisches Landesamt für Landwirtschaft Kassel |
| 8 | | | | Hessisches Landesamt für Straßenbau |
| 9 | | | | Landeskulturamt Wiesbaden |
| 10 | | | Sub-federal state level | Bischofsheim |
| 11 | | | | Berkach |
| 12 | | | | Braunshardt |
| 13 | | | | Buchschlag |
| 14 | | | | Büttelborn |
| 15 | | | | District Comittee Main-Taunus in Frankfurt/M.-Höchst |
| 16 | | | | District Comittee Offenbach |
| 17 | | | | District Comittee Groß-Gerau |
| 18 | | | | District Comittee Darmstadt |
| 19 | | | | Dornheim |
| 20 | | | | Egelsbach |
| 21 | | | | Eddersheim |
| 22 | | | | Flörsheim |
| 23 | | | | Frankfurt am Main |
| 24 | | | | Ginsheim-Gustavsburg |
| 25 | | | | Groß-Gerau |
| 25 | | | | Hattersheim |
| 27 | | | | Heusenstamm |
| 28 | | | | Hochheim |

*Figure A.5: Involved institutions of the RSE Runway 18-West, Part II, based on HMWT (1971: 754).*

| # | Institutional form | | Institution | Institutional Unit |
|---|---|---|---|---|
| | **Non-private** | | | |
| | | **Administrative** | (continuation) | |
| 29 | | | Sub-federal state level | Kelsterbach |
| 30 | | | | Klein-Gerau |
| 31 | | | | Langen |
| 32 | | | | Massenheim |
| 33 | | | | Mörfelden |
| 34 | | | | Nauheim |
| 35 | | | | Neu Isenburg |
| 36 | | | | Offenbach (Main) |
| 37 | | | | Okriftel |
| 38 | | | | Raunheim |
| 39 | | | | Rüsselsheim |
| 40 | | | | Sprendlingen |
| 41 | | | | Treburg |
| 42 | | | | Walldorf |
| 43 | | | | Wasserwirtschaftsamt Darmstadt |
| 44 | | | | Weilbach |
| 45 | | | | Weiterstadt |
| 46 | | | | Wicker |
| 47 | | | | Wiesbaden |
| 48 | | | | Worfelden |
| 49 | | | | Zeppelinheim |
| | **Non-private** | | | |
| | | **Metropolitan** | - | |
| | **Private** | | | |
| | | **Metropolitan** | - | |
| | | | | |
| | | **Economic** | - | |
| | **Further** | | - | |

*Figure A.6: Involved Institutions of the RSE Runway 18-West, Part II, based on HMWT (1971: 754).*

## A.4 Members of Gesprächskreis Flughafen 1998

| # | Institutional form | | Institution | Person |
|---|---|---|---|---|
| | **Non-private** | | | |
| | | **Administrative** | | |
| 1 | | | Federal State of Hesse | Hans Eichel (Minister-President of Hesse) |
| 2 | | | | Klaus Peter Möller (President of Hessian Parliament) |
| 3 | | | | Roland Kern (Judge at the Constitutional Court) |
| 4 | | | | Lothar Klemm (Hessian Economic Minister) |
| 5 | | | | Rupert von Plottnitz (Hessian Attorney General) |
| 6 | | | | Karl Starzacher (Hessian Finance Minister) |
| 7 | | | | Hans Joachim Suchan (Hessian State Secretary) |
| 8 | | | Sub-federal state level | Dr.-Ing. E.h. Kurt Oeser (Principal Councilor of Mörfelden-Waldorf) |
| 9 | | | | Petra Roth (Lord Mayor of Frankfurt am Main) |
| | | **Metropolitan** | | |
| | | | - | |
| | **Private** | | | |
| | | **Metropolitan** | | |
| | | | - | |
| | | **Economic** | | |
| 10 | | | Fraport AG | Dr. Wilhelm Bender (CEO Fraport AG) |
| | **Further** | | | |
| 11 | | Church | Protestant | Jörn Dulige (Delegate of the Protestant Church) |
| 12 | | | Catholics | Prelat Dr. Franz Kasper (Commissioner of Catholics Bishops) |
| 13 | | Civil Society | Author | Peter Härtling |
| 14 | | | | Dr. Hartmut Johnsen |
| 15 | | Union | Deutscher Gewerkschaftsbund (Hessen) | Dieter Hooge (Chairman of DGB (Hesse) |
| 16 | | Chambers | Chamber of Commerce and Industry (Hesse) | Dr. Frank Niethammer (Chairman of Hessian Chamber of Commerce and Industry) |

*Figure A.7: Members of 'Gesprächskreis Flughafen' and institutional characteristics, based on Hochfeld et al. (2000: 146).*

## A.5 Members of "Mediationsgruppe" 1998–2000

| # | Institutional form | | Institution | Person |
|---|---|---|---|---|
| | **Non-private** | | | |
| | | **Administrative** | | |
| 1 | | | Federal Republic of Germany | Gernot Riediger<br>(Deputy of Department LS 11, BMVBW) |
| 2 | | | Federal State of Hesse | Klaus-Peter Güttler<br>(Chief of Traffic Department, HMWVL) |
| 3 | | | | Wenzel Mayer<br>(Chief of Emission Department, HMULF) |
| 4 | | | Sub-federal state level | Peter Benz<br>(Lord Mayor of Darmstadt) |
| 5 | | | | Ingrid Borretty<br>(Councilor of Offenbach) |
| 6 | | | | Bernhard Brehl<br>(Mayor of Mörfelden-Walldorf) |
| 7 | | | | Hildebrand Diehl<br>(Lord Mayor of Wiesbaden) |
| 8 | | | | Erhard Engisch<br>(Mayor of Kelsterbach) |
| 9 | | | | Herbert Haas<br>(Mayor of Raunheim) |
| 10 | | | | Dirk-Oliver Quilling<br>(Mayor of Neu-Isenburg) |
| 11 | | | | Joachim Vandreike<br>(Mayor of Frankfurt) |
| 12 | | | | Dieter Wolf<br>(Mayor Stadt Flörsheim) |
| | | **Metropolitan** | | |
| | | | - | |
| | **Private** | | | |
| | | **Metropolitan** | | |
| | | | - | |
| | | **Economic** | | |
| 13 | | | Fraport AG | Herbert Becker<br>(Board Representative) |
| 14 | | | DFS Deutsche Flugsicherung GmbH | Rüdiger Bonneß<br>(Chief Officer Company Relations) |
| 15 | | | DLH | Viktor Pompe<br>(Coordinator Airport Issues) |

*Figure A.8: Members of "Mediationsgruppe Frankfurt Flughafen" and institutional characteristics, Part I, based on Hochfeld et al. (2000: 147).*

| # | Institutional form | | Institution | Person |
|---|---|---|---|---|
| | **Further** | | | |
| 16 | | Association | Vereinigung der hessischen Unternehmerverbände | Dr. Klaus Bernhard Dott (Member of Executive Committee) |
| 17 | | | BARIG Board of Airline Representatives in Germany e.V. | Martin Gaebges (General Secretary) |
| 18 | | | Offenbacher Vereinigung gegen den Fluglärm e.V. | Richard Müller (Chairman) |
| 19 | | Union | Deutsche Angestellten-Gewerkschaft (Hesse) | Karl Eugen Becker (Regional State Officer) |
| 20 | | | ötv Hesse | Gerold Schraub (District Leader) |
| 21 | | Chambers | Chamber of Commerce and Industry (Frankfurt) | Ewald Vollrath (Committee of Traffic, Tourism, Foreign Trade) |

*Figure A.9: Members of Mediation Group Frankfurt Flughafen and institutional characteristics, Part II, based on 2000 (2000: 147).*

# A.6 Members of task forces

## Mediationsverfahren:
## Mitglieder der Arbeitskreise, Arbeits- und Expertengruppen

### Arbeitskreis „Ökologie, Gesundheit und Soziales"

| | |
|---|---|
| Auer, Inge | Stadt Mörfelden-Walldorf, Büro des Bürgermeisters |
| Becker, Herbert | FAG, Vorstandsbeauftragter externe Kontakte |
| Bolze, Jens | Stadt Darmstadt, Umweltamt |
| Borretty, Ingrid | Stadt Offenbach, Stadträtin |
| Brehl, Bernhard | Stadt Mörfelden-Walldorf, Bürgermeister |
| Bruinier, Johann | Fluglärmschutzbeauftragter der Hessischen Landesregierung |
| Büchen Dr., Matthias | HLfU |
| Engisch, Erhard | Stadt Kelsterbach, Bürgermeister |
| Fenn, Heike | Stadt Mainz, Umweltamt |
| Haas, Herbert | Stadt Raunheim, Bürgermeister |
| Heer, Hermann-Josef | HMWVL, Abt. VI, Ref. B 3 |
| Hermann, Günther E. | HMWVL, PG Mediation |
| Heudorf Dr., Ursula | Stadt Frankfurt, Stadtgesundheitsamt |
| Laubereau Dr., Peter G. | HLfU, AL Umweltplanung |
| Mayer, Wenzel | HMULF, Leiter der Immissionsschutzabteilung |
| Müller, Kurt | HLfU |
| Müller, Richard | Offenbacher Vereinigung gegen den Fluglärm e.V. |
| Neumann-Opitz Dr., P. | FAG, USR-U |
| Ott, Dietrich | Magistrat der Stadt Rüsselsheim, Umwelt- und Grünflächenamt |
| Pompe, Viktor | Deutsche Lufthansa AG, Koordinator Flughafenangelegenheiten |
| Schmitt Dr., Stefan | Stadt Frankfurt, Umweltamt |
| Schönegge Dr., Peter | Stadt Neu-Isenburg, Umweltamt |
| Schuch, Hans J. | Stadt Hattersheim, Stadtrat |
| Schumacher, Renate | Stadt Obertshausen, Stadträtin |
| Stöcker Dr., Ulrich | Bundesministerium für Verkehr, Bau- u. Wohnungswesen, LS 15 |
| Tögel, Maximilian | FAG, Leiter Kapazitäts- und Umfeldentwicklung |
| Urban, Wolfgang | Stadt Damstadt, Stadtplanungsamt |
| Viebrock, Hans-Heinrich | Stadt Mörfelden-Walldorf, Stadtplanungsamt |
| Walle Dr.-Ing., Frank | Deutsche Lufthansa AG, Leiter Konzern Umweltfragen |
| Werner, Michael | Stadt Frankfurt, Dez. VIII Umwelt, Energie und Brandschutz |
| Wolf, Dieter | Stadt Flörsheim, Bürgermeister |

*Figure A.10: Members of task force ‘, based on Mediationsgruppe (2000a).*

**Expertengruppe „Fluglärm" im Arbeitskreis „Ökologie, Gesundheit und Soziales"**

| | |
|---|---|
| Auer, Inge | Stadt Mörfelden-Walldorf, Büro des Bürgermeisters |
| Bolze, Jens | Stadt Darmstadt, Umweltamt |
| Bonneß, Rüdiger | DFS, Leiter Unternehmenskontakte |
| Bruinier, Johann | Fluglärmschutzbeauftragter der Hessischen Landesregierung |
| Eberle Dr., Wolfgang | HMULF, Immissionsschutzabteilung |
| Fenn, Heike | Stadt Mainz, Umweltamt |
| Heer, Hermann-Josef | HMWVL, Abt. VI, Ref. B 3 |
| Heudorf Dr., Ursula | Stadt Frankfurt, Gesundheitsamt |
| Mayer, Wenzel | HMULF, Leiter der Immissionsschutzabteilung |
| Mönch, Rudolf | Deutsche Lufthansa AG, Operationelle Standards |
| Müller, Kurt | HLfU |
| Müller, Richard | Offenb. Vereinigung gegen den Fluglärm |
| Ott, Dietrich | Magistrat der Stadt Rüsselsheim, Umwelt- und Grünflächenamt |
| Riediger, Gernot | BMVBW, Leiter LS 11 |
| Schaab, Jochen | Stadt Kelsterbach, Büro des Bürgermeisters |
| Schmitt Dr., Stefan | Stadt Frankfurt, Umweltamt |
| Tögel, Maximilian | FAG, Leiter Kapazitäts- und Umfeldentwicklung |
| Urban, Wolfgang | Stadt Darmstadt, Stadtplanungsamt |
| Viebrock, Hans-Heinrich | Stadt Mörfelden-Walldorf, Stadtplanungsamt |
| Walle Dr.-Ing., Frank | Deutsche Lufthansa AG, Leiter Konzern Umweltfragen |
| Wetterling Dr., Klaus | Stadt Frankfurt, Amt f. komm. Gesamtentwicklung/Stadtplanung |

*Figure A.11: Members of sub-task force "Fluglärm" of task force "Ökologie, Gesundheit und Soziales," based on Mediationsgruppe (2000a).*

## Arbeitsgruppe „Siedlung und Soziales" im Arbeitskreis „Ökologie, Gesundheit und Soziales"

| | |
|---|---|
| Altenburger, Peter | Stadt Frankfurt, Amt f. komm. Gesamtentwicklung/Stadtplanung |
| Auer, Inge | Stadt Mörfelden-Walldorf, Büro des Bürgermeisters |
| Bergmeier, Matthias | HMWVL, Abt. Infrastruktur, Landesplanung, Regionalentwicklung |
| Brehl, Bernhard | Stadt Mörfelden-Walldorf, Bürgermeister (Leiter der AG) |
| Engisch, Erhard | Stadt Kelsterbach, Bürgermeister |
| Feld, Ingeborg | Kreis Groß-Gerau, KAG |
| Feßenmayr, Herbert | Stadtplanungsamt Mainz, Ltd. Baudirektor |
| Fercher, Leo | Stadt Flörsheim, Liegenschaftsamt |
| Franger Dr., Michael | Stadt Offenbach, Jugendhilfeplaner |
| Goßmann, Holger | Stadt Wiesbaden, Amt für Stadtplanung, Bürgermeister |
| Hartz, Bernd | RP Darmstadt, Leiter Dez. Siedlungswesen und Bauleitplanung |
| Heer, Hermann-Josef | HMWVL, Abt. VI, Ref. 3 b |
| Hermann, Günther E. | HMWVL, PG Mediation |
| Hofmann, Doris | Landkreis Darmstadt-Dieburg, KAG, VS-Mitglied |
| Müller, Robert | Stadt Mühlheim am Main, Stadtrat |
| Müller, Kurt | HLfU |
| Ott, Dietrich | Magistrat der Stadt Rüsselsheim, Umwelt- und Grünflächenamt |
| Pfalzgraf, Klaus | Hess. Städte- u. Gemeindebund, Ref. Stadtplanung / Baurecht |
| Rauschnabel Dr., Kurt | HMWVL, Abt. Bauwesen, Städtebau, Wohnungswesen |
| Riediger, Gernot | BMVBW, Leiter LS 11 |
| Rohrmann, Bernd | Regierungspräsidium Darmstadt, Leiter Dez. Wirtschaft, Verkehr |
| Schmitt, Baldur | Kreis Groß-Gerau, Erster Kreisbeigeordneter |
| Schönegge Dr., Peter | Stadt Neu-Isenburg, Umweltamt |
| Sturm Dr., Peter | Umlandverband Frankfurt/M. |
| Tögel, Maximilian | FAG, Leiter Kapazitäts- und Umfeldentwicklung |
| Urban, Wolfgang | Stadt Darmstadt, Stadtplanungsamt |
| Viebrock, Hans-H. | Stadt Mörfelden-Walldorf, Stadtplanungsamt |
| Werner, Michael | Stadt Frankfurt, Dezernat VIII Umwelt, Energie und Brandschutz |
| Wichert, Klaus | Stadt Frankfurt, Leiter des Umweltamtes |
| Wolf, Dieter | Stadt Flörsheim, Bürgermeister |

*Figure A.12: Members of sub-task force 'Siedlung und Soziales' of task force 'Ökologie, Gesundheit und Soziales', based on Mediationsgruppe (2000a).*

## Arbeitskreis „Ökonomie"

| | |
|---|---|
| Auer, Inge | Stadt Mörfelden-Walldorf, Büro des Bürgermeisters |
| Becker, Herbert | FAG, Vorstandsbeauftragter externe Kontakte |
| Becker, Karl-Eugen | DAG, Landesverbandsleiter |
| Brehl, Bernhard | Stadt Mörfelden-Walldorf, Bürgermeister |
| Dehn, Klaus | FAG, FIF-P 1 |
| Diehl, Hildebrand | Stadt Wiesbaden, Oberbürgermeister |
| Dott Dr., Klaus Bernhard | Vereinigung der hessischen Unternehmerverbände |
| Gretz Dr., Wendelin | HLT |
| Güttler, Klaus-Peter | HMWVL, Leiter der Verkehrsabteilung |
| Hermann, Günther E. | HMWVL, PG Mediation |
| Kania, Peter | Wirtschaftsförderung Frankfurt/M. |
| Lang, Andreas | FAG, PSL-E 2 |
| Neidel, Hans-Günter | Stadt Offenbach, Leiter des Sozialamtes |
| Pompe, Viktor | Deutsche Lufthansa AG, Koordinator f. Flughafenangelegenheiten |
| Riediger, Gernot | BMVBW, Stellv. Leiter LS 11 |
| Ringhoffer, Franz | Stadt Mainz, Beigeordneter |
| Robanus, Gerd | Stadt Maintal, Erster Stadtrat |
| Schaab, Jochen | Stadt Kelsterbach, Büro des Bürgermeisters |
| Schwesinger Dr., H. | Wirtschaftsförderung Frankfurt/M. |
| Urban, Wolfgang | Stadt Darmstadt, Stadtplanungsamt |
| Vandreike, Joachim | Stadt Frankfurt/M, Bürgermeister |
| Vauth, Werner | Stadt Darmstadt, Amt für Wirtschaftsförderung |
| Viebrock, Heinrich | Stadt Mörfelden-Walldorf, Stadtplanungsamt |
| Vollrath, Ewald | IHK Frankfurt/M. |

*Figure A.13: Members of task force 'Ökonomie', based on Mediationsgruppe (2000a).*

**Arbeitskreis „Verkehr"**

| | |
|---|---|
| Auer, Inge | Stadt Mörfelden-Walldorf, Büro des Bürgermeisters |
| Becker, Herbert | FAG, Vorstandsbeauftragter externe Kontakte |
| Bonneß, Rüdiger | DFS, Leiter Unternehmenskontakte |
| Borretty, Ingrid | Stadt Offenbach, Stadträtin |
| Brehl, Bernhard | Stadt Mörfelden-Walldorf, Bürgermeister |
| Gaebges, Martin | BARIG, Generalsekretär |
| Gläser, Winfried | FAG, FIF-P 1 |
| Güttler, Klaus-Peter | HMWVL, Leiter der Verkehrsabteilung |
| Hermann, Günther E. | HMWVL, PG Mediation |
| Mauel, Stefan | FAG, VHM-K 2 |
| Mayer, Wenzel | HMULF, Leiter der Immissionsschutzabteilung |
| Müller, Richard | Offenbacher Vereinigung gegen den Fluglärm e.V. |
| Nymbach, Jürgen | Offenbacher Vereinigung gegen den Fluglärm e.V. |
| Ott, Dietrich | Magistrat der Stadt Rüsselsheim, Umwelt- und Grünflächenamt |
| Pompe, Viktor | Deutsche Lufthansa AG, Koordinator f. Flughafenangelegenheiten |
| Quilling, Dirk-Oliver | Stadt Neu-Isenburg, Bürgermeister |
| Ringhoffer, Franz | Stadt Mainz, Beigeordneter |
| Schaab, Jochen | Stadt Kelsterbach, Büro des Bürgermeisters |
| Schönegge Dr., Peter | Stadt Neu-Isenburg, Umweltamt |
| Schöttler, Ulrich | Stadt Frankfurt/M., Ordnungsamt |
| Schuch, Hans J. | Stadt Hattersheim, Stadtrat |
| Urban, Wolfgang | Stadt Darmstadt, Stadtplanungsamt |
| Viebrock, Heinrich | Stadt Mörfelden-Walldorf, Stadtplanungsamt |
| Vollrath, Ewald | IHK Frankfurt/M. |
| Weber, Frank | Stadt Offenbach, stellv. Leiter des Straßenverkehrsamtes |
| Wetterling Dr., Klaus | Stadt Frankfurt/M., Amt f. komm. Gesamtentwicklung/Stadtplanung |
| Wolf, Dieter | Stadt Flörsheim, Bürgermeister |

*Figure A.14: Members of task force "Verkehr," based on Mediationsgruppe (2000a).*

## Arbeitsgruppe „Flugsicherung und Navigation" im Arbeitskreis „Verkehr"

| | |
|---|---|
| Auer, Inge | Stadt Mörfelden-Walldorf, Büro des Bürgermeisters |
| Bonneß, Rüdiger | DFS, Leiter Unternehmenskontakte |
| Bruinier, Johann | Fluglärmschutzbeauftragter der Hessischen Landesregierung |
| Fongern, Georg | Vereinigung Cockpit e.V. |
| Gebauer, Peter | DFS |
| Grösslein, Egon Franz | HMWVL |
| Heer, Hermann-Josef | HMWVL, Abt. VI, Ref. 3 b |
| Hermann, Günther E. | HMWVL, PG Mediation |
| Knorr, Werner | Deutsche Lufthansa AG, Sicherheitspilot |
| Mauel, Stefan | FAG, VHM-K 2 |
| Menz, Dennys | FAG, VHM-K 2 |
| Müller, Richard | Offenbacher Vereinigung gegen den Fluglärm e.V. |
| Nymbach, Jürgen | Offenbacher Vereinigung gegen den Fluglärm e.V. |
| Pompe, Viktor | Deutsche Lufthansa AG, Koordinator Flugangelegenheiten |
| Riediger, Gernot | BMVBW, Leiter LS 11 |
| Schaab, Jochen | Stadt Kelsterbach, Büro des Bürgermeisters |
| Schierle, Volker | Deutsche Lufthansa AG, Operationelle Standards |
| Schwenk, Rüdiger | Deutsche Lufthansa AG, Operationelle Standards |
| Urban, Wolfgang | Stadt Darmstadt, Stadtplanungsamt |
| Viebrock, Hans-Heinrich | Stadt Mörfelden-Walldorf, Stadtplanungsamt |
| Winter, Walter | Stadt Neu-Isenburg, Natur- und Umweltschutz |

*Figure A.15: Members of sub-task force "Flugsicherung und Navigation" of task force "Verkehr," based on Mediationsgruppe (2000a).*

## A.7 Members of Regionales Dialogforum Flughafen Frankfurt

| # | Institutional form | | Institution | Person | Year |
|---|---|---|---|---|---|
| | **Non-private** | | | | |
| | | **Administrative** | | | |
| 1 | | | Federal State of Hesse* | Gerlinde Borchmann (Hessian State Chancellery) | 2000-2001 |
| 2 | | | | Dr. Holger Sewering (Hessian State Chancellery) | |
| 3 | | | Sub-federal state level | Peter Benz (Lord Mayor of Darmstadt) | 2000-2004 |
| 4 | | | | Klaus Feuchtinger (Councilor of Darmstadt) | 2004-2008 |
| 5 | | | | Bernhard Brehl (Mayor of Mörfelden-Walldorf) | 2000-2006 |
| 6 | | | | Heinz-Peter Becker (City of Mörfelden-Walldorf) | 2006-2008 |
| 7 | | | | Erhard Engisch (Mayor of Kelsterbach) | 2000-2006 |
| 8 | | | | Manfred Ockl (City of Kelsterbach) | 2006-2008 |
| 9 | | | | Hans Franssen (Mayor of Hattersheim) | 2000-2008 |
| 10 | | | | Stefan Gieltowski (Lord Mayor Rüsselsheim) | 2000-2002 |
| 11 | | | | Peter Layer (Councilor of Rüsselsheim) | 2002-2008 |
| 12 | | | | Gerhard Grandke (Lord Mayor of Offenbach) | 2000-2004 |
| 13 | | | | Horst Schneider (Mayor of Offenbach) | 2004-2006 |
| 14 | | | | Paul-Gerhard Weiß (City of Offenbach) | 2006-2008 |
| 15 | | | | Thomas Jühn (Mayor Raunheim) | 2000-2004 |
| 16 | | | | Johannes Wolf (Citiy of Raunheim) | 2004-2006 |
| 17 | | | | Hand-Joachim Engelhardt (City of Raunheim) | 2006-2008 |
| 18 | | | | Ulrich Krebs (Mayor of Flörsheim) | 2000-2006 |
| 19 | | | | Michael Antenbrink (City of Flörsheim) | 2006-2008 |
| 20 | | | | Dirk-Oliver Quilling (Mayor of Neu-Isenburg) | 2000-2008 |
| 21 | | | | Wolfgang Reichel (Councillor of Mainz) | 2000-2008 |

*Figure A.16: Members of Regionales Dialogforum Flughafen Frankfurt and institutional characteristics, Part I, based on RDF (2001), RDF (2002b), RDF (2003), RDF (2004), RDF (2005a), RDF (2006a), and RDF (2008b); *Have been listed as members in the first Annual Report.*

| # | Institutional form | | Institution | Person | Year |
|---|---|---|---|---|---|
| | **Non-private** | | | | |
| | | **Administrative** | (continuation) | | |
| 22 | | | Sub-federal state level | Harald Schindler (City of Hochheim) | 2000-2002 |
| 23 | | | | Angelika Munck (Mayor of Hochheim) | 2002-2008 |
| 24 | | | | Edwin Schwarz (Councilor of Frankfurt) | 2000-2008 |
| 25 | | | | Dieter Zimmer (City of Dreieich) | 2006-2008 |
| | **Non-Private** | | | | |
| | | **Metropolitan** | | | |
| | | | - | | |
| | **Private** | | | | |
| | | **Metropolitan** | | | |
| | | | - | | |
| | | **Economic** | | | |
| 26 | | | Fraport AG | Dr. Barbara Jakubeit | 2000-2002 |
| 27 | | | | Dr. Manfred Schölch (Board Member) | 2002-2006 |
| 28 | | | | Dr. Stefan Schule (Board member) | 2006-2008 |
| 29 | | | DFS Deutsche Flugsicherung GmbH | Dr. Heinrich Olbert | 2000-2004 |
| 30 | | | | Michael Kraft | 2004-2008 |
| 31 | | | DLH | Viktor Pompe (Coordinator Airport Issues) | 2000-2002 |
| 32 | | | | Rita Diede-Stützel | 2001-2006 |
| 33 | | | | Gregor Schlüter | 2006-2006 |
| 34 | | | | Mark Johnson | 2006-2008 |

*Figure A.17: Members of Regionales Dialogforum Flughafen Frankfurt and institutional characteristics, Part II, based on RDF (2001), RDF (2002b), RDF (2003), RDF (2004), RDF (2005a), RDF (2006a), and RDF (2008b); *Have been listed as members in the first Annual Report.*

| # | Institutional form | | Institution | Person | Year |
|---|---|---|---|---|---|
| | Further | | | | |
| 35 | | University/ Research | Technische Universität Darmstadt | Dr.-Ing. Johann-Dietrich Wörne (President TU Darmstadt) | 2000-2008 |
| 36 | | | Öko-Institu e.V.* | Christian Hiochfeld | |
| 37 | | | | Regine Barth | |
| 38 | | Association | Vereinigung der hessischen Unternehmerverbände | Dr. Klaus Bernhard Dott (Member of Executive Committee) | 2000-2002 |
| 39 | | | | Lutz Raettig | 2002-2008 |
| 40 | | | Schutzbund Deutscher Wald (SDW) | Christoph von Eisenhart Rothe | 2000-2006 |
| 41 | | | Vereinigung Cockpit e.V. | Cpt. Georg Fongern | 2000-2002; 2005-2006 |
| 42 | | | | Uwe Kröger | 2002-2005 |
| 43 | | | | Burkhart Kaumanns | 2006-2008 |
| 44 | | | BARIG Board of Airline Representatives in Germany e.V. | Martin Gaebges (General Secretary) | 2000-2008 |
| 45 | | | Fluglärmkommission | Richard Müller | 2000-2004 |
| 46 | | | | Thomas Jühe | 2004-2008 |
| 47 | | | Hessischer Handwerktag | Jürgen Heyne (President) | 2000-2006 |
| 48 | | | | Bernd Ehinger | 2006-2008 |
| 49 | | | Arbeitsgemeinschaft Deutscher Verkehrsflughäfen (ADV) | Bernd Nierobisch | 2000-2006 |
| 50 | | | | Ralph Beisel | 2006-2008 |
| 51 | | | Bundesvereinigung gegen Fluglärm | Dirk Treber | 2000-2008 |
| 52 | | | Arbeitsgemeinschaft Deutscher Luftfahrtsunternehmen | Dr. Detlef Winter | 2000-2006 |

*Figure A.18: Members of Regionales Dialogforum Flughafen Frankfurt and institutional characteristics, Part III, based on RDF (2001), RDF (2002b), RDF (2003), RDF (2004), RDF (2005a), RDF (2006a), and RDF (2008b); *Have been listed as members in the first Annual Report.*

| # | Institutional form | | Institution | Person | Year |
|---|---|---|---|---|---|
| | Further | (continuation) | | | |
| 53 | | | Regionales Dialogforum* | Melanie Beiler | |
| 54 | | | | Dr. Christoph Ewen | |
| 55 | | | | Dr. Hans-Peter Meister | |
| 56 | | | | Carla Schönfelder | |
| 57 | | | Fachverband Spedition und Logistik | Willy Korf (Chief Executive Officer) | 2001-2008 |
| 58 | | | Bürgeraktion pro Flughafen | Christel Karesch (Chairman) | 2001-2008 |
| 59 | | | WiDeMa e.V. | Ernst Grewatta | 2001-2002 |
| 60 | | | | Harm Heldmaier | 2002-2008 |
| 61 | | | Bundesverband der Deutschen Fluggesellschaften (BDF) | Wolfgang Dreher | 2006-2006 |
| 62 | | | | Raphael von Heeremann | 2006-2008 |
| 63 | | Church | Protestant | Jörn Dulige (Consistory) | 2000-2008 |
| 64 | | | Catholics | Dr. Franz Kaspar | 2000-2004 |
| 65 | | | | Dr. Guido Amend | 2004-2008 |
| 66 | | Chambers | Chamber of Commerce and industry (Hessen) | Dr. Wolfgang Lindstaedt (Chief Executive Officer) | 2000-2004 |
| 67 | | | | Matthias Gräßel | 2004-2006 |
| 68 | | | | Hans-Peter Debling | 2006-2008 |
| 69 | | | Landesärztekammer Hessen | Dr. Uwe Gieler | 2001-2004 |
| 70 | | | | Dr. Caroline Herr | 2004-2006 |
| 71 | | | | Dr. Thomas Eikmann | 2006-2008 |
| 72 | | Unions | ver.di (former Deutscher Gewerkschaftsbund (DGB)) | Gerold Schaub (DeputyRegional State Officer) | 2000-2008 |
| 73 | | | ver.di (former Deutsche Angestellten Gewerkschaft (DAG)) | Klaus zum Felde | 2000-2008 |

*Figure A.19: Members of Regionales Dialogforum Flughafen Frankfurt and institutional characteristics, Part IV, based on RDF (2001), RDF (2002b), RDF (2003), RDF (2004), RDF (2005a), RDF (2006a), and RDF (2008b); *Have been listed as members in the first Annual Report.*

## A.8 Members of forum flughafen & region

| # | Institutional form | | Institution | Person |
|---|---|---|---|---|
| | Non-private | | | |
| | | Administrative | | |
| 1 | | | Federal republic of Germany | Hr. Schmidt (Bundesministerium für Verkehr, Bau und Stadtentwicklung) |
| 2 | | | | N.N. Luftfahrtbundesamt |
| 3 | | | Federal State of Hesse | Hr. Lanz (Hessisches Ministerium für Umwelt, Energie, Landwirtschaft und Verbraucherschutz) |
| 4 | | | | N.N. (Hessisches Ministerium für Wirtschaft, Verkehr und Landesentwicklung) |
| 5 | | | | Hr. Dr. Sewering (Hessische Staatskanzlei) |
| 6 | | | | |
| 7 | | | Sub-federal state level | Hr. Quilling (Mayor of the City of Neu-Isenburg) |
| 8 | | | | Hr. Gölzenleuchter (Mayor of Büttelborn) |
| 9 | | | | Hr. Hoffmann (Mayor of the City of Darmstadt) |
| 10 | | | | Hr. Zimmer (Mayor of the City of Dreieich) |
| 11 | | | | Hr. Antenbrink (Mayor of the City Flörsheim am Main) |
| 12 | | | | Fr. Dr. Rottmann (Councilor Frankfurt am Main) |
| 13 | | | | Hr. Leber (Mayor of the City of Griesheim) |
| 14 | | | | Hr. Franssen (Mayor of the City of Hattersheim am Main) |
| 15 | | | | Hr. Jakoby (Mayor of the City of Heusenstamm) |
| 16 | | | | Hr. Munck (Mayor of the City of Hochheim am Main) |
| 17 | | | | Hr. Ockel (Mayor of the City of Kelsterbach) |
| 18 | | | | Hr. Beutel (Lord Mayor of the City of Mainz) |
| 19 | | | | Hr. Rohrbach (Mayor of the City of Maintal) |

*Figure A.20: Members of Regionales Dialogforum Flughafen Frankfurt and institutional characteristics, Part I, based on FFR (2009).*

| # | Institutional form | | Institution | Person |
|---|---|---|---|---|
| | Non-private | | | |
| | | Administrative | (continuation) | |
| 20 | | | Sub-federal state level | Hr. Becker (Mayor of the City of Mörfelden-Walldorf) |
| 21 | | | | Hr. Müller (Mayor of the City of Mühlheim) |
| 22 | | | | Hr. Schneider (Lord Mayor of the City of Offenbach am Main) |
| 23 | | | | Hr. Jühe (Lord Mayor of the City of Raunheim) |
| 24 | | | | Hr. Gieltowski (Lord Mayor of the City of Rüsselsheim) |
| 25 | | | | Hr. Kohl (Mayor of Walluf) |
| 26 | | | | Hr. Rohrbach (Mayor of the City of Weiterstadt) |
| 27 | | | | Hr. Dr. Müller (Lord Mayor of the City of Wiesbaden) |
| 28 | | | | Hr. Siehr (Head of a District Kreis Groß-Gerau) |
| 29 | | | | Hr. Gall (Head of a District Main-Taunus-Kreis) |
| 30 | | | | Hr. Schick (Head of a District Mainz-Bingen) |
| 31 | | | | Hr. Walter (Head of a District Kreis Offenbach) |
| 32 | | | | Hr. Jakoubek (Head of a District Darmstadt Dieburg) |
| | | Metropolitan | | |
| 33 | | | Metropolitan Governance Bodies | Hr. Kasseckert (Planungsverband Ballungsraum Frankfurt/ Rhein-Main) |
| | Private | | | |
| | | Metropolitan | | |
| | | | - | |
| | | Economic | | |
| 34 | | | Fraport AG | Dr. Schulte |
| 35 | | | Deutsche Lufthansa AG | Hr. Pauly |
| 36 | | | DFS Deutsche Flugsicherung GmbH | Hr. Kraft |

*Figure A.21: Members of Regionales Dialogforum Flughafen Frankfurt and institutional characteristics, Part II, based on FFR (2009).*

| # | Institutional form | | Institution | Person |
|---|---|---|---|---|
| | Further | | | |
| 37 | | Association | Deutsches Institut für Luft- und Raumfahrt e.V. | Prof. Dr. Wörner |
| 38 | | | Hessischer Handwerkstag | Hr. Ehinger |
| 39 | | | Arbeitsgemeinschaft deutscher Verkehrsflughäfen e.V. ADV | Hr. Beisel |
| 40 | | | Fachverband Güterkraftverkehr und Logistik Hessen e.V. | Hr. Poppe |
| 41 | | | Vereinigung der hessischen Unternehmerverbände e. V. | Hr. Dr. Raettig |
| 42 | | | Vereinigung Cockpit e.V. - Leiter des Ressorts Berufspolitik/ Arbeitsgruppen | Hr. Kaumanns |
| 43 | | | Landesärztekammer Hessen | Hr. Dr. Popović |
| 44 | | | Bund für Umwelt und Naturschutz Deutschland BUND | Fr. Martin |
| 45 | | | Bürgerinitiative WIDEMA | N.N. |
| 46 | | | Deutscher Fluglärmdienst e.V. DFLD | Hr. Weise |
| 47 | | | Bürgervereinigung Pro Flughafen | Fr. Karesch |
| 48 | | | Deutsches Institut für Luft- und Raumfahrt e.V. | Hr. Reichmuth |
| 49 | | | Board of Airline Representatives e.V. | Hr. Gaebges |
| 50 | | | Bundesvereinigung gegen Fluglärm e.V. | Hr. Treber |
| 51 | | | Kommunale Arbeitsgemeinschaft Flughafen Frankfurt am Main | Hr. Will |
| 52 | | Union | Deutscher Gewerkschaftsbund Landesbezirk Hessen | Hr. Körzell |
| 53 | | | ver.di Hessen | Hr. Schaub |
| 54 | | Church | Evangelische Kirche in Hessen | Hr. Dulige |
| 55 | | | Katholische Kirche (Leiter des Kommissariats der katholischen Bischöfe im Lande Hessen) | Hr. Dr. Amend |
| 56 | | Chambers | Industrie- und Handelskammer Frankfurt, Arbeitsgemeinschaft hessischer Industrie- und Handelskammern | Hr. Dr. Debling |
| 57 | | | Industrie- und Handelskammer Offenbach | Hr. Achenbach |

*Figure A.22: Members of Regionales Dialogforum Flughafen Frankfurt and institutional characteristics, Part III, based on FFR (2009).*

| # | Institutional form | | Institution | Person |
|---|---|---|---|---|
| | Further | | | |
| 58 | | Party | Fraktion der CDU im Hessischen Landtag | Hr. Boddenberg |
| 59 | | | Fraktion der SPD im Hessischen Landtag | Hr. Frankenberger |
| 60 | | | Fraktion der FDP im Hessischen Landtag | Hr. Posch |
| 61 | | | Fraktion Bündnis90/Die Grünen im Hessischen Landtag | Hr. Kaufmann |
| 62 | | | Fraktion „Die Linken" im Hessischen Landtag | Hr. van Oyen |
| 63 | | No affiliation mentioned | Fluglärmbeauftragter | Hr. Müller |

*Figure A.23: Members of Regionales Dialogforum Flughafen Frankfurt and institutional characteristics, Part IV, based on FFR (2009).*

# A.9 Involved institutions of the RPP Frankfurt Airport

| # | Institutional form | | Institution | Institutional Unit |
|---|---|---|---|---|
| | Non-private | | | |
| | | Administrative | | |
| 1 | | | Federal Republic of Germany | Bundesvermögensamt Frankfurt |
| 2 | | | | Eisenbahn-Bundesamt Frankfurt |
| 3 | | | | Wehrbereichsverwaltung IV, Wiesbaden |
| 4 | | | | Deutscher Wetterdienst RMG Offenbach |
| 5 | | | | Bundesministerium für Verkehr, Bau und Wohnungswesen, Berlin |
| 6 | | | only for somebodies attention | Bundesministerium für Umwelt, Naturschutz und Reaktorsicherheit, Berlin |
| 7 | | | Federal State of Hesse | Hessisches Landesamt für Umwelt und Geologie, Abt. Hydrogeologie |
| 8 | | | | Hessisches Landesamt für Umwelt und Geologie, Abt. Lärm |
| 9 | | | | Hessisches Landesamt für Umwelt und Geologie, Abt. Luftreinhaltung |
| 10 | | | | Hessisches Landesamt für Umwelt und Geologie, Abt. Bodenschutz |
| 11 | | | | Landesamt für Denkmalpflege, Schloss Biebrich, Wiesbaden |
| 12 | | | | Landesamt für Denkmalpflege, Abt. Archäologie und Denkmalpflege, Darmstadt |
| 13 | | | | Landesarbeitsamt Hessen, Frankfurt |
| 14 | | | | Hessisches Landesamt für Straßen- und Verkehrswesen, Wiesbaden |
| 15 | | | | Amt für Straßen- und Verkehrswesen Frankfurt |
| 16 | | | | Amt für Straßen- und Verkehrswesen Darmstadt |
| 17 | | | | Amt für Straßen- und Verkehrswesen Gelnhausen |
| 18 | | | | Amt für Straßen- und Verkehrswesen Wiesbaden |
| 19 | | | | Hessisches Forstamt Darmstadt |
| 20 | | | | Hessisches Forstamt Groß-Gerau |
| 21 | | | | Hessisches Forstamt Hofheim |
| 22 | | | | Hessisches Forstamt Königstein |
| 23 | | | | Hessisches Forstamt Langen |
| 24 | | | | Hessisches Forstamt Mörfelden-Walldorf |

*Figure A.24: Involved institutions of RPP Frankfurt Airport and institutional characteristics, Part I, based on RPDA (2002b: 18–24).*

| # | Institutional form | | Institution | Institutional Unit |
|---|---|---|---|---|
| | **Non-private** | | | |
| | | **Administrative** | | |
| 25 | | | Federal State of Hesse (continuation) | Hessisches Forstamt Rodgau |
| 26 | | | | Hessisches Forstamt Wolfgang |
| 27 | | | | Hessisches Forstamt Babenhausen |
| 28 | | | | Landesbetrieb Hessenforst, Kassel |
| 29 | | | | Landesbetrieb Hessenforst, Abt. FIV, Gießen |
| 30 | | | | Landrat des Landkreises Darmstadt-Dieburg, Hauptabteilung LFN, Darmstadt |
| 31 | | | | Landrat des Hochtaunuskreises, Hauptabteilung LFN, Usingen |
| 32 | | | | Landrat des Main-Kinzig-Kreises, Hauptabteilung LFN, Gelnhausen |
| 33 | | | | Landrat des Wetteraukreises, Hauptabteilung LFN, Friedberg |
| 34 | | | | Hessisches Immobilienmanagement, Zentrale Wiesbaden |
| 35 | | | | Landrat des Landkreises Limburg-Weilburg |
| 36 | | | | Landrat des Landkreises Aschaffenburg |
| 37 | | | | Hessische Staatskanzlei, Wiesbaden |
| 38 | | | | Hessisches Ministerium für Umwelt, Landwirtschaft und Forsten, Wiesbaden |
| 39 | | | | Hessisches Sozialministerium, Wiesbaden |
| 40 | | | only for somebodies attention | Hessisches Ministerium des Innern und für Sport, Wiesbaden |
| 41 | | | | Hessisches Ministerium für Wirtschaft, Verkehr und Landesentwicklung, Wiesbaden |
| 42 | | | | Hessisches Ministerium für Wirtschaft, Verkehr und Landesentwicklung - Fluglärmbeauftragter |
| 43 | | | Federal State of Rhineland-Palatinate | Struktur- und Genehmigungsdirektion Nord, Koblenz |
| 44 | | | | Struktur- und Genehmigungsdirektion Süd, Neustadt an der Weinstraße |

*Figure A.25: Involved institutions of RPP Frankfurt Airport and institutional characteristics, Part II, based on RPDA (2002b: 18–24).*

| # | Institutional form | | Institution | Institutional Unit |
|---|---|---|---|---|
| | **Non-private** | | | |
| | | **Administrative** | (continuation) | |
| 45 | | | Federal State of Rhineland-Palatinate | Ministerium für Umwelt und Forsten des Landes Rheinland-Pfalz, Mainz |
| 46 | | only for somebodies attention | | Ministerium für Wirtschaft, Verkehr, Landwirtschaft und Weinbau des Landes Rheinland-Pfalz, Mainz |
| 47 | | | | Ministerium des Innern und für Sport des Landes Rheinland-Pfalz, Mainz |
| 48 | | | | Staatskanzlei des Landes Rheinland-Pfalz, Mainz |
| 49 | | | Regional Council Darmstadt | Bezirksforstausschuss, Darmstadt |
| 50 | | | | Abteilung I, Darmstadt |
| 51 | | | | Dezernat II 23.1, Darmstadt |
| 52 | | | | Dezernat III 22.1, Darmstadt |
| 53 | | | | Dezernat III 22.2, Darmstadt |
| 54 | | | | Dezernat IV/Da 41.1, Darmstadt |
| 55 | | | | Dezernat IV/Da 43.3, Darmstadt |
| 56 | | | | Dezernat IV/Da 43.1, Darmstadt |
| 57 | | | | Dezernat IV/Da 41.5, Darmstadt |
| 58 | | | | Dezernat IV/Da 41.2, Darmstadt |
| 59 | | | | Dezernat IV/Da 41.3, Darmstadt |
| 60 | | | | Dezernat IV/Da 43.2, Darmstadt |
| 61 | | | | Dezernat IV/F 43.4, Darmstadt |
| 62 | | | | Dezernat IV/F 41.3, Darmstadt |
| 63 | | | | Dezernat IV/F 41.5, Darmstadt |
| 64 | | | | Dezernat IV/F 41.2, Darmstadt |
| 65 | | | | Dezernat IV/F 41.1, Darmstadt |
| 66 | | | | Dezernat IV/Hu 43.1, Darmstadt |
| 67 | | | | Dezernat IV/Wi 41.2, Darmstadt |
| 68 | | | | Dezernat V 31.1, Darmstadt |
| 69 | | | | Dezernat V 31.2, Darmstadt |
| 70 | | | | Dezernat V 31.3, Darmstadt |
| 71 | | | | Dezernat V 33.3, Darmstadt |
| 72 | | | | Dezernat VI 51.1, Darmstadt |
| 73 | | | | Dezernat VI 51.2, Darmstadt |

*Figure A.26: Involved Institutions of RPP Frankfurt Airport and institutional characteristics, Part III, based on RPDA (2002b: 18–24).*

| # | Institutional form | | Institution | Institutional Unit |
|---|---|---|---|---|
| | Non-private | | | |
| | | Administrative | (continuation) | |
| 74 | | | Regional Council Darmstadt | Dezernat VI 52, Darmstadt |
| 75 | | | | Dezernat VI 53.1, Darmstadt |
| 76 | | | Regional Council Gießen | Abteilung III, Regierungspräsidium Gießen |
| 77 | | | Sub-federal state level | Magistrat der Wissenschaftsstadt Darmstadt |
| 78 | | | | Magistrat der Stadt Frankfurt am Main |
| 79 | | | | Magistrat der Stadt Offenbach am Main |
| 80 | | | | Magistrat der Landeshauptstadt Wiesbaden |
| 81 | | | | Magistrat der Stadt Babenhausen |
| 82 | | | | Magistrat der Stadt Bad Homburg v.d.H. |
| 83 | | | | Magistrat der Stadt Bad Soden am Taunus |
| 84 | | | | Magistrat der Stadt Bad Vilbel |
| 85 | | | | Magistrat der Stadt Bruchköbel |
| 86 | | | | Magistrat der Stadt Dietzenbach |
| 87 | | | | Magistrat der Stadt Dreieich |
| 88 | | | | Magistrat der Stadt Eltville am Rhein |
| 89 | | | | Magistrat der Stadt Eppstein |
| 90 | | | | Magistrat der Stadt Eschborn |
| 91 | | | | Magistrat der Stadt Flörsheim am Main |
| 92 | | | | Magistrat der Stadt Friedrichsdorf |
| 93 | | | | Magistrat der Stadt Gernsheim |
| 94 | | | | Magistrat der Stadt Griesheim |
| 95 | | | | Magistrat der Stadt Groß-Gerau |
| 96 | | | | Magistrat der Stadt Hanau |
| 97 | | | | Magistrat der Stadt Hattersheim am Main |
| 98 | | | | Magistrat der Stadt Heusenstamm |
| 99 | | | | Magistrat der Stadt Hochheim am Main |
| 100 | | | | Magistrat der Stadt Hofheim am Taunus |
| 101 | | | | Magistrat der Stadt Idstein |
| 102 | | | | Magistrat der Stadt Karben |
| 103 | | | | Magistrat der Stadt Kelkheim (Taunus) |
| 104 | | | | Magistrat der Stadt Kelsterbach |

*Figure A.27: Involved Institutions of RPP Frankfurt Airport and institutional characteristics, Part IV, based on RPDA (2002b: 18–24).*

| # | Institutional form | | Institution | Institutional Unit |
|---|---|---|---|---|
| | Non-private | | | |
| | | Administrative | (continuation) | |
| 105 | | | Sub-federal state level | Magistrat der Stadt Königstein im Taunus |
| 106 | | | | Magistrat der Stadt Kronberg im Taunus |
| 107 | | | | Magistrat der Stadt Langen |
| 108 | | | | Magistrat der Stadt Langenselbold |
| 109 | | | | Magistrat der Stadt Maintal |
| 110 | | | | Magistrat der Stadt Mörfelden-Walldorf |
| 111 | | | | Magistrat der Stadt Mühlheim am Main |
| 112 | | | | Magistrat der Stadt Neu-Isenburg |
| 113 | | | | Magistrat der Stadt Obertshausen |
| 114 | | | | Magistrat der Stadt Oberursel |
| 115 | | | | Magistrat der Stadt Pfungstadt |
| 116 | | | | Magistrat der Stadt Raunheim |
| 117 | | | | Magistrat der Stadt Rödermark |
| 118 | | | | Magistrat der Stadt Rodgau |
| 119 | | | | Magistrat der Stadt Rüsselsheim |
| 120 | | | | Magistrat der Stadt Schwalbach am Taunus |
| 121 | | | | Magistrat der Stadt Seligenstadt |
| 122 | | | | Magistrat der Stadt Steinbach (Taunus) |
| 123 | | | | Magistrat der Stadt Taunusstein |
| 124 | | | | Magistrat der Stadt Weiterstadt |
| 125 | | | | Stadtforstamt Wiesbaden |
| 126 | | | | Stadtforstamt Frankfurt |
| 127 | | | | Gemeindevorstand der Gemeinde Bischofsheim |
| 128 | | | | Bezirksnaturschutzbeirat, Regierungspräsidium Darmstadt |
| 129 | | | | Gemeindevorstand der Gemeinde Büttelborn |
| 130 | | | | Gemeindevorstand der Gemeinde Egelsbach |
| 131 | | | | Gemeindevorstand der Gemeinde Eppertshausen |
| 132 | | | | Gemeindevorstand der Gemeinde Erlensee |
| 133 | | | | Gemeindevorstand der Gemeinde Erzhausen |

*Figure A.28: Involved institutions of RPP Frankfurt Airport and institutional characteristics, Part V, based on RPDA (2002b: 18–24).*

| # | Institutional form | | Institution | Institutional Unit |
|---|---|---|---|---|
| | Non-private | | | |
| | | Administrative | (continuation) | |
| 134 | | | Sub-federal state level | Gemeindevorstand der Gemeinde Ginsheim-Gustavsburg |
| 135 | | | | Gemeindevorstand der Gemeinde Glashütten |
| 136 | | | | Gemeindevorstand der Gemeinde Großkrotzenburg |
| 137 | | | | Gemeindevorstand der Gemeinde Groß-Zimmern |
| 138 | | | | Gemeindevorstand der Gemeinde Hainburg |
| 139 | | | | Gemeindevorstand der Gemeinde Hünstetten |
| 140 | | | | Gemeindevorstand der Gemeinde Kiedrich |
| 141 | | | | Gemeindevorstand der Gemeinde Kriftel |
| 142 | | | | Gemeindevorstand der Gemeinde Liederbach am Taunus |
| 143 | | | | Gemeindevorstand der Gemeinde Mainhausen |
| 144 | | | | Gemeindevorstand der Gemeinde Messel |
| 145 | | | | Gemeindevorstand der Gemeinde Nauheim |
| 146 | | | | Gemeindevorstand der Gemeinde Niederdorfelden |
| 147 | | | | Gemeindevorstand der Gemeinde Niedernhausen |
| 148 | | | | Gemeindevorstand der Gemeinde Riedstadt |
| 149 | | | | Gemeindevorstand der Gemeinde Rodenbach |
| 150 | | | | Gemeindevorstand der Gemeinde Roßdorf |
| 151 | | | | Gemeindevorstand der Gemeinde Sulzbach (Taunus) |
| 152 | | | | Gemeindevorstand der Gemeinde Trebur |
| 153 | | | | Gemeindevorstand der Gemeinde Waldems |
| 154 | | | | Gemeindevorstand der Gemeinde Walluf |
| 155 | | | | Kreisausschuss des Landkreises Darmstadt-Dieburg |

*Figure A.29: Involved institutions of RPP Frankfurt Airport, Part VI, based on RPDA (2002b: 18–24).*

| # | Institutional form | | Institution | Institutional Unit |
|---|---|---|---|---|
| | Non-private | | | |
| | | Administrative | (continuation) | |
| 156 | | | Sub-federal state level | Kreisausschuss des Landkreises Groß-Gerau |
| 157 | | | | Kreisausschuss des Hochtaunuskreises |
| 158 | | | | Kreisausschuss des Main-Kinzig-Kreises |
| 159 | | | | Kreisausschuss des Main-Taunus-Kreises |
| 160 | | | | Kreisausschuss des Landkreises Offenbach |
| 161 | | | | Kreisausschuss des Rheingau-Taunus-Kreises |
| 162 | | | | Kreisausschuss des Wetteraukreises |
| 163 | | | | Magistrat der Stadt Bad Camberg |
| 164 | | | | Kreisausschuss des Landkreises Limburg-Weilburg |
| 165 | | | | Magistrat der Stadt Alzenau in Unterfranken |
| 166 | | | | Gemeindevorstand der Gemeinde Kahl am Main |
| 167 | | | | Gemeindevorstand der Gemeinde Karlstein am Main |
| 168 | | | | Gemeindevorstand der Gemeinde Kleinostheim |
| 169 | | | | Gemeindevorstand der Gemeinde Stockstadt am Main |
| 170 | | | | Regierung von Unterfranken, Würzburg |
| 171 | | | | Kreisverwaltung Mainz-Bingen, Ingelheim |
| 172 | | | | Stadtverwaltung Mainz |
| 173 | | | | Gemeindeverwaltung der Gemeinde Budenheim |
| 174 | | | | Verbandsgemeinde Bodenheim |
| 175 | | | | Verbandsgemeinde Heidesheim am Rhein |
| 176 | | | | Verbandsgemeinde Nierstein/Oppenheim |

*Figure A.30: Involved institutions of RPP Frankfurt Airport, Part VII, based on RPDA (2002b: 18–24).*

| # | Institutional form | | Institution | Institutional Unit |
|---|---|---|---|---|
| | Non-private | | | |
| 177 | | Metropolitan | | Planungsverband Frankfurt/Rhein-Main, Frankfurt |
| 178 | | | | Regionaler Planungsverband Bayrischer Untermain, Aschaffenburg |
| 179 | | | | Planungsgemeinschaft Rheinhessen-Nahe, Mainz |
| 180 | | | | Regionalversammlung Südhessen |
| | Private | | | |
| | | Metropolitan | | |
| | | | - | |
| 181 | | Economic | | DB Netz AG, Niederlassung Mitte, Immobilienmanagement Frankfurt |
| 182 | | | | DB Station & Services AG, Recht (GRI), Frankfurt |
| 183 | | | | De-TE Immobilien Frankfurt |
| 184 | | | | Deutsche Post AG, Bauen GmbH, Karlsruhe |
| 185 | | | | Deutsche Flugsicherung GmbH, Offenbach |
| 186 | | | | Rhein-Main-Verkehrsverbund, Hofheim |
| 187 | | | | Stadtwerke Verkehrsgesellschaft Frankfurt am Main GmbH |
| 188 | | | | Lokale Nahverkehrsgesellschaft mbH des Kreises Groß-Gerau |
| 189 | | | | Lokale Nahverkehrsgesellschaft Frankfurt am Main GmbH |
| 190 | | | | Riedwerke Kreis Groß-Gerau |
| 191 | | | | Hessenwasser GmbH, Frankfurt |
| 192 | | | | Fernleitungsbetriebs GmbH, Bonn |
| 193 | | | | Rhein-Main-Rohrleitungstransport GmbH, Köln |
| 194 | | | | Preussen Elektra AG, Borken |
| 195 | | | | RWE Net AG, Dortmund |
| 196 | | | | WINGAS GmbH, Kassel |
| 197 | | | | Ruhrgas AG, Essen |
| 198 | | | | Pipeline Engineering GmbH, Essen |
| 199 | | | | Energieversorgung Offenbach AG |

*Figure A.31: Involved institutions of RPP Frankfurt Airport and institutional characteristics, Part VIII, based on RPDA (2002b: 18–24).*

| # | Institutional form | | Institution | Unit |
|---|---|---|---|---|
| | **Private** | | | |
| 200 | | **Economic** | (continuation) | Gas-Union GmbH, Frankfurt |
| 201 | | | | Hessische Elektrizitäts AG, Darmstadt |
| 202 | | | | Kraftwerke Mainz-Wiesbaden AG, Mainz |
| 203 | | | | Süwag Energie AG, Niederlassung MKW/EWH, Bad Homburg v. d. H. |
| 204 | | | | Mainova AG, Frankfurt |
| 205 | | | | Süwag Energie AG, Niederlassung MKW, Frankfurt |
| 206 | | | | Infraserv GmbH & Co Höchst KG, Frankfurt |
| 207 | | | | Stadtwerke Mörfelden-Walldorf |
| 208 | | | | Stadtwerke Groß-Gerau |
| 209 | | | | Stadtwerke Langen GmbH |
| 210 | | | | Stadtwerke Mainz AG |
| 211 | | | | Stadtwerke Neu-Isenburg |
| 212 | | | | Stadtwerke Wiesbaden |
| 213 | | | | Südhessische Gas- und Wasser AG, Darmstadt |
| 214 | | | | Überlandwerk Groß-Gerau GmbH |
| 215 | | | | E.ON Netz GmbH, Borken |
| 216 | | | | Hessische Flugplatz GmbH Egelsbach |
| | **Further** | | | |
| 217 | | Association | | Pro Bahn, Landesverband Hessen, Darmstadt |
| 218 | | | | VCD Hessen e.V., Kassel |
| 219 | | | | Schutzgemeinschaft Deutscher Wald, Landesverband Hessen e.V., Wiesbaden |
| 220 | | | | Hessische Gesellschaft für Ornithologie und Naturschutz Hessen, Echzell |
| 221 | | | | Landesjagdverband Hessen e.V., Bad Nauheim |
| 222 | | | | Bund für Umwelt und Naturschutz, Landesverband Hessen, Frankfurt |
| 223 | | | | Deutsche Gebirgs- und Wandervereine, Landesverband Hessen, Groß-Umstadt |

*Figure A.32: Involved institutions of RPP Frankfurt Airport and institutional characteristics, Part IX, based on RPDA (2002b: 18–24).*

| # | Institutional form | | Institution | Institutional Unit |
|---|---|---|---|---|
| | Private | | | |
| 224 | | Association | (continuation) | Verband Hessischer Sportfischer, Wiesbaden |
| 225 | | | | Naturschutzbund Deutschland, Landesverband Hessen, Wetzlar |
| 226 | | | | Botanische Vereinigung für Naturschutz in Hessen e.V., Marburg |
| 227 | | | | Staatliche Vogelschutzwarte für Hessen, Rheinland-Pfalz und Saarland, Frankfurt |
| 228 | | | | Wasserverband Kinzig, Frankfurt |
| 229 | | | | Wasserbeschaffungsverband Rheingau-Taunus, Wiesbaden |
| 230 | | | | Arbeitsgruppe Wasserwerk Hessisches Ried, Einhausen-Jägersburg |
| 231 | | | | Zweckverband Wasserversorgung Stadt und Kreis Offenbach, Seligenstadt |
| 232 | | | | Wasserverband Schwarzbachgebiet-Ried, Groß-Gerau |
| 233 | | | | Hessischer Waldbesitzerverband e.V., Friedrichsdorf |
| 234 | | | | Hessischer Bauernverband e.V., Friedrichsdorf |
| 235 | | | | Arbeitsgemeinschaft Deutscher Verkehrsflughäfen, Geschäftsführung Stuttgart |
| 236 | | | | Bundesvereinigung gegen Fluglärm e.V., Mörfelden-Walldorf |
| 237 | | | | Vorsitzender der Kommission zur Abwehr des Fluglärms für Flughafen Frankfurt Main |
| 238 | | | | Vorsitzender der Kommission zur Abwehr des Fluglärms für den Verkehrslandeplatz Egelsbach |
| 239 | | | | ARIG e.V., Frankfurt |
| 240 | | | | Hessischer Landkreistag, Wiesbaden |
| 241 | | | | Hessischer Städtetag e.V., Wiesbaden |
| 242 | | | | Hessischer Städte- und Gemeindebund, Mühlheim |
| 243 | | | | Regionales Dialogforum Flughafen Frankfurt Main, Rüsselsheim |
| 244 | | | | Hessischer Städte- und Gemeindebund, Mühlheim |
| 245 | | | Union | Deutscher Gewerkschaftsbund, Landesbezirk Hessen, Frankfurt |

*Figure A.33: Involved institutions of RPP Frankfurt Airport and institutional characteristics, Part X, based on RPDA: (2002b: 18–24).*

| # | Institutional form | | Institution | Institutional Unit |
|---|---|---|---|---|
| | **Further** | | | |
| 246 | | Chruch | Protestant | Evangelische Kirche in Hessen und Nassau, Darmstadt |
| 247 | | | Catholics | Kommissariat der Katholischen Bischöfe, Wiesbaden |
| 248 | | Chamber | Industrie- und Handelskammer | Darmstadt |
| 249 | | | | Frankfurt |
| 250 | | | | Gießen-Friedberg |
| 251 | | | | Hanau-Gelnhausen-Schlüchtern |
| 252 | | | | Offenbach |
| 253 | | | | Wiesbaden |
| 254 | | | | Arbeitsgemeinschaft Hessischer Industrie- und Handelskammern, Frankfurt |
| 255 | | | Handwerkskammer | Rhein-Main, Hauptverwaltung, Frankfurt |
| 256 | | | | Rhein-Main Darmstadt |
| 257 | | | | Wiesbaden |
| 258 | | | | Arbeitsgemeinschaft der Hessischen Handwerkskammern, Wiesbaden |

*Figure A.34: Involved institutions of RPP Frankfurt Airport and institutional characteristics, Part XI, based on RPDA (2002b: 18–24).*

## A.10 Involved institutions of the RSE A380 Shipyard, 1st evaluation

| # | Institutional form | | Institution | Institutional Unit |
|---|---|---|---|---|
| | **Non-private** | | | |
| | | **Administrative** | | |
| 1 | | | Federal Republic of Germany | Bundesministerium für Verkehr, Bau und Wohnungswesen |
| 2 | | | | Bundesministerium für Umwelt, Naturschutz und Reaktorsicherheit |
| 3 | | | | Wehrbereichsverwaltung West - Außenstelle Wiesbaden |
| 4 | | | | Wasser- und Schifffahrtsamt Aschaffenburg |
| 5 | | | Federal State of Hesse | Hessisches Forstamt Mörfelden-Walldorf |
| 6 | | | | Landesbetrieb Hessenforst |
| 7 | | | | Hessisches Landesamt für Umwelt und Geologie |
| 8 | | | | Landesamt für Denkmalpflege |
| 9 | | | | Hessisches Landesamt für Straßen- und Verkehrswesen |
| 10 | | | | Polizeipräsidium Südhessen |
| 11 | | | | Hessisches Ministerium des Innern und für Sport |
| 12 | | | | Hessisches Sozialministerium |
| 13 | | | | Hessisches Ministerium für Umwelt, ländlichen Raum und Verbraucherschutz |
| 14 | | | | Hessisches Ministerium für Umwelt, ländlichen Raum und Verbraucherschutz – Landesforstausschuss |
| 15 | | | | Hessisches Ministerium für Wirtschaft, Verkehr und Landesentwicklung - Fluglärmbeauftragter |
| 16 | | | | Landrat des Landkreises Groß-Gerau |
| 17 | | | | Landrat des Hochtaunuskreises |
| 18 | | | | Landrat des Landkreises Darmstadt-Dieburg |
| 19 | | | Regional Council Darmstadt | Abteilung V, Bezirksforstausschuss |
| 20 | | | Sub-federal state level | Stadtforstamt Frankfurt am Main |
| 21 | | | | Amt für Straßen- und Verkehrswesen Frankfurt |

*Figure A.35: Involved institutions of RSE A380 Shipyard and institutional characteristics, 1st evaluation, Part I, based on HMWVL (2004: 54–57).*

| # | Institutional form | | Institution | Institutional Unit |
|---|---|---|---|---|
| | Non-private | | | |
| | | Administrative | (continuation) | |
| 22 | | | Sub-federal state level | Amt für Straßen- und Verkehrswesen Darmstadt |
| 23 | | | | Polizeipräsidium Frankfurt am Main |
| 24 | | | | Kreisausschuss des Landkreises Groß-Gerau |
| 25 | | | | Oberbürgermeisterin der Stadt Frankfurt am Main |
| 26 | | | | Magistrat der Stadt Frankfurt am Main |
| 27 | | | | Magistrat der Stadt Rüsselsheim |
| 28 | | | | Magistrat der Stadt Kelsterbach |
| 29 | | | | Magistrat der Stadt Mörfelden-Walldorf |
| 30 | | | | Magistrat der Stadt Raunheim |
| | Non-private | | | |
| | | Metropolitan | | |
| 31 | | | | Planungsverband Frankfurt/Rhein-Main |
| | Private | | | |
| | | Metropolitan | | |
| | | | - | |
| | | Economic | | |
| 32 | | | | Staatliche Vogelschutzwarte für Hessen, Rheinland-Pfalz und das Saarland |
| 33 | | | | Deutscher Wetterdienst |
| 34 | | | | Deutsche Telekom AG |
| 35 | | | | Mainova AG |
| 36 | | | | Rhein-Main-Verkehrsverbund GmbH |
| 37 | | | | Lokale Nahverkehrsgesellschaft mbH des Kreises Groß-Gerau |
| 38 | | | | Traffiq Lokale Nahverkehrsgesellschaft Frankfurt am Main mbH |
| 39 | | | | Hessenwasser GmbH |
| 40 | | | | Deutsche Flugsicherung GmbH |
| 41 | | | | Riedwerke Kreis Groß-Gerau |
| | Further | | | |
| 42 | | Association | | Kommission zur Abwehr des Fluglärms Flughafen Frankfurt Main |
| 43 | | | | Arbeitsgruppe Wasserwerk Hessisches Ried |

*Figure A.36: Involved institutions of RSE A380 Shipyard and institutional characteristics, 1st evaluation, Part II, based on HMWVL (2004: 54–57).*

| # | Institutional form | | Institution | Institutional Unit |
|---|---|---|---|---|
| | **Further** | | | |
| 44 | | Association | (continuation) | Wasserverband Schwarzbachgebiet-Ried |
| 45 | | | | Bund für Umwelt und Naturschutz - Landesverband Hessen e.V. - |
| 46 | | | | Botanische Vereinigung für Naturschutz in Hessen e.V. |
| 47 | | | | Deutsche Gebirgs- und Wandervereine - Landesverband Hessen |
| 48 | | | | Hessische Gesellschaft für Ornithologie und Naturschutz Hessen |
| 49 | | | | Landesjagdverband Hessen e.V. |
| 50 | | | | Naturschutzbund Deutschland - Landesverband Hessen e.V. |
| 51 | | | | Schutzgemeinschaft Deutscher Wald - Landesverband Hessen e.V. |
| 52 | | | | Verband Hessischer Sportfischer e.V. |
| 53 | | | | Bund Ökologische Lebensmittelwirtschaft |
| 54 | | | | Hessischer Waldbesitzerverband e.V. |
| 55 | | | | Vereinigung Ökologischer Landbau in Hessen (VÖL) |
| 56 | | | | Landesverband Hessen für Obstbau, Garten und Landschaftspflege e.V. |
| 57 | | | | Hessischer Bauernverband e.V. |
| 58 | | | | Kreisbauernverband Frankfurt am Main |
| 59 | | | | Verband der Landwirte im Nebenberuf e.V. - Landesverband Hessen |
| 60 | | | | Hessenhöfe |
| 61 | | | | Hessischer Grundbesitzerverband e.V. |
| 62 | | | | Ökologischer Jagdverein in Hessen e.V. |
| 63 | | | | Regionalverband Starkenburg e.V. |
| 64 | | | | Verband der Jagdgenossenschaften und Eigenjagdbesitzer in Hessen |
| 65 | | | | Wasser-, Boden- und Landschaftspflegeverband Hessen |
| 66 | | Chambers | IHK | Frankfurt am Main |
| 67 | | | | Darmstadt |
| 68 | | | Handwerkskammer Rhein-Main | Hauptverwaltung Frankfurt |
| 69 | | | | Hauptverwaltung Darmstadt - |

*Figure A.37: Involved institutions of RSE A380 Shipyard and institutional characteristics, 1st evaluation, Part III, based on HMWVL (2004: 54–57).*

## A.11 Involved institutions of the RSE A380 Shipyard, 2nd evaluation

| # | Institutional form | | Institution | Institutional Unit |
|---|---|---|---|---|
| | Non-private | | | |
| | | Administrative | | |
| 1 | | | Federal Republic of Germany | Wehrbereichsverwaltung West - Außenstelle Wiesbaden |
| 2 | | | | Bundesvermögensamt Frankfurt |
| 3 | | | Federal State of Hesse | Hessisches Forstamt Mörfelden-Walldorf |
| 4 | | | | Landesbetrieb Hessenforst |
| 5 | | | | Hessisches Landesamt für Umwelt und Geologie |
| 6 | | | | Hessisches Landesamt für Straßen- und Verkehrswesen |
| 7 | | | | Landrat des Landkreises Groß-Gerau |
| 8 | | | | Landrat des Hochtaunuskreises |
| 9 | | | | Landrat des Landkreises Darmstadt-Dieburg |
| 10 | | | Regional Council Darmstadt | Dez. II 22.1 Brandschutz - Katastrophenschutz, Zivile Verteidigung |
| 11 | | | | Dez. V 52 - Forsten |
| 12 | | | | Dez. V 53.1 - Eingriffsregelung/ Planungsbeiträge |
| 13 | | | | Dez V (Bezirksforstauschuss) |
| 14 | | | | Abt V Naturschutzbeiräte |
| 15 | | | | Staatliches Umweltamt Frankfurt - Dez. 41.1 Grundwasserschutz, Wasserversorgung |
| 16 | | | | Staatliches Umweltamt Frankfurt - Dez. 41.3 Kommunales Abwasser, Schutz der oberirdischen Gewässer |
| 17 | | | | Staatliches Umweltamt Frankfurt - Dez. 41.4 Industrielles Abwasser, Wassergefährdende Stoffe |
| 18 | | | | Staatliches Umweltamt Frankfurt - Dez. 41.5 Altlasten, Grundwasserschadensfälle |
| 19 | | | | Staatliches Umweltamt Frankfurt - Dez. 42.1 |
| 20 | | | | Staatliches Umweltamt Frankfurt - Dez. 42.2 - Abfallwirtschaft |
| 21 | | | | Dez. IV/41.2 Abflussverhältnisse, Hydrologie, flächenbezogene Planung |
| 22 | | | | Regionalversammlung Südhessen |

*Figure A.38: Involved institutions of RSE A380 Shipyard and institutional characteristics, 2nd evaluation, Part I, based on HMWVL (2004: 69–71).*

| # | Institutional form | | Institution | Institutional Unit |
|---|---|---|---|---|
| | **Non-private** | | | |
| | | **Administrative** | (continuation) | |
| 23 | | | Sub-federal state level | Amt für Straßen- und Verkehrswesen Frankfurt |
| 24 | | | | Amt für Straßen- und Verkehrswesen Darmstadt |
| 25 | | | | Kreisausschuss des Landkreises Groß-Gerau |
| 26 | | | | Oberbürgermeisterin der Stadt Frankfurt am Main |
| 27 | | | | Magistrat der Stadt Frankfurt am Main |
| 28 | | | | Magistrat der Stadt Rüsselsheim |
| 29 | | | | Magistrat der Stadt Mörfelden-Walldorf |
| | **Non-private** | | | |
| | | **Metropolitan** | | |
| 30 | | | | Planungsverband Frankfurt/Rhein-Main |
| | **Private** | | | |
| | | **Metropolitan** | | |
| | | | - | |
| | | **Economic** | | |
| 31 | | | | Staatliche Vogelschutzwarte für Hessen, Rheinland-Pfalz und das Saarland |
| 32 | | | | Deutsche Telekom AG |
| 33 | | | | Mainova AG |
| 34 | | | | Deutsche Flugsicherung GmbH |
| | **Further** | | | |
| 35 | | Association | | Botanische Vereinigung für Naturschutz in Hessen e.V. |
| 36 | | | | Deutsche Gebirgs- und Wandervereine - Landesverband Hessen |
| 37 | | | | Hessische Gesellschaft für Ornithologie und Naturschutz Hessen |
| 38 | | | | Naturschutzbund Deutschland - Landesverband Hessen e.V. |
| 39 | | | | Schutzgemeinschaft Deutscher Wald - Landesverband Hessen e.V. |

*Figure A.39: Involved institutions of RSE A380 Shipyard and institutional characteristics, 2nd evaluation, Part II, based on HMWVL (2004: 69–71).*

## A.12 Involved institutions of the RSE Runway North-West/Terminal 3, 1st Evaluation

| # | Institutional form | | Institution | Institutional Unit |
|---|---|---|---|---|
| | **Non-private** | | | |
| | | **Administrative** | | |
| 1 | | | Federal Republic of Germany | Bundesministerium für Verkehr, Bau- und Wohnungswesen |
| 2 | | | | Luftfahrt-Bundesamt |
| 3 | | | | Wehrbereichsverwaltung West, Außenstelle Wiesbaden |
| 4 | | | | Eisenbahn-Bundesamt, Außenstelle Frankfurt |
| 5 | | | | Bundesgrenzschutzamt Flughafen Frankfurt am Main |
| 6 | | | | Wasser- und Schifffahrtsamt Aschaffenburg |
| 7 | | | | Wasser- und Schifffahrtsamt Mannheim |
| 8 | | | Federal State of Hesse | Hessisches Ministerium für Umwelt, ländlichen Raum und Verbraucherschutz |
| 9 | | | | Landesforstausschuss beim Hessischen Ministerium für Umwelt, ländlichen Raum und Verbraucherschutz |
| 10 | | | | Kommission zur Abwehr des Fluglärms Flughafen Frankfurt am Main |
| 11 | | | | Hessisches Sozialministerium |
| 12 | | | | Beauftragter der Hessischen Landesregierung für behinderte Menschen, Herrn Friedel Rinn beim Hessischen Ministerium des Innern und für Sport |
| 13 | | | | Hessisches Ministerium des Innern und für Sport |
| 14 | | | | Ministerium für Wirtschaft, Verkehr, Landwirtschaft und Weinbau des Landes Rheinland-Pfalz |
| 15 | | | | Hessisches Landesamt für Umwelt und Geologie, Hydrogeologie |
| 16 | | | | Hessisches Landesamt für Umwelt und Geologie, G 5 Bodenschutz |
| 17 | | | | Hessisches Landesamt für Umwelt und Geologie, Luftreinhaltung |
| 18 | | | | Hessisches Landesamt für Umwelt und Geologie, Lärmschutz |
| 19 | | | | Hessisches Landesamt für Straßen und Verkehrswesen, Dezernat 22 |
| 20 | | | | Amt für Straßen- und Verkehrswesen Darmstadt |

*Figure A.40: Involved Institutions of the RSE Runway North-West/Terminal 3 and institutional characteristics, 1st evaluation, Part I, based on HMWVL (2007: 207–212).*

| # | Institutional form | | Institution | Institutional Unit |
|---|---|---|---|---|
| | Non-private | | | |
| | | Administrative | (continuation) | |
| 21 | | | Federal State of Hesse | Amt für Straßen- und Verkehrswesen Frankfurt |
| 22 | | | | Landesamt für Denkmalpflege Hessen |
| 23 | | | | Landrat des Landkreises Darmstadt-Dieburg |
| 24 | | | | Landrat des Landkreises Groß-Gerau |
| 25 | | | | Landrat des Landkreises Offenbach |
| 26 | | | | Landrat des Main-Taunus-Kreises |
| 27 | | | | Landrat des Main-Kinzig-Kreises |
| 28 | | | | Landrat des Hochtaunuskreises |
| 29 | | | Regierungspräsidium Darmstadt | Dezernat II 22.1 Kampfmittelräumdienst |
| 30 | | | | Dezernat II 22.2 Brandschutz |
| 31 | | | | Dezernat III 31.1 |
| 32 | | | | Dezernat III 31.3, Regionale Infrastruktur und Umwelt |
| 33 | | | | Dezernat III 33.1, Straßen- und Schienenverkehr |
| 34 | | | | Dezernat III 33.3 |
| 35 | | | | Abteilung IV/Da Umwelt, Dezernat IV/Da 41.1 |
| 36 | | | | Abteilung IV/Da Umwelt, Dezernat IV/Da 41.2 |
| 37 | | | | Abteilung IV/Da Umwelt, Dezernat IV/Da 41.3 |
| 38 | | | | Abteilung IV/Da Umwelt, Dezernat IV/Da 41.4 (Energie) |
| 39 | | | | Abteilung IV/Da Umwelt, Dezernat IV/Da 41.5 |
| 40 | | | | Abteilung IV/Da Umwelt, Dezernat IV/Da 42.1 |
| 41 | | | | Abteilung IV/Da Umwelt, Dezernat IV/Da 42.2 |
| 42 | | | | Abteilung IV/Da Umwelt, Dezernat IV/Da 43.1 |
| 43 | | | | Abteilung IV/Da Umwelt, Dezernat IV/Da 43.2 |
| 44 | | | | Abteilung IV/Da Umwelt, Dezernat IV/Da 43.3 |
| 45 | | | | Abteilung IV/F Umwelt, Dezernat IV/F 41.1 |

*Figure A.41: Involved institutions of the RSE Runway North-West/Terminal 3 and institutional characteristics, 1st evaluation, Part II, based on HMWVL (2007: 207–212).*

| # | Institutional form | | Institution | Institutional Unit |
|---|---|---|---|---|
| | Non-private | | | |
| | | Administrative | (continuation) | |
| 46 | | | Regierungspräsidium Darmstadt | Abteilung IV/F Umwelt, Dezernat IV/F 41.2 |
| 47 | | | | Abteilung IV/F Umwelt, Dezernat IV/F 41.3 |
| 48 | | | | Abteilung IV/F Umwelt, Dezernat IV/F 41.4 |
| 49 | | | | Abteilung IV/F Umwelt, Dezernat IV/F 41.5 |
| 50 | | | | Abteilung IV/F Umwelt, Dezernat IV/F 42 |
| 51 | | | | Abteilung IV/F Umwelt, Dezernat IV/F 43.1 |
| 52 | | | | Abteilung IV/F Umwelt, Dezernat IV/F 43.2 |
| 53 | | | | Abteilung IV/F Umwelt, Dezernat IV/F 43.3 |
| 54 | | | | Abteilung IV/Wi Umwelt, Dezernat IV/Wi 44 (Bergaufsicht) |
| 55 | | | | Dezernat V 51.1 |
| 56 | | | | Dezernat V 52 |
| 57 | | | | Dezernat V 53.1 |
| 58 | | | | Naturschutzbeirat, Dezernat V 53.4 |
| 59 | | | | Bezirksforstausschuss , Abteilung V |
| 60 | | | | Dezernat VII/F 74.2, Herr Röhling |
| 61 | | | | Dezernat VII/Da 71.1 |
| 62 | | | | Regionalversammlung Südhessen |
| 63 | | | Regierungspräsidium Karlsruhe | Herr Görnert |
| 64 | | | Regierungspräsidium Kassel | Obere Jagdbehörde, Außenstelle Darmstadt |
| 65 | | | Sub-federal state level | Kreisausschuss des Landkreises Groß-Gerau |
| 66 | | | | Kreisausschuss des Landkreises Darmstadt-Dieburg |
| 67 | | | | Kreisausschuss des Main-Kinzig-Kreises |
| 68 | | | | Kreisausschuss des Main-Taunus-Kreises |
| 69 | | | | Kreisausschuss des Landkreises Offenbach |
| 70 | | | | Struktur- und Genehmigungsdirektion Süd |

*Figure A.42: Involved institutions of the RSE Runway North-West/Terminal 3 and institutional characteristics, 1st evaluation, Part III, based on HMWVL (2007: 207–212).*

| # | Institutional form | | Institution | Institutional Unit |
|---|---|---|---|---|
| | Non-private | | | |
| | | Administrative | (continuation) | |
| 71 | | | Sub-federal state level | Kreisverwaltung Mainz-Bingen |
| 72 | | | | Forstamt Chausseehaus |
| 73 | | | | Forstamt Dieburg |
| 74 | | | | Forstamt Darmstadt |
| 75 | | | | Forstamt Groß-Gerau |
| 76 | | | | Forstamt Schlüchtern |
| 77 | | | | Forstamt Hanau-Wolfgang |
| 78 | | | | Oberbürgermeister der Stadt Rüsselsheim, Straßenverkehrsbehörde |
| 79 | | | | Oberbürgermeisterin der Stadt Frankfurt am Main, Straßenverkehrsbehörde |
| 80 | | | | Magistrat der Wissenschaftsstadt Darmstadt |
| 81 | | | | Magistrat der Stadt Dietzenbach |
| 82 | | | | Magistrat der Stadt Dreieich |
| 83 | | | | Magistrat der Stadt Eschborn |
| 84 | | | | Magistrat der Stadt Flörsheim am Main |
| 85 | | | | Magistrat der Stadt Frankfurt am Main |
| 86 | | | | Magistrat der Stadt Griesheim |
| 87 | | | | Magistrat der Stadt Groß-Gerau |
| 88 | | | | Magistrat der Stadt Hanau |
| 89 | | | | Magistrat der Stadt Hattersheim am Main |
| 90 | | | | Magistrat der Stadt Heusenstamm |
| 91 | | | | Magistrat der Stadt Hochheim am Main |
| 92 | | | | Magistrat der Stadt Hofheim am Taunus |
| 93 | | | | Magistrat der Stadt Kelsterbach |
| 94 | | | | Magistrat der Stadt Langen |
| 95 | | | | Magistrat der Stadt Langenselbold |
| 96 | | | | Magistrat der Stadt Maintal |
| 97 | | | | Magistrat der Stadt Mörfelden Walldorf |

*Figure A.43: Involved institutions of the RSE Runway North-West/Terminal 3 and institutional characteristics, 1st evaluation, Part IV, based on HMWVL (2007: 207–212).*

| # | Institutional form | | Institution | Institutional Unit |
|---|---|---|---|---|
| | **Non-private** | | | |
| | | **Administrative** | (continuation) | |
| 98 | | | Sub-federal state level | Magistrat der Stadt Mühlheim am Main |
| 99 | | | | Magistrat der Stadt Neu-Isenburg |
| 100 | | | | Magistrat der Stadt Obertshausen |
| 101 | | | | Magistrat der Stadt Offenbach am Main |
| 102 | | | | Magistrat der Stadt Pfungstadt |
| 103 | | | | Magistrat der Stadt Raunheim |
| 104 | | | | Magistrat der Stadt Reinheim |
| 105 | | | | Magistrat der Stadt Rodgau |
| 106 | | | | Magistrat der Stadt Rüsselsheim |
| 107 | | | | Magistrat der Stadt Weiterstadt |
| 108 | | | | Magistrat der Landeshauptstadt Wiesbaden |
| 109 | | | | Gemeindevorstand der Gemeinde Bischofsheim |
| 110 | | | | Gemeindevorstand der Gemeinde Büttelborn |
| 111 | | | | Gemeindevorstand der Gemeinde Egelsbach |
| 112 | | | | Gemeindevorstand der Gemeinde Erlensee |
| 113 | | | | Gemeindevorstand der Gemeinde Erzhausen |
| 114 | | | | Gemeindevorstand der Gemeinde Ginsheim-Gustavsburg |
| 115 | | | | Gemeindevorstand der Gemeinde Groß-Zimmern |
| 116 | | | | Gemeindevorstand der Gemeinde Hasselroth |
| 117 | | | | Gemeindevorstand der Gemeinde Kriftel |
| 118 | | | | Gemeindevorstand der Gemeinde Liederbach am Taunus |
| 119 | | | | Gemeindevorstand der Gemeinde Messel |
| 120 | | | | Gemeindevorstand der Gemeinde Nauheim |
| 121 | | | | Gemeindevorstand der Gemeinde Riedstadt |
| 122 | | | | Gemeindevorstand der Gemeinde Rodenbach |

*Figure A.44: Involved Institutions of the RSE Runway North-West/Terminal 3 and institutional characteristics, 1st evaluation, Part V, based on HMWVL (2007: 207–212).*

| # | Institutional form | | Institution | Institutional Unit |
|---|---|---|---|---|
| | Non-private | | | |
| | | Administrative | (continuation) | |
| 123 | | | Sub-federal state level | Gemeindevorstand der Gemeinde Ronneburg |
| 124 | | | | Gemeindevorstand der Gemeinde Roßdorf |
| 125 | | | | Gemeindevorstand der Gemeinde Sulzbach (Taunus) |
| 126 | | | | Gemeindevorstand der Gemeinde Trebur |
| 127 | | | | Stadtverwaltung der Stadt Mainz |
| 128 | | | | Verbandsgemeindeverwaltung Bodenheim |
| 129 | | | | Verbandsgemeindeverwaltung Gau Algesheim |
| 130 | | | | Verbandsgemeindeverwaltung Nieder Olm |
| 131 | | | | Verbandsgemeindeverwaltung Nierstein-Oppenheim |
| 132 | | | | Amt für Bodenmanagement Heppenheim, Dienststelle Reichelsheim |
| 133 | | | | Amt für Bodenmanagement Limburg |
| 134 | | | | Amt für Bodenmanagement Büdingen |
| 135 | | | | Ortsgemeindeverwaltung Schwabenheim |
| | Non-private | | | |
| 136 | | Metropolitan | | Planungsverband Frankfurt/Rhein-Main |
| 137 | | | | Regionalverband Starkenburg e.V. |
| | Private | | | |
| | | Metropolitan | | |
| | | | - | |
| 138 | | Economic | | Deutsche Flugsicherung GmbH |
| 139 | | | Zu Bundesebene | Deutscher Wetterdienst, RMG Offenbach |
| 140 | | | | Infraserv Höchst GmbH & Co. Höchst KG |
| 141 | | | | Hessenwasser GmbH |
| 142 | | | | NRM Netzdienste Rhein-Main GmbH (Mainova AG) |
| 143 | | | | Stadtwerke Neu-Isenburg |
| 144 | | | | Rhein-Main-Verkehrsverbund GmbH |

*Figure A.45: Involved institutions of the RSE Runway North-West/Terminal 3 and institutional characteristics, 1st evaluation, Part VI, based on HMWVL (2007: 207–212).*

| # | Institutional form | | Institution | Institutional Unit |
|---|---|---|---|---|
| | **Private** | | | |
| 145 | | **Economic** | (continuation) | Kreiswerke Hanau GmbH |
| 146 | | | | Kreisverkehrsgesellschaft Offenbach mbH |
| 147 | | | | Lokale Nahverkehrsgesellschaft mbH Frankfurt am Main |
| 148 | | | | Lokale Nahverkehrsgesellschaft mbH des Kreises Groß-Gerau |
| 149 | | | | Stadtwerke Rüsselsheim Verkehrsbetrieb |
| 150 | | | | Traffiq Lokale Nahverkehrsgesellschaft mbH |
| 151 | | | | Offenbacher Verkehrs-Betriebe GmbH |
| 152 | | | | Deutsche Telekom AG, Technikniederlassung Eschborn, Ressort SBN/Herr Brühl |
| 153 | | | | Deutsche Post AG, Geschäftsbereich 16, Transport und Briefzentrum |
| 154 | | | | DB Netz AG, Niederlassung Mitte |
| 155 | | | | DB Station & Service AG, Regionalbereich Mitte |
| 156 | | | | DB Fernverkehr AG |
| 157 | | | | DB Regio AG, Regionalbereich Rhein-Main |
| 158 | | | | Süwag Energie AG |
| 159 | | | | WINGAS GmbH.ON Energie AG |
| 160 | | | | Ruhrgas AG |
| 161 | | | | Pipeline Engineering GmbH |
| 162 | | | | Fernleitungsbetriebsgesellschaft mbH |
| 163 | | | | Level 3 Communications GmbH, LWL-Kabel |
| 164 | | | | Telia Sonera International Carrier Germany GmbH |
| 165 | | | | Colt Telecom GmbH |
| 166 | | | | Shell Deutschland Oil GmbH |
| 167 | | | | Abwasserverband Flörsheim |
| 168 | | | | Arcor AG & Co. KG |
| 169 | | | | Caltex Oil (Germany) GmbH |
| 170 | | | | DUAL Grundstücksverwaltungs-gesellschaft mbH |

*Figure A.46: Involved institutions of the RSE Runway North-West/Terminal 3 and institutional characteristics, 1st evaluation, Part VII, based on HMWVL (2007: 207–212).*

| # | Institutional form | | Institution | Institutional Unit |
|---|---|---|---|---|
| | **Private** | | | |
| 171 | | **Economic** | (continuation) | Stadtwerke Frankfurt am Main GmbH |
| 172 | | | | Kraftwerke Mainz-Wiesbaden AG |
| 173 | | | | Wasserwerk "Gerauer Land" |
| 174 | | | | Rhein-Main Rohrleitungstransport GmbH Rodenkirchen |
| 175 | | | | Überlandwerk Fulda GmbH |
| 176 | | | | BASF AG |
| 177 | | | | GasLine Telekommunikations-gesellschaft deutscher Gasversorgungs-unternehmen mbH & Co. KG |
| 178 | | | | BT (Germany) GmbH & Co. OHG |
| 179 | | | | E.ON Energie AG |
| | **Further** | | | |
| 180 | | Association | | Bund für Umwelt und Naturschutz Deutschland, Hessen e.V. |
| 181 | | | | Deutsche Gebirgs- und Wandervereine, Landesverband Hessen e.V., Verteilstelle Götz |
| 182 | | | | Hessische Gesellschaft für Ornithologie und Naturschutz Hessen e.V. |
| 183 | | | | Schutzgemeinschaft Deutscher Wald e.V., Landesverband Hessen |
| 184 | | | | Hessischer Gärtnereiverband e.V. |
| 185 | | | | Wasser-, Boden- und Landschaftspflege Verband Hessen |
| 186 | | | | Verband der Jagdgenossenschaften und Eigenjagdbesitzer in Hessen |
| 187 | | | | Astheim-Erfelder Entwässerungsverband, Vorsteher Hans Heinrich Hannemann |
| 188 | | | | Beregnungs- und Bodenverband Geinsheim, Vorsteher Edwin Paul |
| 189 | | | | Sommerdammverband Kornsand - Treburer - Auen sowie Wasserverband Kornsand Ludwigsaue, Vorsteher Reinhold Hennig |
| 190 | | | | Bodenverband Main-Kinzig, Geschäftsführer A. Zeller |
| 191 | | | | Wasserverband Rodau-Bieber |
| 192 | | | | Naturschutzbund Deutschland, Landesverband Rheinland-Pfalz e.V. |
| 193 | | | | Landesjagdverband Rheinland-Pfalz e.V. |

*Figure A.47: Involved institutions of the RSE Runway North-West/Terminal 3 and institutional characteristics, 1st evaluation, Part VIII, based on HMWVL (2007: 207–212).*

| # | Institutional form | | Institution | Institutional Unit |
|---|---|---|---|---|
| | Further | | | |
| 194 | | Association | (continuation) | Verband Deutscher Sportfischer, Landesverband Rheinland-Pfalz e.V. |
| 195 | | | | Die Naturfreunde, Verband für Umweltschutz, sanften Tourismus, Sport und Kultur, Landesverband Rheinland-Pfalz e.V. |
| 196 | | | | Landesverband Rheinland-Pfalz der deutschen Gebirgs- und Wandervereine e.V. |
| 197 | | | | Schutzgemeinschaft Deutscher Wald e.V.,- Landesverband Rheinland-Pfalz |
| 198 | | | | Zweckverband Wasserversorgung Stadt und Kreis Offenbach |
| 199 | | Chambers | | IHK Darmstadt |
| 200 | | | | IHK Frankfurt am Main |
| 201 | | | | IHK Offenbach am Main |
| 202 | | | | IHK Rheinhessen |
| 203 | | | | IHK Wiesbaden |
| 204 | | | | IHK Hanau-Gelnhausen-Schlüchtern |
| 205 | | | | Handwerkskammer Rhein-Main, Hauptverwaltung Darmstadt |
| 206 | | Church | | Evangelische Kirche in Hessen und Nassau |

*Figure A.48: Involved institutions of the RSE Runway North-West/Terminal 3 and institutional characteristics, 1st evaluation, Part IX, based on HMWVL (2007: 207–212).*

# A.13 Involved institutions of the RSE Runway North-West/Terminal 3, 2nd Evaluation

| # | Institutional form | | Institution | Institutional Unit |
|---|---|---|---|---|
| | Non-private | | | |
| | | Administrative | | |
| 1 | | | Federal Republic of Germany | Bundesministerium für Verkehr, Bau und Stadtentwicklung Referat LS 11 |
| 2 | | | | Wehrbereichsverwaltung West Außenstelle Wiesbaden |
| 3 | | | | Eisenbahn-Bundesamt Außenstelle Frankfurt |
| 4 | | | | Bundeseisenbahnvermögen Dienststelle Mitte |
| 5 | | | | Wasser- und Schifffahrtsamt Aschaffenburg |
| 6 | | | | Wasser- und Schifffahrtsamt Mannheim |
| 7 | | | | Bundespolizeiamt Flughafen Frankfurt am Main |
| 8 | | | | Flughafenkoordination Deutschland |
| 9 | | | Federal State of Hesse | Hessisches Ministerium für Umwelt, ländlichen Raum und Verbraucherschutz |
| 10 | | | | Landesnaturschutzbeirat beim Hessischen Ministerium für Umwelt, ländlichen Raum und Verbraucherschutz |
| 11 | | | | Landesforstausschuss beim Hessischen Ministerium für Umwelt, ländlichen Raum und Verbraucherschutz |
| 12 | | | | Hessisches Ministerium für Wirtschaft, Verkehr und Landesentwicklung |
| 13 | | | | Kommission zur Abwehr des Fluglärms Flughafen Frankfurt am Main |
| 14 | | | | Hessisches Sozialministerium |
| 15 | | | | Beauftragter der Hessischen Landesregierung für behinderte Menschen beim Hessischen Ministerium des Innern und für Sport |
| 16 | | | | Hessisches Ministerium des Innern und für Sport |
| 17 | | | | Hessisches Ministerium der Finanzen |
| 18 | | | | Hessisches Landesamt für Umwelt und Geologie |
| 19 | | | | Hessisches Landesamt für Straßen- und Verkehrswesen |
| 20 | | | | Landesamt für Denkmalpflege Hessen |

*Figure A.49: Involved Institutions of the RSE Runway North-West/Terminal 3 and institutional characteristics, 2nd evaluation, Part I, based on HMWVL (2007: 245-257).*

| # | Institutional form | | Institution | Institutional Unit |
|---|---|---|---|---|
| | **Non-private** | | | |
| | | **Administrative** | (continuation) | |
| 21 | | | Federal State of Hesse | Landesamt für Denkmalpflege Hessen Archäologische Denkmalpflege |
| 22 | | | | Obere Jagdbehörde, Außenstelle Darmstadt |
| 23 | | | | Polizeipräsidium Südosthessen Abteilungsstab - E 1 - |
| 24 | | | | Polizeipräsidium Westhessen |
| 25 | | | | Landrat des Landkreises Darmstadt-Dieburg |
| 26 | | | | Landrat des Landkreises Groß-Gerau |
| 27 | | | | Landrat des Landkreises Offenbach |
| 28 | | | | Landrat des Main-Taunus-Kreises |
| 29 | | | | Landrat des Main-Kinzig-Kreises |
| 30 | | | | Landrat des Hochtaunuskreises |
| 31 | | | | Landrat des Wetteraukreises |
| 32 | | | | Landesbetrieb Hessenforst |
| 33 | | | only for somebodies attention | Hessiche Staatskanzleich |
| 34 | | | Federal State of Rhineland-Palatinate | Ministerium des Innern und für Sport des Landes Rheinland-Pfalz |
| 35 | | | | Ministerium für Umwelt, Forsten und Verbraucherschutz |
| 36 | | | | Staatskanzlei Rheinland-Pfalz |
| 37 | | | Regierungspräsidium Darmstadt | Abteilung I, Dezernat I 18 |
| 38 | | | | Abteilung III |
| 39 | | | | Abteilung IV/DA Umwelt |
| 40 | | | | Abteilung IV/F Umwelt |
| 41 | | | | Abteilung IV/WI Umwelt |
| 42 | | | | Abteilung V |
| 43 | | | | Abteilung II |
| 44 | | | | Regionalversammlung Südhessen |
| 45 | | | Regierungspräsidium Karlsruhe | Regierungspräsidium Karlsruhe |
| 46 | | | Regierungspräsidium Kassel | Obere Jagdbehörde, Außenstelle Darmstadt |

*Figure A.50: Involved institutions of the RSE Runway North-West/Terminal 3 and institutional characteristics, 2nd evaluation, Part II, based on HMWVL (2007: 245–257).*

| # | Institutional form | | Institution | Institutional Unit |
|---|---|---|---|---|
| | **Non-private** | | | |
| | | **Administrative** | | |
| 47 | | | Sub-federal state level | Amt für Bodenmanagement Limburg |
| 48 | | | | Amt für Bodenmanagement Büdingen |
| 49 | | | | Polizeipräsidium Frankfurt am Main Abteilung Einsatz E1 |
| 50 | | | | Kreisausschuss des Landkreises Groß-Gerau |
| 51 | | | | Kreisausschuss des Landkreises Darmstadt-Dieburg |
| 52 | | | | Kreisausschuss des Main-Kinzig-Kreises |
| 53 | | | | Kreisausschuss des Main-Taunus-Kreises |
| 54 | | | | Kreisausschuss des Landkreises Offenbach |
| 55 | | | | Kreisausschuss des Wetteraukreises |
| 56 | | | | Struktur- und Genehmigungsdirektion Süd |
| 57 | | | | Kreisverwaltung Mainz-Bingen z. Hd. Frau Leyendecker |
| 58 | | | | Forstamt Wiesbaden-Chausseehaus |
| 59 | | | | Forstamt Dieburg |
| 60 | | | | Forstamt Darmstadt |
| 61 | | | | Forstamt Groß-Gerau |
| 62 | | | | Forstamt Königstein |
| 63 | | | | Forstamt Langen |
| 64 | | | | Forstamt Schlüchtern |
| 65 | | | | Forstamt Hanau-Wolfgang |
| 66 | | | | Bürgermeister der Stadt Mörfelden-Walldorf, Straßenverkehrsbehörde |
| 67 | | | | Bürgermeister der Stadt Kelsterbach, Straßenverkehrsbehörde |
| 68 | | | | Oberbürgermeister der Stadt Rüsselsheim, Straßenverkehrsbehörde |
| 69 | | | | Bürgermeister der Stadt Neu-Isenburg, Straßenverkehrsbehörde |
| 70 | | | | Oberbürgermeisterin der Stadt Frankfurt am Main, Straßenverkehrsamt |
| 71 | | | | Magistrat der Wissenschaftsstadt Darmstadt |
| 72 | | | | Magistrat der Stadt Dietzenbach |

*Figure A.51: Involved institutions of the RSE Runway North-West/Terminal 3 and institutional characteristics, 2nd evaluation, Part III, based on HMWVL (2007: 245–257).*

| # | Institutional form | | Institution | Institutional Unit |
|---|---|---|---|---|
| | **Non-private** | | | |
| | | **Administrative** | (continuation) | |
| 73 | | | Sub-federal state level | Magistrat der Stadt Dreieich |
| 74 | | | | Magistrat der Stadt Eschborn |
| 75 | | | | Magistrat der Stadt Flörsheim am Main |
| 76 | | | | Magistrat der Stadt Frankfurt am Main |
| 77 | | | | Magistrat der Stadt Gernsheim |
| 78 | | | | Magistrat der Stadt Griesheim |
| 79 | | | | Magistrat der Stadt Groß-Gerau |
| 80 | | | | Magistrat der Stadt Hanau |
| 81 | | | | Magistrat der Stadt Hattersheim am Main |
| 82 | | | | Magistrat der Stadt Heusenstamm |
| 83 | | | | Magistrat der Stadt Hochheim am Main |
| 84 | | | | Magistrat der Stadt Hofheim a.Ts. |
| 85 | | | | Magistrat der Stadt Kelsterbach |
| 86 | | | | Magistrat der Stadt Langen |
| 87 | | | | Magistrat der Stadt Langenselbold |
| 88 | | | | Magistrat der Stadt Mörfelden-Walldorfagistrat der Stadt Maintal |
| 89 | | | | Magistrat der Stadt Mühlheim am Main |
| 90 | | | | Magistrat der Stadt Neu-Isenburg |
| 91 | | | | Magistrat der Stadt Obertshausen |
| 92 | | | | Magistrat der Stadt Offenbach am Main |
| 93 | | | | Magistrat der Stadt Pfungstadt |
| 94 | | | | Magistrat der Stadt Raunheim |
| 95 | | | | Magistrat der Stadt Reinheim |
| 96 | | | | Magistrat der Stadt Rodgau |
| 97 | | | | Magistrat der Stadt Rüsselsheim |
| 98 | | | | Magistrat der Stadt Schwalbach a. Ts. |
| 99 | | | | Magistrat der Stadt Steinau an der Straße |
| 100 | | | | Magistrat der Stadt Weiterstadt |
| 101 | | | | Magistrat der Landeshauptstadt Wiesbaden |
| 102 | | | | Gemeindevorstand der Gemeinde Bischofsheim |

*Figure A.52: Involved institutions of the RSE Runway North-West/Terminal 3 and institutional characteristics, 2nd evaluation, Part IV, based on HMWVL (2007: 245–257).*

| # | Institutional form | | Institution | Institutional Unit |
|---|---|---|---|---|
| | **Non-private** | | | |
| | | **Administrative** | (continuation) | |
| 103 | | | Sub-federal state level | Gemeindevorstand der Gemeinde Büttelborn |
| 104 | | | | Gemeindevorstand der Gemeinde Egelsbach |
| 105 | | | | Gemeindevorstand der Gemeinde Erlensee |
| 106 | | | | Gemeindevorstand der Gemeinde Erzhausen |
| 107 | | | | Gemeindevorstand der Gemeinde Ginsheim-Gustavsburg |
| 108 | | | | Gemeindevorstand der Gemeinde Groß-Zimmern |
| 109 | | | | Gemeindevorstand der Gemeinde Hasselroth |
| 110 | | | | Gemeindevorstand der Gemeinde Kriftel |
| 111 | | | | Gemeindevorstand der Gemeinde Liederbach a. Ts. |
| 112 | | | | Gemeindevorstand der Gemeinde Messel |
| 113 | | | | Gemeindevorstand der Gemeinde Nauheim |
| 114 | | | | Gemeindevorstand der Gemeinde Otzberg |
| 115 | | | | Magistrat der Stadt Riedstadt |
| 116 | | | | Gemeindevorstand der Gemeinde Rodenbach |
| 117 | | | | Gemeindevorstand der Gemeinde Ronneburg |
| 118 | | | | Gemeindevorstand der Gemeinde Roßdorf |
| 119 | | | | Gemeindevorstand der Gemeinde Sulzbach (Taunus) |
| 120 | | | | Gemeindevorstand der Gemeinde Trebur |
| 121 | | | | Stadtverwaltung der Stadt Mainz |
| 122 | | | | Verbandsgemeindeverwaltung Bodenheim |
| 123 | | | | Verbandsgemeindeverwaltung Gau-Algesheim |
| 124 | | | | Verbandsgemeindeverwaltung Nieder-Olm |
| 125 | | | | Verbandsgemeindeverwaltung Nierstein-Oppenheim |

*Figure A.53: Involved Institutions of the RSE Runway North-West/Terminal 3 and institutional characteristics, 2nd evaluation, Part V, based on HMWVL (2007: 245–257).*

| # | Institutional form | | Institution | Institutional Unit |
|---|---|---|---|---|
| | **Non-private** | | | |
| | | **Administrative** | (continuation) | |
| 126 | | | Sub-federal state level | Magistrat der Stadt Bad Vilbel |
| 127 | | | | Gemeindevorstand der Gemeinde Bickenbach |
| 128 | | | | Magistrat der Stadt Dieburg |
| 129 | | | | Gemeindevorstand der Gemeinde Niederdorfelden |
| 130 | | | | Gemeindevorstand der Gemeinde Seeheim-Jugenheim |
| 131 | | | | Gemeindevorstand der Gemeinde Mühltal |
| 132 | | | | Gemeindevorstand der Gemeinde Gründau |
| 133 | | | only for somebodies attention | Amt für Straßen- und Verkehrswesen Darmstadt |
| 134 | | | | Amt für Straßen- und Verkehrswesen Frankfurt |
| | | **Metropolitan** | | |
| 135 | | | | Planungsverband Ballungsraum Frankfurt/Rhein-Main |
| 136 | | | | Planungsgemeinschaft Rheinhessen-Nahe |
| 137 | | | | Regionalverband Starkenburg e.V. |
| | **Private** | | | |
| | | **Metropolitan** | | |
| | | | - | |
| | | **Economic** | | |
| 138 | | | | Deutsche Flugsicherung GmbH |
| 139 | | | Zu Bundesebene | Deutscher Wetterdienst Abt. Flugmeteorologie |
| 140 | | | | Infraserv GmbH & Co. Höchst KG Geschäftsfeld Umwelt/Sicherheit/ Gesundheit und Genehmigungen |
| 141 | | | | Hessenwasser GmbH |
| 142 | | | | NRM Netzdienste Rhein-Main GmbH, Asset Management |
| 143 | | | | Stadtwerke Mörfelden-Walldorf |
| 144 | | | | Stadtwerke Langen GmbH |
| 145 | | | | Stadtwerke Neu-Isenburg GmbH |

*Figure A.54: Involved institutions of the RSE Runway North-West/Terminal 3 and institutional characteristics, 2nd evaluation, Part VI, based on HMWVL (2007: 245–257).*

| # | Institutional form | | Institution | Institutional Unit |
|---|---|---|---|---|
| | **Private** | | | |
| 146 | | **Economic** | (continuation) | Kreiswerke Hanau GmbH |
| 147 | | | | Rhein-Main-Verkehrsverbund GmbH |
| 148 | | | | Kreisverkehrsgesellschaft Offenbach mbH |
| 149 | | | | Traffiq Lokale Nahverkehrsgesellschaft mbH Frankfurt am Main |
| 150 | | | | Lokale Nahverkehrsgesellschaft mbH des Kreises Groß-Gerau |
| 151 | | | | Stadtwerke Rüsselsheim GmbH, Verkehrsbetrieb |
| 152 | | | | Offenbacher Verkehrs-Betriebe GmbH |
| 153 | | | | Deutsche Telekom AG, Technikniederlassung Eschborn |
| 154 | | | | Deutsche Post AG SP 161, Transport und Briefzentrum |
| 155 | | | | DB Netz AG, Niederlassung Mitte, c/o Services Immobilien GmbH, Kundenteam DB Netz (FRI-Ffm-N) |
| 156 | | | | DB Station & Service AG, Leiter Bauherrenvertretung Frankfurt Flughafen (I. SBE) |
| 157 | | | | DB Fernverkehr AG P.TVN 4 |
| 158 | | | | DB Regio AG Regionalbereich Rhein-Main |
| 159 | | | | DB Energie GmbH, Niederlassung West |
| 160 | | | | DB Telematik GmbH |
| 161 | | | | DB Services Immobilien GmbH, Niederlassung Frankfurt |
| 162 | | | | Süwag Netz GmbH |
| 163 | | | | RWE Transportnetz Strom GmbH, Abt. ETE-A-AG |
| 164 | | | | WINGAS GmbH |
| 165 | | | | PLEdoc Gesellschaft für Dokumentationserstellung und -pflege mbH |
| 166 | | | | Fernleitungs-Betriebsgesellschaft mbH |
| 167 | | | | MCI Deutschland GmbH Verizon |
| 168 | | | | Level 3 Communications GmbH, LWL-Kabel |
| 169 | | | | TeliaSonera International Carrier-Germany GmbH |

*Figure A.55: Involved institutions of the RSE Runway North-West/Terminal 3 and institutional characteristics, 2nd evaluation, Part VII, based on HMWVL (2007: 245–257).*

| # | Institutional form | | Institution | Institutional Unit |
|---|---|---|---|---|
| | **Private** | | | |
| 170 | | **Economic** | (continuation) | Colt Telecom GmbH, Abt. Fibre/Construction |
| 171 | | | | Hydrantenbetriebsgesellschaft Flughafen Frankfurt am Main |
| 172 | | | | Shell Deutschland Oil GmbH, Abt. LFMB Rechtsabteilung |
| 173 | | | | 3T Telekommunikationsgesellschaft mbH |
| 174 | | | | Biomasse Rhein Main GmbH |
| 175 | | | | Abwasserverband Flörsheim |
| 176 | | | | HEAG Südhessische Energie AG |
| 177 | | | | Arcor AG & Co. KG, Region Mitte, Abt. TRN-M |
| 178 | | | | AGIP Deutschland AG |
| 179 | | | | Deutsche BP AG |
| 180 | | | | ESSO Deutschland GmbH |
| 181 | | | | Exxon Mobil Central, Europe Holding GmbH |
| 182 | | | | TOTAL Deutschland GmbH |
| 183 | | | | TLR Tanklager Raunheim GmbH |
| 184 | | | | Mc Stay Luby |
| 185 | | | | DUAL Grundstücksverwaltungs-gesellschaft mbH |
| 186 | | | | Stadtwerke Verkehrsgesellschaft Frankfurt am Main mbH |
| 187 | | | | O2 Germany GmbH & Co. OHG, Projektbüro Frankfurt |
| 188 | | | | Kraftwerke Mainz-Wiesbaden AG |
| 189 | | | | Überlandwerk Groß-Gerau GmbH |
| 190 | | | | Wasserwerk „Gerauer Land" |
| 191 | | | | NRM Netzdienste Rhein-Main GmbH, Technische Verwaltung Gas-Union, Abteilung N 1-RT 8 |
| 192 | | | | Rhein-Main Rohrleitungstransport GmbH |
| 193 | | | | Süwag Netz GmbH |
| 194 | | | | BASF AG, WLS/IB-C 13 |
| 195 | | | | BT (Germany) GmbH & Co. OHG |
| 196 | | | | Mc Stay Luby (Ireland) |

*Figure A.56: Involved institutions of the RSE Runway North-West/Terminal 3 and institutional characteristics, 2nd evaluation, Part VIII, based on HMWVL (2007: 245–257).*

| # | Institutional form | | Institution | Institutional Unit |
|---|---|---|---|---|
| | Further | | | |
| 197 | | Association | | Botanische Vereinigung für Naturschutz in Hessen e.V. |
| 198 | | | | Bund für Umwelt und Naturschutz, Landesverband Hessen e.V. |
| 199 | | | | Deutsche Gebirgs- und Wandervereine, Landesverband Hessen e.V., Verteilerstelle |
| 200 | | | | Hessische Gesellschaft für Ornithologie und Naturschutz Hessen |
| 201 | | | | Landesjagdverband Hessen e.V. |
| 202 | | | | Naturschutzbund Deutschland -Landesverband Hessen e.V.- |
| 203 | | | | Schutzgemeinschaft Deutscher Wald, -Landesverband Hessen e.V.- |
| 204 | | | | Verband Hessischer Sportfischer e.V. |
| 205 | | | | Bioland e.V. Landesverband Hessen-Thüringen-Sachsen-Anhalt |
| 206 | | | | Deutscher Bundesverband der Landwirte im Nebenberuf e.V. |
| 207 | | | | Gäa e.V., Landesverband Thüringen-Hessen |
| 208 | | | | Hessischer Gärtnereiverband |
| 209 | | | | Hessischer Grundbesitzerverband e.V. |
| 210 | | | | Hessischer Landesverband für Erwerbsobstbau |
| 211 | | | | Landesverband Hessen für Obstbau, Garten und Landschaftspflege e. . |
| 212 | | | | Hessischer Pächterverband e.V. |
| 213 | | | | Hessischer Waldbesitzerverband e.V. |
| 214 | | | | Kreisbauernverband Hochtaunus e.V. |
| 215 | | | | Regionalbauernverband Wetterau und Frankfurt e.V. |
| 216 | | | | Kreisbauernverband Main-Kinzig e.V. |
| 217 | | | | Landesverband Hessen im Bund Deutscher Baumschulen |
| 218 | | | | Ökologischer Jagdverein in Hessen e.V. |
| 219 | | | | Verband der Jagdgenossenschaften und Eigenjagdbesitzer in Hessen |
| 220 | | | | Hessischer Bauernverband Geschäftsstelle - Referat VII |
| 221 | | | | Naturschutzbund Deutschland Landesverband Rheinland-Pfalz e.V. |

*Figure A.57: Involved institutions of the RSE Runway North-West/Terminal 3 and institutional characteristics, 2nd evaluation, Part IX, based on HMWVL (2007: 245–257).*

| # | Institutional form | | Institution | Institutional Unit |
|---|---|---|---|---|
| | Further | | | |
| 222 | | Association | (continuation) | Bund für Umwelt und Naturschutz Deutschland (BUND), Landesverband Rheinland-Pfalz e.V. |
| 223 | | | | POLLICHIA Verein für Naturforschung und Landespflege e.V. |
| 224 | | | | Gesellschaft für Naturschutz und Ornithologie Rheinland- Pfalz e.V. (GNOR) |
| 225 | | | | Landesjagdverband Rheinland- Pfalz e.V. |
| 226 | | | | Verband Deutscher Sportfischer, Landesverband Rheinland-Pfalz e.V. |
| 227 | | | | Landesaktionsgemeinschaft Natur und Umwelt Rheinland-Pfalz e.V. (LAG) |
| 228 | | | | Die Naturfreunde, Verband für Umweltschutz, sanften Tourismus, Sport und Kultur, Landesverband Rheinland-Pfalz e.V. |
| 229 | | | | Schutzgemeinschaft Deutscher Wald, Landesverband Rheinland-Pfalz e.V. |
| 230 | | | | Arbeitsgruppe Wasserwerke Hessisches Ried |
| 231 | | | | Zweckverband Wasserversorgung Stadt und Kreis Offenbach |
| 232 | | | | Wasserverband Schwarzbachgebiet-Ried |
| 233 | | | | Wasserversorgungsverband Main-Taunus-West |
| 234 | | | only for somebodies attention | Regionales Dialogforum Flughafen Frankfurtm, c/o IFOK GmbH |
| 235 | | | | Regionales Dialogforum Flughafen Frankfurt, c/o Öko-Institut |
| 236 | | Chamber | | IHK Darmstadt |
| 237 | | | | IHK Frankfurt am Main, Geschäftsfeld Standortpolitik |
| 238 | | | | IHK Offenbach am Main |
| 239 | | | | IHK Rheinhessen |
| 240 | | | | IHK Wiesbaden |
| 241 | | | | IHK Hanau-Gelnhausen-Schlüchtern |
| 242 | | | | IHK Gießen-Friedberg |
| 243 | | | | Handwerkskammer Rhein-Main, Hauptverwaltung Darmstadt |
| 244 | | | | Handwerkskammer Rheinhessen-Mainz |

*Figure A.58: Involved institutions of the RSE Runway North-West/Terminal 3 and institutional characteristics, 2nd evaluation, Part X, based on HMWVL (2007: 245–257).*

| # | Institutional form | | Institution | Institutional Unit |
|---|---|---|---|---|
| | **Further** | | | |
| 245 | | Church | | Evangelische Kirche in Hessen und Nassau |
| 246 | | | | Kommissariat der Katholischen Bischöfe |
| 247 | | Legal representative | | Anwaltskanzlei Dr. Heribert Fislake |
| 248 | | | | Anwaltskanzlei Schmitz |
| 249 | | | | Anwaltskanzlei Möller- Meinecke |
| 250 | | | | Anwaltskanzlei Nörr, Stiefenhofer, Lutz |
| 251 | | | | Anwaltskanzlei Dr. Reiner Geulen |
| 252 | | | | Anwaltskanzlei Dr. Ebner/Falk/ Dr. Berghäuser/Dr. Albach/Dr. Landzettel/Dr. Wieland/Dr. Berg |
| 253 | | | | Anwaltskanzlei Haldenwang |
| 254 | | | | Anwaltskanzlei Wurster, Wirsing & Schotten |
| 255 | | | | Anwaltskanzlei Baumann |
| 256 | | | | Anwaltskanzlei Philipp-Gerlach |
| 257 | | | | Anwaltskanzlei Nickel |
| 258 | | | | Anwaltskanzlei Redeker, Sellner, Dahs & Widmaier |
| 259 | | | | Anwaltskanzleien Mayer, Brown, Rowe und Maw |
| 260 | | | | Arnecke & Siebold |
| 261 | | | | Gleiss Lutz |

*Figure A.59: Involved institutions of the RSE Runway North-West/Terminal 3 and institutional characteristics, 2nd evaluation, Part XI, based on HMWVL (2007: 245–257).*

## A.14 Responding Institutions of the RSE Runway North-West/Terminal 3, 2nd Evaluation

| # | Institutional form | | Institution | Institutional Unit |
|---|---|---|---|---|
| | Non-private | | | |
| | | Administrative | | |
| 1 | | | Federal Republic of Germany | Wehrbereichsverwaltung West Außenstelle Wiesbaden |
| 2 | | | | Eisenbahn-Bundesamt Außenstelle Frankfurt |
| 3 | | | | Bundeseisenbahnvermögen Dienststelle Mitte |
| 4 | | | | Wasser- und Schifffahrtsamt Aschaffenburg |
| 5 | | | | Wasser- und Schifffahrtsamt Mannheim |
| 6 | | | | Bundespolizeiamt Flughafen Frankfurt am Main |
| 7 | | | | Deutscher Wetterdienst Abt. Flugmeteorologie |
| 8 | | | Federal State of Hesse | Hessisches Ministerium für Umwelt, ländlichen Raum und Verbraucherschutz |
| 9 | | | | Hessisches Ministerium für Wirtschaft, Verkehr und Landesentwicklung |
| 10 | | | | Kommission zur Abwehr des Fluglärms Flughafen Frankfurt am Main |
| 11 | | | | Hessisches Sozialministerium |
| 12 | | | | Beauftragter der Hessischen Landesregierung für behinderte Menschen beim Hessischen Ministerium des Innern und für Sport |
| 13 | | | | Hessisches Ministerium des Innern und für Sport |
| 14 | | | | Hessisches Landesamt für Umwelt und Geologie |
| 15 | | | | Hessisches Landesamt für Straßen- und Verkehrswesen |
| 16 | | | | Landesamt für Denkmalpflege Hessen |
| 17 | | | | Landesamt für Denkmalpflege Hessen Archäologische Denkmalpflege |
| 18 | | | | Landrat des Hochtaunuskreises |
| 19 | | | | Landesbetrieb Hessenforst |
| 20 | | | Regierungspräsidium Darmstadt | Abteilung III |
| 21 | | | | Abteilung IV/DA Umwelt |
| 22 | | | | Abteilung IV/F Umwelt |
| 23 | | | | Abteilung IV/WI Umwelt |

*Figure A.60: Responding Institutions of the RSE Runway North-West/Terminal 3 and institutional characteristics, 2nd evaluation, Part I, based on HMWVL (2007: 260–265).*

| # | Institutional form | | Institution | Institutional Unit |
|---|---|---|---|---|
| | Non-private | | | |
| | | Administrative | (continuation) | |
| 24 | | | Regierungspräsidium Darmstadt | Abteilung V |
| 25 | | | | Abteilung II |
| 26 | | | | Regionalversammlung Südhessen |
| 27 | | | Sub-federal state level | Amt für Bodenmanagement Limburg |
| 28 | | | | Kreisausschuss des Landkreises Groß-Gerau |
| 29 | | | | Kreisausschuss des Landkreises Darmstadt-Dieburg |
| 30 | | | | Kreisausschuss des Main-Kinzig-Kreises |
| 31 | | | | Kreisausschuss des Main-Taunus-Kreises |
| 32 | | | | Kreisausschuss des Landkreises Offenbach |
| 33 | | | | Kreisverwaltung Mainz-Bingen z. Hd. Frau Leyendecker |
| 34 | | | | Forstamt Wiesbaden-Chausseehaus |
| 35 | | | | Forstamt Groß-Gerau |
| 36 | | | | Forstamt Langen |
| 37 | | | | Magistrat der Wissenschaftsstadt Darmstadt |
| 38 | | | | Magistrat der Stadt Dietzenbach |
| 39 | | | | Magistrat der Stadt Dreieich |
| 40 | | | | Magistrat der Stadt Eschborn |
| 41 | | | | Magistrat der Stadt Flörsheim am Main |
| 42 | | | | Magistrat der Stadt Frankfurt am Main |
| 43 | | | | Magistrat der Stadt Griesheim |
| 44 | | | | Magistrat der Stadt Hanau |
| 45 | | | | Magistrat der Stadt Hattersheim am Main |
| 46 | | | | Magistrat der Stadt Heusenstamm |
| 47 | | | | Magistrat der Stadt Hochheim am Main |
| 48 | | | | Magistrat der Stadt Hofheim a.Ts. |
| 49 | | | | Magistrat der Stadt Kelsterbach |
| 50 | | | | Magistrat der Stadt Langen |
| 51 | | | | Magistrat der Stadt Langenselbold |
| 52 | | | | Magistrat der Stadt Mühlheim am Main |

*Figure A.61: Responding Institutions of the RSE Runway North-West/Terminal 3 and institutional characteristics, 2nd Evaluation, Part II, based on HMWVL (2007: 260–265).*

| # | Institutional form | | Institution | Institutional Unit |
|---|---|---|---|---|
| | **Non-private** | | | |
| | | **Administrative** | (continuation) | |
| 53 | | | Sub-federal state level | Magistrat der Stadt Neu-Isenburg |
| 54 | | | | Magistrat der Stadt Offenbach am Main |
| 55 | | | | Magistrat der Stadt Raunheim |
| 56 | | | | Magistrat der Stadt Rodgau |
| 57 | | | | Magistrat der Stadt Rüsselsheim |
| 58 | | | | Magistrat der Stadt Weiterstadt |
| 59 | | | | Magistrat der Landeshauptstadt Wiesbaden |
| 60 | | | | Gemeindevorstand der Gemeinde Bischofsheim |
| 61 | | | | Gemeindevorstand der Gemeinde Egelsbach |
| 62 | | | | Gemeindevorstand der Gemeinde Erzhausen |
| 63 | | | | Gemeindevorstand der Gemeinde Ginsheim-Gustavsburg |
| 64 | | | | Gemeindevorstand der Gemeinde Groß-Zimmern |
| 65 | | | | Gemeindevorstand der Gemeinde Hasselroth |
| 66 | | | | Gemeindevorstand der Gemeinde Kriftel |
| 67 | | | | Gemeindevorstand der Gemeinde Liederbach a. Ts. |
| 68 | | | | Gemeindevorstand der Gemeinde Otzberg |
| 69 | | | | Magistrat der Stadt Riedstadt |
| 70 | | | | Gemeindevorstand der Gemeinde Rodenbach |
| 71 | | | | Gemeindevorstand der Gemeinde Ronneburg |
| 72 | | | | Gemeindevorstand der Gemeinde Roßdorf |
| 73 | | | | Stadtverwaltung der Stadt Mainz |
| 74 | | | | Verbandsgemeindeverwaltung Bodenheim |
| 75 | | | | Verbandsgemeindeverwaltung Gau-Algesheim |
| 76 | | | | Verbandsgemeindeverwaltung Nieder-Olm |

*Figure A.62: Responding Institutions of the RSE Runway North-West/Terminal 3 and institutional characteristics, 2nd Evaluation, Part III, based on HMWVL (2007: 260–265).*

| # | Institutional form | | Institution | Institutional Unit |
|---|---|---|---|---|
| | Non-private | | | |
| | | Administrative | (continuation) | |
| 77 | | | Sub-federal state level | Verbandsgemeindeverwaltung Nierstein-Oppenheim |
| 78 | | | | Magistrat der Stadt Dieburg |
| 79 | | | | Gemeindevorstand der Gemeinde Niederdorfelden |
| 80 | | | | Gemeindevorstand der Gemeinde Seeheim-Jugenheim |
| 81 | | | | Gemeindevorstand der Gemeinde Mühltal |
| | | Metropolitan | | |
| 82 | | | | Planungsverband Ballungsraum Frankfurt/Rhein-Main |
| 83 | | | | Regionalverband Starkenburg e.V. |
| | Private | | | |
| | | Metropolitan | | |
| | | | - | |
| | | Economic | | |
| 84 | | | | Deutsche Flugsicherung GmbH |
| 85 | | | | Infraserv GmbH & Co. Höchst KG Geschäftsfeld Umwelt/Sicherheit/ Gesundheit und Genehmigungen |
| 86 | | | | Hessenwasser GmbH |
| 87 | | | | NRM Netzdienste Rhein-Main GmbH, Asset Management |
| 88 | | | | Stadtwerke Neu-Isenburg GmbH |
| 89 | | | | Rhein-Main-Verkehrsverbund GmbH |
| 90 | | | | Kreisverkehrsgesellschaft Offenbach mbH |
| 91 | | | | Lokale Nahverkehrsgesellschaft mbH des Kreises Groß-Gerau |
| 92 | | | | Deutsche Telekom AG, Technikniederlassung Eschborn |
| 93 | | | | Deutsche Post AG SP 161, Transport und Briefzentrum |
| 94 | | | | DB Netz AG, Niederlassung Mitte, c/o Services Immobilien GmbH, Kundenteam DB Netz (FRI-Ffm-N) |

*Figure A.63: Responding Institutions of the RSE Runway North-West/Terminal 3 and institutional characteristics, 2nd Evaluation, Part IV, based on HMWVL (2007: 260–265).*

| # | Institutional form | | Institution | Institutional Unit |
|---|---|---|---|---|
| | **Private** | | | |
| 95 | | **Economic** | (continuation) | |
| 96 | | | | DB Energie GmbH, Niederlassung West |
| 97 | | | | RWE Transportnetz Strom GmbH, Abt. ETE-A-AG |
| 98 | | | | PLEdoc Gesellschaft für Dokumentationserstellung und -pflege mbH |
| 99 | | | | Abwasserverband Flörsheim |
| 100 | | | | TLR Tanklager Raunheim GmbH |
| 101 | | | | NRM Netzdienste Rhein-Main GmbH, Technische Verwaltung Gas-Union, Abteilung N 1-RT 8 |
| 102 | | | | Rhein-Main Rohrleitungstransport GmbH |
| 103 | | | | Süwag Netz GmbH |
| | **Further** | | | |
| 104 | | Association | | Bund für Umwelt und Naturschutz, Landesverband Hessen e.V. |
| 105 | | | | Hessische Gesellschaft für Ornithologie und Naturschutz Hessen |
| 106 | | | | Naturschutzbund Deutschland -Landesverband Hessen e.V.- |
| | | | | Schutzgemeinschaft Deutscher Wald, -Landesverband Hessen e.V.- |
| 107 | | | | Bund für Umwelt und Naturschutz Deutschland (BUND), Landesverband Rheinland-Pfalz e. V. |
| 108 | | | | Zweckverband Wasserversorgung Stadt und Kreis Offenbach |
| 109 | | | | Wasserverband Schwarzbachgebiet-Ried |
| 110 | | | | Wasserversorgungsverband Main-Taunus-West |
| 111 | | Chamber | | IHK Frankfurt am Main, Geschäftsfeld Standortpolitik |
| 112 | | | | IHK Offenbach am Main |
| 113 | | | | IHK Rheinhessen |
| 114 | | | | IHK Gießen-Friedberg |
| 115 | | | | Handwerkskammer Rhein-Main, Hauptverwaltung Darmstadt |
| 116 | | Church | | Evangelische Kirche in Hessen und Nassau |

*Figure A.64: Responding Institutions of the RSE Runway North-West/Terminal 3 and institutional characteristics, 2nd Evaluation, Part V, based on HMWVL (2007: 260–265).*

## A.15 Clarification letters of Fraport AG

- Aufklärungsschreiben vom 16.12.2005
- Aufklärungsschreiben vom 13.02.2006
- Aufklärungsschreiben vom 06.04.2006
- Aufklärungsschreiben vom 24.05.2006
- Aufklärungsschreiben vom 29.06.2006
- Aufklärungsschreiben vom 21.09.2006
- Aufklärungsschreiben vom 27.11.2006
- Aufklärungsschreiben vom 18.01.2007
- Aufklärungsschreiben vom 25.01.2007
- Aufklärungsschreiben vom 29.01.2007
- Aufklärungsschreiben vom 08.02.2007
- Aufklärungsschreiben vom 09.02.2007
- Aufklärungsschreiben vom 02.03.2007
- Aufklärungsschreiben vom 26.03.2007
- Aufklärungsschreiben vom 02.04.2007
- Aufklärungsschreiben vom 17.04.2007
- Aufklärungsschreiben vom 23.04.2007
- Aufklärungsschreiben vom 18.05.2007
- Aufklärungsschreiben vom 28.06.2007
- Aufklärungsschreiben vom 04.07.2007
- Aufklärungsschreiben vom 18.07.2007
- Aufklärungsschreiben vom 27.07.2007
- Aufklärungsschreiben vom 07.08.2007
- Aufklärungsschreiben vom 17.08.2007
- Aufklärungsschreiben vom 27.08.2007
- Aufklärungsschreiben vom 05.09.2007
- Aufklärungsschreiben vom 10.09.2007
- Aufklärungsschreiben vom 12.09.2007
- Aufklärungsschreiben vom 20.09.2007
- Aufklärungsschreiben vom 26.09.2007
- Aufklärungsschreiben vom 05.10.2007
- Aufklärungsschreiben vom 16.10.2007
- Aufklärungsschreiben vom 17.10.2007
- Aufklärungsschreiben vom 22.10.2007

- Aufklärungsschreiben vom 23.10.2007
- Aufklärungsschreiben vom 25.10.2007
- Aufklärungsschreiben vom 31.10.2007
- Aufklärungsschreiben vom 01.11.2007
- Aufklärungsschreiben vom 07.11.2007
- Aufklärungsschreiben vom 08.11.2007
- Aufklärungsschreiben vom 09.11.2007
- Aufklärungsschreiben vom 15.11.2007
- Aufklärungsschreiben vom 22.11.2007
- Aufklärungsschreiben vom 23.11.2007
- Aufklärungsschreiben vom 27.11.2007
- Aufklärungsschreiben vom 29.11.2007
- Aufklärungsschreiben vom 30.11.2007

HMWVL (2007: 289–291).

# Bibliography


ACI (2014). *2013 ACI Annual World Airport Traffic Report*. Gütersloh: Airports Council International (ACI).

Adamascheck, Bernd and Marga Pöhl (2003). *Regionen erfolgreich steuern. Regional Governance – von der kommunalen zur regionalen Strategie*. Gütersloh: Verlag Bertelsmann Stiftung.

Ahrend, Rüdiger, Emily Farchy, et al. (2014). "What Makes Cities More Productive? Evidence on the Role of Urban Governance from Five OECD Countries". In: *OECD Regional Development Working Papers* 2014.V, pp. 1–32.

Ahrend, Rüdiger, Catherine Gamper, and Abel Schumann (2014). "The OECD Metropolitan Governance Survey. A Quantitative Description of Governance Structure in Large Urban Agglomerations". In: *OECD Regional Development Working Papers* 2014.IV, pp. 1–49.

Ahrend, Rüdiger and Abel Schumann (2014a). "Approaches to Metropolitan Area Governance: A Country Overview". In: *OECD Regional Development Working Papers* 2014.III, pp. 1–21.

— (2014b). "Does Regional Economic Growth Depend on Proximity to Urban Centres?" In: *OECD Regional Development Working Papers* 2014.VII, pp. 1–23.

Alemann, Ulrich von (1981). *Neokorporatismus*. Frankfurt am Main & New York: Campus Verlag.

Alemann, Ulrich von and Rolf G. Heinze (1981). *Verbände und Staat. Vom Pluralismus zum Korporatismus. Analysen, Positionen, Dokumente*. 2nd ed. Opladen: Westdeutscher Verlag.

Amann, Horst (2012). "Ausbau Flughafen Frankfurt. Ausbauprogramm der Superlative – Fit for Future". In: "Neunzehntes Kolloquium Luftverkehr an der Technischen Universität Darmstadt. August-Euler-Luftfahrtpreis-Verleihung 2012, Luftverkehr – Intelligente Infrastrukturen für das 21. Jahrhundert". Ed. by TUD. Darmstadt: Technische Universität Darmstadt (TUD), pp. 1–16.

Anderson, James E. (1975). *Public Policymaking*. New York: Praeger Publishers.

Andrews, Matt (2014). *The Limits of Institutional Reform in Development: Changing Rules for Realistic Solutions*. Cambridge: Cambridge University Press.

Andrews, Matt, Lant Pritchett, and Michael Woolcock (2012). "Escaping Capability Traps through Problem-Driven Iterative Adaptation (PDIA)." In: *CGD Working Papers* June 2012.CCXCIX, pp. 1–26.

Arndt, Rudi (1975). "Eine dritte Startbahn für den Flughafen Frankfurt/Main". In: "Chronik Flughafen Frankfurt am Main". Ed. by Rudolph Lange. Köln: Digital River, pp. 327–336.

Ascher, François (2003). "Urban Homogenisation and Diversificatin in West Europe". In: "Globalism and Local Democracy. Challenge and Change in Europe and North America". Ed. by Robin Hambleton, Hank V. Savitch, and Murray Stewart. Reprint. Houndsmill & New York: Palgrave Macmillan, pp. 52–66.

Ashford, Norman J., Saleh Mumayiz, and Paul H. Wright (2011). "Airport City". In: "Airport Engineering: Planning, Design and Development of 21st Century Airports". Ed. by Norman J. Ashford, Saleh Mumayiz, and Paul H. Wright. 4th ed. Hoboken: John Wiley & Sons, pp. 659–703.

ATSC (1980). *Untersuchung des 'praktischen Betriebswertes' der geplanten Startbahn 18*. Kronberg im Taunus: Air Traffic Service Consulting.

Audretsch, Andreas (2009). *Zur Entstehung von Good Governance. Gründe, Muster und Bedingungen einer afrikanischen Entwicklung. Das Beispiel Ghana*. Potsdam: Universität Potsdam.

Aulich, Chris, Graham Sansom, and Peter McKinlay (2014). "A Fresh Look at Municipal Consolidation in Australia". In: *Local Government Studies* 40.I, pp. 1–20.

AZ (2002). *"Verdoppelung der Kapazität ist unzumutbar". Bürgermeister der Mainspitze zum Flughafenausbau: Kein öffentliches Interesse mehr vorhanden, 4. Februar 2002*. Allgemeine Zeitung.

— (2005). *"Erörterung ist unrechtmäßig". Kommunen fordern erneut Abbruch des Ausbau-Termins / Fraport gelassen, 14. September 2005*. Allgemeine Zeitung, Lars Hennemann.

Barber, Bejamin R. (2014). *If Mayors Ruled the World. Dysfunctional Nations, Rising Cities*. New Haven: Yale University Press.

Barber, Bernard (1995). "All Economies Are "Embedded": The Career of a Concept, and Beyond." In: *Social Research* 62.II, pp. 378–413.

BARIG (1981). *Brief an den Vorsitzenden des Haushaltsausschusses, Herrn Dr. Erwin Lang, Hessischer Landtag, 26. Januar 1981.*

— (1997). *Resolution – Kapazitätenentwicklung des Flughafen Frankfurts, 4. Dezember 1997.* Frankfurt am Main: Board of Airline Representatives in Germany e.V. (BARIG).

— (2007). *"Pressemitteilung: BARIG Position zum Thema "Planfeststellungsbeschluss zum Ausbau des Frankfurter Flughafens"", 18. Dezember 2007.* Board of Airline Representatives in Germany e.V. (BARIG).

— (2014). *Vorstand des "Air Cargo Community Frankfurt e.V." konstituiert sich und nimmt die Geschäfte auf.* Board of Airline Representatives in Germany e.V. (BARIG), Website: http://www.barig.org – rev. January 9th, 2015.

Bartels, Dietrich (1970). "Einleitung". In: *Wirtschafts- und Sozialgeographie.* Ed. by Dietrich Bartels. Neue Wissenschaftliche Bibliothek - Wirtschaftswissenschaften. Köln & Berlin: Kiepenhauer & Witsch, pp. 13–45.

Bathelt, Harald and Johannes Glückler (2002). "Wirtschaftsgeographie in relationaler Perspektive: Das Argument der zweiten Transition." In: *Geographische Zeitschrift* 90.I, pp. 20–39.

— (2003). "Towards a Relational Economic Geography". In: *Journal of Economic Geography* 3.II, pp. 117–144.

Bathelt, Harald and Johannes Glückler (2011). *The Relational Economy. Geographies of Knowing and Learning.* Oxford etc.: Oxford University Press.

— (2012). *Wirtschaftsgeographie. Ökonomische Beziehungen in räumlicher Perspektive.* 3rd ed. Stuttgart: Eugen Ulmer, UTB.

Bauer, Thomas and Florian Baumann (2008). "Politische Führung und Differenzierung in Europa". In: "Lissabon in der Analyse. Der Reformvertrag der Europäischen Union". Ed. by Werner Weidenfeld. Münchner Beiträge zur Europäischen Einigung, 20. Baden-Baden: Nomos, pp. 189–212.

Baumheier, Ralph (2007). *Metropolregionen in Nordwestdeutschland. Metropolregionen und Raumentwicklung Teil 1*. Hannover: Akademie für Raumforschung und Landesplaung.

BBI (2004). *Pressemitteilung, 14. Januar 2004*. Bündnis der Bürgerinitiativen (BBI).

— (2013). *Pressemitteilung: Planergänzungsbeschluss zu Wirbelschleppenschäden panisch und wenig durchdacht, 16. Mai 2013*. Bündnis der Bürgerinitiativen (BBI).

Beatty, Paul (2003). "Understanding the Standardized\Non-Standardized Interviewing Controversy". In: *Interviewing, Vol I*. Ed. by Nigel Fielding. Sage Bench-marks in Social Research Methods. London etc.: Sage Publications, pp. 109–123.

Beck, Ulrich, Anthony Giddens, and Scott Lash (1994). *Reflexive Modernization: Politics, Tradition and Aesthetics in the Modern Social Order*. Cambridge & Oxford: Polity Press.

Becker-Ritterspach, Luisa (2015). *Steuerung der Implementation von Flughäfen. Eine Untersuchung des Flughafens in Frankfurt a.M. und Berlin-Brandenburg*. Wiesbaden: Springer SV.

Beckert, Jens (1996). "What is Sociological About Economic Sociology? Uncertainty and the Embeddedness of Economic Action". In: *Theory and Society* 25.VI, pp. 803–840.

— (2007). "The Great transformation of Embeddedness. Karl Polanyi and the New Economic Sociology". In: *MPIfG Discussion Papers* 2007.01, pp. 1–25.

Beetham, David (2012). "Evaluating New vs Old Forms of Citizens Engagment and Participation". In: *Evaluating Democratic Innovations*. Ed. by Brigitte Geißel and Kenneth Newton. London & New York: Routledge, pp. 56–67.

Benz, Arthur (1997). "Kooperativer Staat? Gesellschaftliche Einflußnahme auf staatliche Steuerung". In: *Politische Beteiligung und Bürgerengagement in Deutschland*. Ed. by Ansgar Klein and Rainer Schmalz-Bruns. Schriftenreihe der Bundeszentrale für politische Bildung. Bonn: Bundeszentrale für politische Bildung, pp. 88–113.

— (2003a). "Governance – Modebegriff oder nützliches sozialwissenschaftliches Konzept?" In: *Governance. Eine Einführung.*

*Dreifachkurseinheit.* Ed. by Fachbereich Kultur-und Sozialwissenschaften FernUniversität in Hagen. 9th ed. Hagen: FernUniversität in Hagen, pp. 13–31.

— (2003b). "Regional Governance mit organisatorischem Kern. Das Beispiel der Region Stuttgart". In: *Informationen zur Raumentwicklung* 2003.VIII/IX, pp. 505–512.

— (2006). "Governance in Mehrebenensystemen". In: *Governance-Forschung. Vergewisserung über Stand und Entwicklungslinien.* Ed. by Gunnar Folke Schuppert. 2nd ed. Baden-Baden: Nomos, pp. 95–120.

— (2010). "Multilevel Governance – Governance in Mehrebenensystemen". In: *Governance – Regieren in komplexen Regelsystemen. Eine Einführung.* Ed. by Arthur Benz and Nicolai Dose. 2nd ed. Wiesbaden: VS Verlag für Sozialwissenschaften, pp. 111–135.

Benz, Arthur and Dietrich Fürst (2003). "Region – 'Regional Governance' – Regionalentwicklung". In: *Regionen erfolgreich steuern. Regional Governance – von der kommunalen zur regionalen Strategie.* Ed. by Bernd Adamaschek and Marga Pröhl, Gütersloh: Bertelsmann Stiftung, pp. 11–66.

Benz, Arthur, Dietrich Fürst, et al. (1999). *Regionalisierung. Theorie – Praxis – Perspektiven.* Opladen: Leske + Budrich.

Berelson, Bernhard (1952). *Content Analysis in Communication Research.* Foundations of Communications Research. New York: Hafner.

BfF (1981). *Brief der Bundesanstalt für Flugsicherung z.Hd. des Vorsitzenden des Haushaltsausschusses Herrn Dr. Lang, 15. Januar 1981.* Bundesanstalt für Flugsicherung (BfF).

Biersteker, Thomas J. (1992). "The "Triumph" of Neoclassical Economics in the Developing World: Policy Convergance in the International Economic Order." In: *Governance without Government: Order and Change in World Politics.* Ed. by James N. Rosenau and Ernst-Otto Czempiel. Cambridge: Cambridge University Press, pp. 102–131.

BIMW (2004). *Pressemitteilung, 14. Januar 2004.* Bürgerinitiative Mörfelden-Walldorf (BIMW).

Bish, Robert L. (1971). *The Public Economy of Metropolitan Areas.* Markham Series in Public Policy Analysis. Chicago: Markham.

— (2001). "Local Government Amalgamations: Discredited Nineteenth-Century Ideals Alive in the Twenty-First." In: *The Urban Papers* 2001.150, pp. 1–35.

Blättel-Mink, Birgit and Alexander Ebner (2009). *Innovationssysteme. Technologie, Institutionen und die Dynamik der Wettbewerbsfähigkeit*. Vol. 3. Modernes Regieren. Baden-Baden: Nomos.

Blatter, Joachim (2007). *Governance – theoretische Formen und historische Transformation. Politische Steuerung und und Integration in den Metropolregionen der USA (1850-2000)*. Heidelberg etc.: Springer.

Blotevogel, Hans Heinrich (1999). "Einführung. Zur Neubewertung der Region für Regionalentwicklung und Regionalpolitik". In: *Europäische Einflüsse auf die Raum- und Regionalentwicklung am Beispiel des Naturschutzes, der Agenda 2000 und des regionalen Milieus*. Ed. by ARL. Hannover: Verlag der Akademie für Raumforschung und Landesplanung (ARL), pp. 44–60.

— (2002). "Deutsche Metropolregionen in der Vernetzung". In: *Informationen zur Raumentwicklung* 2002.VI/VII, pp. 345–351.

— (2005). "Metropolregionen". In: *Handwörterbuch der Raumordnung*. Ed. by ARL. 4th ed. Hannover: Verlag der Akademie für Raumforschung und Landesplanung (ARL), pp. 642–647.

— (2010). "Raumordnung und Metropolregionen". In: *Geographische Rundschau* 2010.XI, pp. 4–12.

Blum, Sonja and Klaus Schubert (2011). *Politikfeldanalye*. 2nd ed. Wiesbaden: VS Verlag für Sozialwissenschaften.

Blumenthal, Julia von (2005). "Governance – eine kritische Zwischenbilanz." In: *Zeitschrift für Politikwissenschaft* 15.IV, pp. 1149–1180.

Blumenthal, Julia von and Stephan Bröchler (2006). *Von Government zu Governance. Analysen zum Regieren im modernen Staat*. Hamburg: LIT Verlag.

BMVBW (2000). *Flughafenkonzept der Bundesregierung, 30. August 2000*. Berlin: Bundesministerium für Verkehr, Bau- und Wohningswesen (BMVBW).

BMZ (2009). *Förderung von Good Governance in der deutschen Entwicklungspolitik*. BMZ Konzept 172. Bonn: Bundesministerium für wirtschaftliche Zusammenarbeit und Entwicklung (BMZ).

— (2015). "Was wir machen. Arbeitsweise. Bilaterale Zusammenarbeit. Akteure. Nichtregierungsorganisationen (Private Träger und Sozialstrukturträger)." Bundesministerium für wirtschaftliche Zusammenarbeit und Entwicklung (BMZ), Website: http://www.bmz.de – rev. January 5th, 2015.

Böderlin, Ruth (2000). "Die neue Institutionalisierung der Region. Das Beispiel Rhein-Main". In: *Informationen zur Raumentwicklung* 2000.9/10, pp. 537–548.

Boger, Peter and Mette Zølner (2007). *Methods in Democratic Network Governance*. Houndsmills & New York: Palgrave Macmillan.

Borrás, Susana and Thomas Conzelmann (2007). "Democracy, Legitimacy and Soft Modes of Governance in the EU: The Empirical Turn." In: *Journal of European Integration* 29.V, pp. 531–548.

Börzel, Tanja A. (2008). "Der "Schatten der Hierarchie" – Ein Governance-Paradox?" In: *Governance in einer sich wandelenden Welt*. Ed. by Gunnar Folke Schuppert and Michael Zürn. PVS – Politische Vierteljahresschrift, Sonderheft 41/2008. Wiesbaden: VS Verlag für Sozialwissenschaften, pp. 118–131.

— (2012). "Experimentalist governance in the EU: The emperor's new clothes?" In: *Regulation & Governance* 6.III, pp. 378–384.

Boswell, Christina (2008). "The Political Functions of Expert Knowledge: Knowledge and Legitimation in European Union Immigration Policy". In: *Journal of European Public Policy* 15.IV, pp. 471–488.

Boud, David (2001). "Introduction: Making the Move to Peer Learning". In: *Peer Learning in Higher Education: Learning from and With Each Other*. Ed. by David Boud, Ruth Cohen, and Jane Sampson. Milton Park & New York: Routledge, pp. 1–17.

BPB (2015). "Handwörterbuch des politischen Systems der Bundesrepublik Deutschland. Bürgerinitiativen". Bundeszentrale für politische Bildung (BPB), Website: http://www.bpb.de – rev. January 5th, 2015.

Brandt, Willy (1969). *Regierungserklärung, 28. Oktober 1969*.

Brassett, James, Ben Richardson, and William Smith (2010). *Experimentalist Governance, Deliberation and Democracy: A Case Study of Primary Commodity Roundtables*. Vol. 270/10. CSGR Working Paper. Warwick: Centre for the Study of Globalisation, Regionalisation, Department of Politics, and International Studies, University of Warwick.

Breuer, Christina B. (2012). *The Regional Puzzle. How Regions and Encompassed Actors are Involved in EU Regional Policy*. Vol. 4. Regionen in Europa / Euro-pean Regions. Münster: LIT Verlag.

Brewer, Garry D. (1974). "Editorial: The Policy Sciences Emerge: To Nurture and Structure a Discipline". In: *Policy Sciences* 5.III, pp. 239–244.

Brezzi, Monica and Paolo Veneri (2014). "Assessing Polycentric Urban Systems in the OECD: Country, Regional and Metropolitan Perspectives". In: *OECD Regional Development Working Papers* 2014.I, pp. 1–20.

Brikinshaw, Patrick, Ian Harden, and Norman Lewis (1990). *Government by Moonlight. The Hybrid Parts of the State*. London etc.: Unwin Hyman.

Bröchler, Stephan and Julia von Blumenthal (2006). "Von Government zu Gov-ernance – Analysen zu einem schwierigen Verhältnis". In: *Von Government zu Governance. Analysen zum Regieren im modernen Staat*. Ed. by Julia von Blu-menthal and Stephan Bröchler. Hamburg: LIT Verlag, pp. 7–21.

Brown, Dan (2009). *The Lost Symbol*. New York: Anchor Books.

Brühl, Tanja et al. (2001). *Die Privatisierung der Weltpolitik. Entstaatlichung und Kommerzialisierung im Globalisierungsprozess*. Bonn: Verlag J.H.W. Dietz Nachf.

Bryant, Anthony and Kathy Charmaz (2013). *The SAGE Handbook of Grounded Theory*. Reprint. London etc.: Sage Publications.

Bryman, Alan (2012). *Social Research Methods*. 4th ed. Oxford etc.: Oxford University Press.

Bryman, Alan and Robert G. Burgess (1999). "Qualitative Research Methodology – A Review". In: *Qualitative Research. Vol. I*. Ed. by Alan Bryman and Robert G. Burgess. London etc.: Sage Publications, pp. IX–XLVI.

Buhr, Daniel and Josef Schmid (2012). "Wirtschaft und Gesellschaft: Politische Wirtschaftslehre". In: *Politikwissenschaft: Eine Einführung*. Ed. by Hans-Joachim Lauth and Christian Wagner. 7th ed. Paderborn: UTB Schöningh, pp. 233–262.

Bulmer, Simon J. (1993). "The Governance of the European Union: A New Institutionalist Approach". In: *Journal of Public Policy* 13.IV, pp. 351–380.

BUND (1998). *Flugverkehr schädigt Mensch, Klima und Umwelt – kein Ausbau des Frankfurter Flughafens Resolution der Landesdelegiertenversammlung des BUND Hessen e.V. am 21. März 1998*. Bund für Umwelt und Naturschutz Deutschland (BUND) Hessen, Website: http://www.bundhessen.de – rev. February 4th, 2015.

— (2000a). *Pressemitteilung: Initiativantrag des Landesvorstandes zur Erweiterung des Frankfurter Flughafens: Rückkehr zur Aufrichtigkeit in der Flughafendebatte!, 25.03.2000*. Bund für Umwelt und Naturschutz Deutschland (BUND) Hessen, Website: http://www.bund-hessen.de – rev. February 4th, 2015.

— (2000b). *Pressemitteilung: Zur Zukunftsfähigkeit des Frankfurter Flughafens – Forderungen, Fakten und Thesen des BUND Hessen, Januar 2000*. Bund für Umwelt und Naturschutz Deutschland (BUND) Hessen, Website: http://www.bund-hessen.de – rev. February 4th, 2015.

— (2001). *Pressemitteilung: Die Zukunft des Waldes im Rhein-Main-Gebiet ist akut gefährdet! Ein Vortrag von Thomas Norgall für das 3. Umweltsymposium Rhein-Main am 24. Januar 2001 in Mörfelden-Walldorf*. Bund für Umwelt und Naturschutz Deutschland (BUND) Hessen, Website: http://www.bund-hessen.de – rev. February 5th, 2015.

— (2002a). *Pressemitteilung: "Diese Entscheidung ist ein vorsätzlicher Rechtsbruch", 12. Juni 2002*. Bund für Umwelt und Naturschutz Deutschland (BUND) Hessen.

— (2002b). *Pressemitteilung: Kein Ausbau des Frankfurter Flughafens – Kein Flughafensystem "Frankfurt – Hahn", gemeinsame Position BUND LV Hessen & Rheinland-Pfalz, 17. September 2002*. Bund für Umwelt und Naturschutz Deutschland (BUND) Hessen, Website: http://www.bund-hessen.de – rev. February 5th, 2015.

— (2002c). *Pressemitteilung: Laudatio zur Verleihung des BUND – Umwelthammers an die Fraport AG am 27. 02. 2002, Der Frankfurter Flughafen würde nach der Erweiterung größer sein als Hessens größtes Naturschutzgebiet, das Europareservat "Kühkopf-Knoblochsaue"*. Bund für Umwelt und Naturschutz Deutschland (BUND) Hessen, Website: http://www.bund-hessen.de – rev. February 5th, 2015.

— (2003a). *Pressemitteilung, 10. April 2003*. Bund für Umwelt und Naturschutz Deutschland (BUND) Hessen.

— (2003b). *Pressemitteilung, 8. April 2003*. Bund für Umwelt und Naturschutz Deutschland (BUND) Hessen.

— (2004a). *Pressemitteilung: BUND: Verhandlungsführer im A380-Verfahren ist befangen, 15. Januar 2004*. Bund für Umwelt und Naturschutz Deutschland (BUND) Hessen.

— (2004b). *Pressemitteilung: BUND wird gegen die A380-Werft klagen, 27. November 2004*. Bund für Umwelt und Naturschutz Deutschland (BUND) Hessen.

— (2004c). *Pressemitteilung: Erneutes Planungschaos bei der Fraport, 3. März 2004*. Bund für Umwelt und Naturschutz Deutschland (BUND) Hessen.

— (2004d). *Pressemitteilung: Regionalplan Südhessen 2000 nichtig - Schnelle Planfeststellung der A380-Werft nicht mehr vorstellbar, 30. Juli 2004*. Bund für Umwelt und Naturschutz Deutschland (BUND) Hessen.

— (2005a). *Pressemitteilung: BUND wird Revision gegen A380-Urteil prüfen, 28. Juni 2005*. Bund für Umwelt und Naturschutz Deutschland (BUND) Hessen.

— (2005b). *Pressemitteilung: Rodung kann noch gestoppt werden, 4. August 2005*. Bund für Umwelt und Naturschutz Deutschland (BUND) Hessen.

— (2007). *Pressemitteilung: BUND kündigt Verbandsklage an, 18. Dezember 2007*. Bund für Umwelt und Naturschutz Deutschland (BUND) Hessen.

BUND and NABU (2000). *Pressemitteilung: Mehr Aufrichtigkeit in der Flughafen Debatte: Stellungnahme für das Landtags-Hearing zum Frankfurter Flughafen, 10. - 12. Mai 2000*. Bund für Umwelt und Naturschutz Deutschland (BUND) Hessen, Website: http://www.bund-hessen.de – rev. February 4th, 2015.

Bundestag (1998). *Abschlußbericht der Enquete-Kommission "Schutz des Menschen und der Umwelt - Ziele und Rahmenbedingungen einer nachhaltig zukunftsverträglichen Entwicklung". Konzept Nachhaltigkeit. Vom Leitbild zur Umsetzung*. Vol. 13/11200. *Drucksache*. Bonn: Bundesanzeiger Verlagsgesellschaft mbH.

— (2000). *Unterrichtung durch die Bundesregierung: Umweltgutachten 2000 des Rates von Sachverständigen für Umweltfragen. Schritte ins nächste Jahrtausend.* Vol. 14/3363. *Drucksache.* Bonn: Bundesanzeiger Verlagsgesellschaft mbH.

Bunte (1982). *Interview zum 1. Mai 1982.* Bunte Illustrierte.

Busch, Per-Olaf (2000). *Konfliktfall Flughafenerweiterung. Eine kritische Würdigung des Verfahrens "Mediation – Eine Zukunftsregion im offenen Dialog" zum Flughafen Frankfurt/Main*, volume 8 of *HSFK-Report.* Frankfurt am Main: Hessische Stiftung Friedens- und Konfliktforschung (HSFK).

BVerwG (2005). *Pressemitteilung: Kein Baustopp für die Airbus 380-Wartungshalle am Flughafen Frankfurt a.M. (Nr. 45/2005), 8. September 2005.* Bundesverwaltungsgericht (BVerwG).

— (2012). *Pressemitteilung: Planmäßige Flüge in der Mediationsnacht weiterhin unzulässig – Kontingent für die Gesamtnacht auf durchschnittlich 133 Flüge beschränkt - Schallschutz für gewerbliche Grundstücke nachbesserungsbedürftig (Nr. 33/2012), 4. April 2012.* Bundesverwaltungsgericht (BVerwG).

BVF (2003). *Pressemitteilung: Das von der Fraport AG im Scoping-Termin vorgelegte Lärmschutzkonzept bietet keinen wirksamen Schutz der Bevölkerung vor Fluglärm – das angekündigte Nachtflugverbot ist unverbindlich!, 14. April 2003.* Bundesvereinigung gegen Fluglärm e.V. (BVG).

Campbell, John L., Leon N. Lindberg, and J. Rogers Hollingsworth (1991). *Governance of the American Economy.* Cambridge etc.: Canbridge University Press.

Campbell-Verduyn, Malcom and Tony Porter (2014). "Experimentalism in European Union and Global Financial Governance: Interactions, Contrasts, and Implications." In: *Journal of European Public Policy* 20.III, pp. 408–429.

CDU (1998). "Pressemitteilung, 16. Juni 1998." Christlich Demokratische Union Deutschlands, Hessen.

— (2006). "Pressemitteilung: Ausbau Frankfurter Flughafen. Michael Boddenberg: "Keine Verzögerung im Anhörungsverfahren" / Offenbach geht es nur um Klamauk und nicht um die Sache, 11. Januar 2006." Christlich Demokratische Union Deutschlands, Hessen.

— (2007). "Pressemitteilung: Die CDU hat Wort gehalten. Ausbau und Nachtflugverbot kommen!, 18. Dezember 2007." Christlich Demokratische Union Deutschlands, Hessen.

Cerny, Philip G. (2006). "Restructuring the State in a Globalizing World: Capital Accumulation, Tangled Hierarchies and the Search for a New Spatio-Temporal Fix." In: *Review of International Political Economy* 13.IV, pp. 679–695.

Chandler, Alfred Dupont (2002). *The Visible Hand: The Managerial Revolution in American Business*. 16th ed. Cambridge: Belknap Press of Harvard University.

Chattopadhyay, Rupak (2011). "Why Metropolitan Regions Matter?" In: "The Governance of Metropolitan Regions – European and Global Experience". Ed. by EU. Workshop on "Governance of Metropolitan Regions in Federal Systems", Brussels, 20–21 June 2011. Brussels & Ottawa: European Union (EU), Committee of the Regions / Forum of Federations, pp. 21–27.

Cheffins, Brain R. (2013). "The History of Corporate Governance." In: "The Oxford Handbook of Corporate Governance". Ed. by Mike Wright et al. Oxford: Oxford University Press, pp. 46–64.

Cho, Eun Meeh (2014). *Privatisierung und Differenzierung im Strafvollzug*. Vol. 5582. Europäische Hochschulschriften. Rechtswissenschaft. Frankfurt am Main etc.: Peter Lang.

Christiansen, Thomas and Simona Piattoni (2003). *Informal Governance in the European Union*. Cheltenham & Northampton: Edward Elgar Publishing Limited.

Clinton, Bill (1993). *State of the Union Address. Speech to the Congress. Presentation of the Economic Plan, Joint Session of Congress on 17th February 1993*. Washington: United States Congress.

Coase, Ronald Harry (1937). "The Nature of the Firm." In: *Economica, New Series* 4.XVI, pp. 386–405.

Cockpit, Vereinigung (1981). "*Brief der Vereinigung Cockpit e.V. an den Hessischen Landtag, 19. Januar 1981.*"

Cohen, Joschua and Charles F. Sabel (2006). "Global Democracy?" In: *NYU Journal of International Law and Politics* 37.IV, pp. 763–797.

Combes, Pierre-Phillipe, Gilles Duranton, and Laurent Gobillon (2011). "The Identification of Agglomeration Economies." In: *Journal of Economic Geography* 11.II, pp. 253–266.

Copper, Richard N. (1972). "Economic Interdependence and Foreign Policy in the Seventies." In: *World Politics* 24, pp. 159–181.

Crouch, Colin (2005). "Die Bedeutung von Governance für die Vielfallt und Wandel im modernen Kapitalismus." In: "*Welten des Kapitalismus. Institutionelle Alternativen in der globalisierten Ökonomie*". Ed. by Max Miller. Frankfurt am Main & New York: Campus Verlag, pp. 101–126.

Crouch, Colin and Henry Farrell (2002). "Breaking the Path of Institutional Development? Alternatives to the New Determinism." In: *MPIfG Discussion Papers* 2002.05, pp. 49–67.

Crouch, Colin, Patrick Le Galés, et al. (2003). *Local Production Systems in Europe: Rise or Demise?* Reprint. Oxford etc.: Oxford University Press.

— (2004). *Changing Governance of Local Economies: Responses of European Local Production Systems*. Oxford etc.: Oxford University Press.

Crouch, Colin and Wolfgang Streeck (2006). *The Diversity of Democracy. Corporatism, Social Order and Political Conflict*. Chaltenham & Northamton: Edward Elgar Publishing.

Crouch, Colin and Carlo Trigilia (2003). "Conclusion: Still Local Economies in Global Capitalism?" In: *Local Production Systems in Europe. Rise or Demise?*. Ed. by Crouch Colin et al. Reprint. Oxford etc.: Oxford University Press, pp. 212–237.

Dąbrowska, Patrycja (2010). "EU Governance of GMOs: Political Struggles and Experimentalist Solutions?" In: *Experimentalist Governance in the European Union. Towards a New Architecture*. Ed. by Charles F. Sabel and Jonathan Zeitlin, Oxford etc.: Oxford University Press, pp. 177–214.

Dacin, M. Tina, Marc J. Ventresca, and Brent D. Beal (1999). "The Embeddedness of Organizations: Dialogue & Directions." In: *Journal of Management* 25.III, pp. 317–356.

DAG (1976). *Brief des DAG-Landesverbandes Hessen an den Hessischen Minister für Wirtschaft und Technik, 13. April 1976.*

— (1979). *Brief des DAG-Landesverbandes Hessen an die Hessische Landesregierung über den Hessischen Ministerpräsidenten Holger Börner, 11. Juni 1979.*

— (1980). *Brief des DAG-Landesverbandes Hessen an die Hessische Landesregierung über den Hessischen Ministerpräsidenten Holger Börner, 20. November 1980.*

Dahl, Robert A. and Edward R. Tufte (1973). *Size and Democracy. The Policies of the Smaller European Democracies*. Stanford: Stanford University Press.

Dale, Gareth (2012). *Karl Polanyi. The Limits of the Market*. Key Contemporary Thinkers. Cambridge & Malden: Polity Press.

Dangschat, Jens S. and Astrid Segert (2011). "Nachhaltige Alltagsmobilität – soziale Ungleichheiten und Milieus." In: *Österreichische Zeitschrift für Soziologie* 36.II, pp. 55–73.

Danielzyk, Rainer and Jürgen Oßenbrügge (2003). "Globale Bedingungen und lokale Perspektiven für Wettbewerbs- und Kooperationsformen in Metropolregionen." In: "Stadt - Raum - Natur. Die Metropolregion als politisch konstruierter Raum". Ed. by Martin Döring et al. Hamburg: Hamburg University Press, pp. 17–38.

Danielzyk, Rainer and Rüdiger Reitzel (2003). "Regionalplanung als Motor regionaler Kooperation. Das Beispiel Oderland-Spree." In: *Informationen zur Raumentwicklung* 2003.VIII/IX, pp. 513–521.

Dawson, Mark (2011). *New Governance and the Transformation of European Law*. Studies on European Law and Policy. Cambridge & New York: Cambridge University Press.

De Búrca, Gráinne (2010). "Strumbling into Experimentalism: The EU Anti-Discrimination Regime." In: "Experimentalist Governance in the European Union. Towards a New Architecture". Ed. by Charles F. Sabel and Jonathan Zeitlin. Oxford etc.: Oxford University Press, pp. 215–235.

De Búrca, Gráinne, Robert O. Keohane, et al. (2014). "Global Experimentalist Governance." In: *Britsh Journal of Political Science* 44.III, pp. 477–486.

de Schutter, Olivier (2010). "The Role of Evaluation in Experimentalist Governance: Learning by Monitoring in the Establishment of the Area of Freedom, Security and Justice." In: "Experimentalist Governance in the European Union. Towards a New Architecture". Ed. by Charles F. Sabel and Jonathan Zeitlin. Oxford etc.: Oxford University Press, pp. 261–296.

Deeg, Richard (2006). "Governance and the Nation-State in a Global Era." In: "Governance in der politischen Ökonomie. Struktur und Wandel des modernen Kapitalismus". Ed. by Susanne Lütz. Wiesbaden: VS Verlag für Sozialwissenschaften, pp. 57–106.

Destatis (2014). "Städte nach Fläche, Bevölkerung und Bevölkerungsdichte am 31.12.2012." Destatis, Website: https://www.destatis.de – rev. December 28th, 2014.

Dewey, John (2012). *The Public and its Problems: An Essay in Political Inquiry*. Pennsylvania: Pennsylvania State University Press.

DFLD (2015). *Aktuelles, 24.03.2015*. Deutscher Fluglärmdienst e.V. (DFLD), Website: http://www.dfld.de – rev. July 6th, 2015.

DGB (1981). *Brief des DGB-Landesbezirk Hessen an des Vorsitzenden der Anörungsvorstandes "Flughafenausbau" im Hessischen Landtag Herrn Dr. Erwin Lang, 19. Januar 1981.*

DIL (1955). *Der Flughafen Frankfurt am Main in Gegenwart und Zukunft. Betrachtungen über die Probleme eines Grossflughafens*. 2nd ed. Frankfurt am Main: Deutsches Institut für Luftverkehrsstatistik e.V. (DIL).

— (1965). *Der Flughafen Frankfurt (Main) und seine wirtschaftliche Bedeutung für die nähere Umgebung*. 2nd ed. Frankfurt am Main: Deutsches Institut für Luftverkehrsstatistik e.V. (DIL).

Dingeldey, Irene (2008). "Governance und Sozialpolitik: Der aktivierende Wohlfahrtsstaat als Gewährleistungsstaat." In: *Governance in einer sich wandelenden Welt*. Ed. by Gunnar Folke Schuppert and Michael Zürn. Wiesbaden: VS Verlag für Sozialwissenschaften, PVS – Politische Vierteljahresschrift, Sonderheft 41/2008, pp. 313–329.

DLH (1980). *Konzernperspektiven – bis i die 90er Jahre, CGN CZ*. Deutsche Lufthansa (DLH), intern.

— (1981). *Brief an den Hessischen Landtag, den Vorsizenden des Haushaltsausschusses, 9. Januar 1981.*

— (2000). *Pressemitteilung, 31. Januar 2000*. Deutsche Lufthansa (DLH).

DM (2011). *'I have my own army in the NYPD - the seventh largest army in the world': Bloomberg's bizarre boast about city's police force'*. Daily Mail, Website: http://www.dailymail.co.uk – rev. June 3rd, 2015.

Dobbin, Frank, Beth Simmons, and Geoffrey Garrett (2007). "The Global Diffusion of Public Policies: Social Construction, Coercion, Competition, or Learning?" In: *Annual Review of Sociology* 33, pp. 449–472.

Dolzer, Rudolf, Matthias Herdegen, and Bernhard Vogel (2009). *Good Governance – Gute Regierungsführung im 21. Jahrhundert*. Freiburg: Herder Verlag.

Dorf, Michael C. and Charles F. Sabel (1998). "A Constitution of Democratic Experimentalism." In: *Columbia Law Review* 98.II, pp. 267–473.

Dornier (1980). *Prognose des Verkehrsaufkommens für den Flughafen Frankfurt 1990*, Vol. Studienbericht Nr. 253. Friedrichshafen: Dornier.

Dose, Nicolai (2006). "Governance im Geflecht von Problemen, Ebenen und Akteuren." In: *Von Government zu Governance. Analysen zum Regieren im modernen Staat*. Ed. by Julia von Blumenthal and Stephan Bröchler. Hamburg: LIT Verlag, pp. 23–56.

Dunleavy, Patrick and O'Leary Brendan (1988). *Theories of the State. The Politics of Liberal Democracy*. Reprint. Houndmills: Macmillan.

Durth, Werner (1987). "Urbanität und Stadtplanung." In: "Die Mentaliät des Städtischen. Standentwicklung und Urbanität im gesellschaftlichen Umbruch". Ed. by Walter Prigge. Stadtforschung aktuell. Basel & Boston: Birkhäuser Verlag, Stadtforschung aktuell, pp. 155–165.

DVZ (2002). *Frankfurt: Neue Landebahn soll 2006 fertig sein*. Deutsche Verkehrs Zeitung.

DW (1965). *Gutachten über die mögliche klimatische Auswirkung eines beim Flughafen/Main vorgesehenen Waldeinschlages*. Frankfurt: Deutscher Wetterdienst.

— (1980). *Verkehrswissenschaftliches Institut*. Stuttgart: Universiät Stuttgart.

Eberle, Dagmar (2010). "Governance in der politischen Ökonomie II: Corporate Governance." In: *Governance – Regieren in komplexen Regelsystemen. Eine Einführung*. Ed. by Arthur Banz and Nikolai Dose. 2nd ed. Wiesbaden: VS Verlag für Sozialwissenschaften, pp. 155–173.

Eberlein, Burkard (2010). "Experimentalist Governance in the European Energy Sector." In: "Experimentalist Governance in the European Union. Towards a New Architecture". Ed. by Charles F. Sabel and Jonathan Zeitlin. Oxford etc.: Oxford University Press, pp. 61–78.

Ebner, Alexander (2002). *Institutions and innovation in economic development: Schumpeterian perspectives*. Frankfurt am Main: Johann Wolfgang Goethe-Universität, Universitätsverlag.

— (2004). "Innovation Policies and Locational Competitiveness: Lessons from Singapore." In: *Journal of Technology Innovation* 12.II, pp. 47–66.

— (2006a). "Institutions, Entrepreneurship and the Rationale of Government: An Outline of the Schumpeterian Theory of the State." In: *Journal of Economic Behavior and Organization* 59.IV, pp. 497–515.

— (2006b). "Schumpeterian Entrepreneurship Revisited: Historical Specificity and the Phases of Capitalist Development." In: *Journal of the History of Economic Thought* 28.III, pp. 315–332.

— (2007). "Public Policy, Governance, and Innovation: Entrepreneurial States in East Asian Economic Development." In: *International Journal of Technology and Globalisation* 3.I, pp. 103–124.

— (2008). "Introduction: The Institutions of the Market." In: *The Institutions of the Market. Organizations, Social Systems, and Governance*. Ed. by Alexander Ebner and Nikolaus Beck. New York etc.: Oxford University Press, pp. 1–20.

— (2009a). "Entrepreneurial State: The Schumpeterian Theory of Industrial Policy and the East Asian "Miracle"." In: *Schumpeterian Perspectives on Innovation, Competition and Growth*. Ed. by Uwe Cantner, Jean-Luc Gaffard, and Lionel Nesta. Berlin & Heidelberg: Springer, pp. 369–390.

— (2009b). "Governance von Innovationssystemen und die politische Ökonomie der Wettbewerbsfähigkeit." In: *Innovationssysteme. Technologie, Institutionen und die Dynamik der Wettbewerbsfähigkeit*. Ed. by Birgit Blättel-Mink and Alexander Ebner. Wiesbaden: VS Verlag für Sozialwissenschaften, pp. 119–141.

— (2013). "Cluster Policies and Entrepreneurial States in East Asia." In: "Clusters and Economic Growth in Asia". Ed. by Sören Eriksson. Cheltenham & Northhampton: Edward Elgar, pp. 1–20.

— (2014). "Die industriepolitische Transformation der ostasiatischen Entwicklungsstaaten." In: *Die großen Schwellenländer. Ursachen und Folgen ihres Aufstiegs in der Weltwirtschaft*. Ed. by Andreas Nölke Christian May, and Simone Claar. Globale Politische Ökonomie. Wiesbaden: Springer VS, Globale Politische Ökonomie, pp. 135–151.

— ((forthcoming) 2015[a]). *Embedded Entrepreneurship. The Institutional Dynamics of Innovation*. London: Routledge.

— ((forthcoming) 2015[b]). *Innovationsstrategien und Regionalentwicklung. Theorie und Empirie regionaler Innovationsprozesse*. Wiesbaden: Springer VS.

— ((forthcoming) 2016). "Institutional Transformation. Technology Policy in East Asia: The Rise of the Entrepreneurial State." In: "Routledge Handbook of Politics and Technology". Ed. by Alexander Ebner, Klaus Heine, and Jan Schnellenbach. Neue Studien zur Politischen Ökonomie, 2. Baden-Baden: Nomos.

Ebner, Alexander and Falk Raschke (2012). *Clusterstudie FrankfurtRhein-Main. Wettbewerbsvorteile durch Vernetzung*. Frankfurt am Main: Schumpeter Center for Clusters, Innovation and Public Policy.

EC (1993). *Growth, Competitiveness, Employment. The Challenges and Ways Forward Into the 21st Century*. Bulletin of the European Communities, Supplement 6/93. Luxembourg: Office for Official Publications of the European Communities, European Commission (EC).

— (1999). *EUREK. Europäisches Raumentwicklungskonzept: Auf dem Wege zu einer räumlich ausgewogenen und nachhaltigen Entwicklung der EU*. Luxembourg: Amt für amtliche Veröffentlichungen der Europäischen Gemeinschaften, European Commission (EC).

— (2001). "European Governance – A White Paper." In: *Official Journal of the European Communities* 2001.C 287/01, pp. 1–29.

— (2003). "Governance and Development." In: *Communication from the Commission to the Council, the European Parliament and the European Economic and Social Committee* COM(2003)(615 final), pp. 1–34.

Eckert, Sandra and Tanja A. Börzel (2012). "Experimentalist Governance: An Introduction." In: *Regulation & Governance* 6.III, pp. 371–377.

Eising, Rainer and Beate Kohler-Koch (2002). "Introduction: Network Governance in the European Union." In: *The Transformation of*

*Governance in the European Union*. Ed. by Rainer Eising and Beate Kohler-Koch. Reprint. London & New York: Routledge, pp. 3–13.

Eschner, A.R. (1971). *Stellungnahme zum Gutachten des Deutschen Wetterdienstes (Wettera Frankfurt) vom Okt. 1965 und Nov. 1967 über die möglichen klimatischen Auswirkungen eines beim Flzghafen Frankfurt/Main vorgesehenen Waldeinschlags, 28. August 1971*.

Esser, Josef (1975). *Einführung in die materialistische Staatsanalyse*. Frankfurt am Main: Campus Verlag.

— (1999). "Der kooperative Nationalstaat im Zeitalter der "Globalisierung"." In: *Sozialstaat in der Globalisierung*. Ed. by Diether Döring. Frankfurt am Main: Suhrkamp Verlag, pp. 117–144.

Esser, Josef, Frieder Naschold, and Werner Väth (1972). *Gesellschaftsplanung in kapitalistischen und sozialistischen Systemen. 11 Beiträge*. Gütersloh: Bertelsmann Fachverlag.

Esser, Josef and Eike W. Schamp (2001). "Einführung." In: *Metropolitane Region in der Vernetzung. Der Fall Frankfurt Rhein/Main*. Ed. by Josef Esser and Eike W. Schamp. Frankfurt am Main & New York: Campus Verlag, pp. 9–24.

Esteva, Gustavo (1993). "Entwicklung." In: *Wie im Westen so auf Erden. Ein polemisches Handbuch zur Entwicklungspolitik*. Ed. by Wolfgang Sachs. reinbeck: Rowohlt Taschenbuchverlag, pp. 89–121.

Eurostats (2015a). *European Cities – the EU-OECD functional Urban Area Definition*. Eurostats, Website: http://ec.europa.eu – rev. April 22th, 2015.

— (2015b). *Nuts – Nomenclature of Territorial Units for Statistics*. Eurostats, Website: http://ec.europa.eu – rev. June 25th, 2015.

Evans, Peter, Dietrich Rueschemeyer, and Theda Skocpol (1997). *Bring the State Back In*. 10th Reprint. Cambridge: Cambridge University Press.

FAG (1971). *Flughafen Frankfurt, Main, Planungsabteilung. Airport Frankfurt, Main, Planning Dep., Aéroport Frankfurt, Majn, Planovyj Otdel*. Frankfurt am Main: Flughafen Frankfurt Main AG (FAG).

— (1978). *Airconsult*. Frankfurt am Main: Flughafen Frankfurt Main AG (FAG).

— (1990). *Ausbau 2000*. Frankfurt am Main: Flughafen Frankfurt Main AG (FAG).

— (1996). *FRA*. Frankfurt am Main: Flughafen Frankfurt Main AG (FAG).

— (1999). *FRA 2000 Plus*. Frankfurt am Main: Flughafen Frankfurt Main AG (FAG).

FAZ (1996). *Das allgegenwärtige Schreckgespenst der Regionalstadt. Vor 25 Jahren stellte Frankfurts Oberbürgermeister Walter Möller sein Modell für Rhein-Main vor.* Frankfurter Allgemeine Zeitung, Detlev Janik.

— (1997). *18. November 1997*. Frankfurter Allgemeine Zeitung.

— (1998). *5. Juni 1998*. Frankfurter Allgemeine Zeitung.

— (2002a). *"Nach der Planfeststellung folgt der Gang vor die Gerichte." Unverständnis und Kritik in den Anrainerkommunen nach der Stellungnahme des Regierungspräsidiums / Klagen über Schwächen in der Verfahrensführung, 16. Januar 2002*. Frankfurter Allgemeine Zeitung.

— (2002b). *Flughafen: Jetzt sind Städte und Kreise am Zug. Land hofft, daß Regionalversammlung Ausbau nicht verzögert, 13. Juni 2002*. Frankfurter Allgemeine Zeitung, Mechthild Harting.

— (2002c). *Kein bedingungsloses Ja. CDU Bad Vilbel für Nordostvariante des Flughafenausbaus, 14. Januar 2002*. Frankfurter Allgemeine Zeitung.

— (2002d). *Noch keine Festlegung. Kreis nur gegen Einschränkung der Siedlungsentwicklung, 30. Januar 2002*. Frankfurter Allgemeine Zeitung.

— (2002e). *SPD besteht weiter auf Klage. Union will nicht gegen "eigenen Flughafen" vorgehen / Roth prüft, 12. Juni 2002*. Frankfurter Allgemeine Zeitung, Mechthild Harting.

— (2002f). *SPD gegen Flughafenausbau. Bürgermeister und Fraktionsspitze einig / "Nullsummenspiel", 25. Januar 2002*. Frankfurter Allgemeine Zeitung.

— (2002g). *Stopp des Verfahrens gefordert. Fraport-Unterlagen zum Ausbau als unvollständig kritisiert, 17. Januar 2002*. Frankfurter Allgemeine Zeitung.

— (2003). *Fraport sieht sich wegen Ticona-Untersuchungen nicht unter Zeitdruck, 17. April 2003*. Frankfurter Allgemeine Zeitung, jk.

— (2004a). *Auswertung dauert Monate. A 380: Regierungspräsident zufrieden mit der Anhörung, 9. März 2004*. Frankfurter Allgemeine Zeitung, Peter Zitzmann.

— (2004b). *BUND will klagen. Die geplante Werft für den Airbus 380 ist aus Sicht der Naturschützer eindeutig rechtswidrig, 28. November 2004*. Frankfurter Allgemeine Zeitung, lhe.

— (2004c). *Erörterung beginnt mit drei Stunden Verspätung, 16. Januar 2004*. Frankfurter Allgemeine Zeitung, Hanns Mattes.

— (2004d). *Werft für A 380: Klage zu Anhörung, 17. Januar 2004*. Frankfurter Allgemeine Zeitung, lhe.

— (2005a). *A380-Werft darf gebaut werden. Bundesverwaltungsgericht hat keine Bedenken / Koch: Historische Entscheidung für Hessen, 9. September 2005*. Frankfurter Allgemeine Zeitung, Helmut Schwan.

— (2005b). *Am Montag beginnt der Bau der A-380-Werft. Das Bundesverwaltungsgericht macht den Weg für Airbus-Wartungshalle am Frankfurter Flughafen frei, 9. September 2005*. Frankfurter Allgemeine Zeitung, Hans-Christoph Noack.

— (2005c). *Fluglärm stärker thematisiert. Ausbaugegner werfen Fraport Ignoranz vor, 26. November 2005*. Frankfurter Allgemeine Zeitung, Helmut Schwan.

— (2005d). *Halle für Airbus A380 darf gebaut werden. Gericht weist Klagen ab / Neue Arbeitsplätze am Flughafen Frankfurt, 29. Juni 2005*. Frankfurter Allgemeine Zeitung, Helmut Schwan & Hans-Christoph Noack.

— (2005e). *Leserbrief: An einem Arbeitstag, 21. September 2005*. Frankfurter Allgemeine Zeitung, Wolfgang Roblick.

— (2005f). *Leserbrief: Beängstigendes Polizeiaufgebot, 20. September 2005*. Frankfurter Allgemeine Zeitung, Iris Papenfus.

— (2005g). *Offenbach: Fraport soll in einen Dialog treten, 17. September 2005*. Frankfurter Allgemeine Zeitung, Anton Jakob Weinberger.

— (2005h). *Signale für ein Urteil zugunsten der A-380-Werft. Verwaltungsgerichtshof entscheidet am Dienstag, 27. Juni 2005*. Frankfurter Allgemeine Zeitung, Helmut Schwan.

— (2005i). *Vor leeren Stühlen, 18. September 2005*. Frankfurter Allgemeine Zeitung, Anton Jakob Weinberger.

— (2006). *Noch keine Entscheidung über Abbruch des Erörterungstermins. Regierungspräsidium läßt Offenbacher Antrag prüfen, 13. Januar 2006*. Frankfurter Allgemeine Zeitung, Anton Jakob Weinberger & Ralf Euler.

— (2008). *Frankfurt muss gegen Ausbaubeschluss klagen. Stadtverordnete heben Roths Widerspruch auf, 17. Januar 2008*. Frankfurter Allgemeine Zeitung, Helmut Schwan.

— (2010). *Die Waldbürger, 5. Novmember 2012*. Frankfurter Allgemeine Zeitung, Katharina Wagner.

— (2014a). *Neue Dachschäden in Raunheim, 18. Mai 2014*. Frankfurter Allgemeine Zeitung.

— (2014b). *Wirbelschleppen: Größere "Vorsorgezone" möglich, 14. April 2014*. Frankfurter Allgemeine Zeitung.

— (2015). *Fraport will Terminal 3 noch in diesem Jahr bauen, 15. April 2015*. Frankfurter Allgemeine Zeitung, bee./dpa/Reuters.

FDP (1998). *Pressemitteilung, 04. März 1998*. Freie Demokratische Partei, Hessen.

— (2006). *Pressemitteilung: Dieter Posch, Staatsminister a. D.: Natürlich muss das Land Hessen VGH-Entscheidung beachten "Jetzt kommt es auf besonders umsichtiges Handeln an", 12. Januar 2006*. Freie Demokratische Partei, Hessen.

— (2007). *Pressemitteilung: Dieter Posch: "Baurecht für Flughafen ist wichtigster Beschluss und Meilenstein in dieser Legislaturperiode – jedoch kein Grund zum Jubeln für die FDP.", 18. Dezember 2007*. Freie Demokratische Partei, Hessen.

Federwisch, Tobias (2012). *Metropolregion 2.0: Konsequenzen einer neoliberalen Raumentwicklungspolitik*, Vol. 15. Sozialgeographische Bibliothek. Stuttgart: Steiner.

FFR (2009). *Forum Flughafen und Region (FFR). Geschäftsordnung, Stand: 29. Mai 2009*. forum flughafen & region (ffr), Website: http://www.forum-flughafen-region.de – rev. July 3rd, 2015.

— (2014). *Zahlen, Daten Fakten des Flughafens Frankfurt – Entwicklungsgeschichte*. forum flughafen & region (FFR), Website: http://www.forum-flughafen-region.de – rev. December 23rd, 2014.

— (2015). *Über uns*. forum flughafen & region (FFR), Website: http://www.forum-flughafen-region.de – rev. March 6th, 2015.

Fishkin, James S. (2012). "Deliberative Polling: Reflections on an Ideal Made Practical." In: *Evaluating Democratic Innovations*. Ed. by Brigitte Geißel and Kenneth Newton. London & New York: Routledge, pp. 69–89.

Flick, Uwe (2011). *Qualitative Sozialforschung. Ein Einführung*. 5th ed. Reinbek bei Hamburg: Rowohlt Taschenbuch Verlag.

FMG (2014). *Zahlen, Daten, Perspektiven. Konzern-Jahresabschluss 2013*. München: Flughafen München GmbH (FMG).

FNP (2002a). *Flughafen-Ausbau nimmt erste Verfahrenshürde, 11. Juni 2002*. Frankfurter Neue Presse, dpa.

— (2002b). *Flughafenausbau I: Nicht nur die SPD ist in Argumentationsnöten, 28. Januar 2002*. Frankfurter Neue Presse, Joachim Geiger.

— (2002c). *Initiativen: Flughafen-Beschluss "Schlechter Tag für Südhessen", 11. Juni 2002*. Frankfurter Neue Presse, dpa.

— (2002d). *Viele Zweifel an Verfahren zum Landebahn-Bau, 28. Januar 2002*. Frankfurter Neue Presse, Georg Haupt.

— (2003a). *Fraport erwartet keine Überraschungen nach Flughafen-Anhörung, 12. April 2003*. Frankfurter Neue Presse, Christoph Barkewitz.

— (2003b). *Scoping-Termin stößt auf Kritik. Fraport AG ignoriert die Sicherheit, 12. April 2003*. Frankfurter Neue Presse, Bernd Diefenbach.

— (2004a). *Ausbaugegner wollen die Tagesordnung selbst bestimmen, 7. Februar 2004*. Frankfurter Neue Presse, val.

— (2004b). *Kreis bezweifelt Nutzen der A380-Werft, 5. Oktober 2004*. Frankfurter Neue Presse, Bernd Diefenbach.

— (2004c). *Kreis lehnt die neuen Pläne für A380-Halle ab, 4. Oktober 2004*. Frankfurter Neue Presse.

— (2004d). *SDW kritisiert Bau der A380-Halle, 2. Dezember 2004*. Frankfurter Neue Presse, meh.

— (2004e). *Streit um Flughafen nimmt zu. Gutachter für Umbau bei Ticona, 16. Januar 2004*. Frankfurter Neue Presse.

— (2004f). *Verfahren kostet zwei Millionen. Stadt moniert die Mängel, 9. März 2004*. Frankfurter Neue Presse, Bernd Diefenbach.

— (2004g). ""Wartungshalle darf nicht gebaut werden", 6. März 2004." Frankfurter Neue Presse, meh.

— (2004h). "Widema kritisiert Anhörung. "Fiasko für den Veranstalter", 20. Januar 2004." Frankfurter Neue Presse, Manfred Becht.

— (2004i). "Zankapfel Flughafen Anhörung zu A380 begann mit Tumult, 16. Januar 2004." Frankfurter Neue Presse, Bernd Diefenbach.

— (2005a). "... und die Anwälte werfen Fraport Verfahrensfehler vor, 14. September 2005." Frankfurter Neue Presse, dib.

— (2005b). "5 Klagen gegen A380-Halle – Ausbau-Akten liegen aus, 7. Januar 2005." Frankfurter Neue Presse.

— (2005c). "A380-Halle: Erste Bäume sind gefällt, 13. September 2005." Frankfurter Neue Presse.

— (2005d). *Anhörung ohne Bürger. Gegner des Flughafenausbaus kritisieren das Erörterungsverfahren in Offenbach, 19. Oktober 2005.* Frankfurter Neue Presse, Ud.

— (2005e). *Auch wenig Lärm macht krank. Sachsenhäuser Bürgerinitiative bezweifelt Fraport-Gutachten zum Ausbau, 25. November 2005.* Frankfurter Neue Presse, Patricia C. Borna.

— (2005f). *Aufruf zur Teilnahme am Anhörungstermin. Flughafenausbau: CDU appelliert an Betroffene, 28. Oktober 2005.* Frankfurter Neue Presse, Bernd Diefenbach.

— (2005g). *Ausbau-Gegner von Resonanz enttäuscht. 500 demonstrieren zu Beginn des Erörterungstermins in Offenbach / Viele Stühle leer, 13. September 2005.* Frankfurter Neue Presse, fs.

— (2005h). *BUND: Das Kasseler Urteil zur A 380-Werft ist ein Skandal, 9. September 2005.* Frankfurter Neue Presse, dib.

— (2005i). *Bundesverwaltungsgericht lehnt Klagen ab – Montag Baubeginn. A380-Halle kommt, 9. September 2005.* Frankfurter Neue Presse.

— (2005j). *Erörterung mit Mängeln, 24. September 2005.* Frankfurter Neue Presse, dib.

— (2005k). *Gericht statt Demonstrationen. Der Widerstand gegen den Ausbau des Frankfurter Flughafens hat sich verändert, 22. Oktober 2005.* Frankfurter Neue Presse, Bernd Diefenbach.

— (2005l). *Quilling warnt vor Desaster der Regionalplanung, 17. September 2005.* Frankfurter Neue Presse, dib.

FNP (2005m). *Rechtsanwalt begründet die Klage gegen den Bau der A-380-Wartungshalle. Es geht um den tatsächlichen Lärm, 25. Februar 2005.* Frankfurter Neue Presse, Ud.

— (2006). *"Kleinkrieg" um Flughafen-Akten, 14. Januar 2006*. Frankfurter Neue Presse, Anne-Katrin Einfeldt.

— (2008a). *Stadt tendiert in Richtung Klage, 24. Januar 2008*. Frankfurter Neue Presse, dib.

— (2008b). *Stadtparlament will mit Klage absolutes Nachtflugverbot durchsetzen. Hier stimmen sie gegen Roth, 17. Januar 2008*. Frankfurter Neue Presse, mu.

— (2013a). *Das Theater geht weiter, 3. Juli 2013*. Frankfurter Neue Presse, Sascha Kröner.

— (2013b). *Wirbelschleppe im Anflug, 9. April 2013*. Frankfurter Neue Presse, Sascha Kröner.

Fossum, John Erik (2012). "Reflections on Experimentalist Governance." In: *Regulation & Governance* 6.III, pp. 394–400.

FR (2000). *31. März 2000*. Frankfurter Rundschau.

— (2002a). *Ausbaugegner lassen kein gutes Haar an der Stellungnahme des RP. Plädoyer für die Nordwestbahn provoziert Kommentare wie Rechtsbruch, Affront, Ziel verfehlt / Fraport ist zufrieden, 12. Juni 2002*. Frankfurter Rundschau.

— (2002b). *Die Botschaft lautet: So nicht. Kommunen fordern Stopp des Raumordnungsverfahrens / Einwendungen und Klagen, 26. Januar 2002*. Frankfurter Rundschau, Peter Hanack.

— (2002c). *Einig nur in der Kritik am Lufthansa-Chef. Landtag debattiert Flughafen-Ausbau: Bökel verlangt Konkretes zum Thema Nachtflugverbot, 14. Juni 2002*. Frankfurter Rundschau.

— (2002d). *Grundsätzlich für den Ausbau des Flughafens. Mehrheit im Bad Vilbeler Stadtparlament stimmte zu / SPD fordert ein Nachflugverbot, 31. Januar 2002*. Frankfurter Rundschau.

— (2002e). *RP soll Akten an Fraport zurückgeben, 31. Januar 2002*. Frankfurter Rundschau.

— (2002f). *SPD-Kommunen gegen Ausbau, 25. Januar 2002*. Frankfurter Rundschau.

— (2002g). *Von Lügnern und Landtagskandidaten. In der Sondersitzung der Stadtverordnetenversammlung zum Flughafenausbau gerät die SPD unter Druck, 28. Januar 2002*. Frankfurter Rundschau, Claus-Jürgen Göpfert.

— (2003a). *Bedenken der Flughafengegner nicht zerstreut. "Augenwischerei" kontra "konstruktive Beratung": Nach einer Woche Scoping zum Ausbau gehen die Meinungen auseinander, 15. April 2002.* Frankfurter Rundschau, Stephan Börnecke.

— (2003b). *Die Stadt soll gegen den Flughafenausbau noch stärker Flagge zeigen, 17. April 2002.* Frankfurter Rundschau, Walter Keber.

— (2003c). *Siehr und Gieltowski üben heftig Kritik an der Fraport, 15. April 2002.* Frankfurter Rundschau, cas.

— (2004a). *Anwalt klagt Freiheit in der Berufsausübung ein, 25. Februar 2004.* Frankfurter Rundschau, ke & Jutta Rippegather.

— (2004b). *Barsche Töne im Streit um Stellungnahme zu A 380, 17. Februar 2004.* Frankfurter Rundschau, Peter Hanack.

— (2004c). *Eine Region im Widerstand, 12. April 2004.* Frankfurter Rundschau, Arne Löffel.

— (2004d). *In Schwarz zum Start des ersten Flugzeugs, 13. April 2004.* Frankfurter Rundschau, Georg Leppert.

— (2004e). *Kirchenvorstand will Widerstand unterstützen, 16. Dezember 2004.* Frankfurter Rundschau, Arne Löffel.

— (2004f). *Protest gegen Werft: "Wir sind das Volk". Funk und Fernsehen werden bei Erörterung ausgeschlossen, 16. Januar 2004.* Frankfurter Rundschau, Jürgen Gelis.

— (2005a). *A 380-Werft kann gebaut werden. Bundesverwaltungsgericht gibt grünes Licht für Airbus-Werft in Frankfurt, 8. September 2005.* Frankfurter Rundschau, Jörg Reckmann.

— (2005b). *Angst vor Fliegern. Beim Anhörungsmarathon zum Ausbau des Frankfurter Flughafens geht es nicht nur um Zahlen und Verfahrenstricks, 24. September 2005.* Frankfurter Rundschau, Stephan Börnecke.

— (2005c). *Anwälte der Städte nennen Fraport-Unterlagen unvollständig. Erneut Forderung nach Abbruch des Verfahrens für den Ausbau des Frankfurter Flughafens / 150 Menschen bei Erörterungstermin, 14. September 2005.* Frankfurter Rundschau, Stephan Börnecke.

— (2005d). "Ausbaugegner sind entrüstet. Kritik an Verfahrensablauf, 14. Oktober 2005." Frankfurter Rundschau, Stephan Börnecke.

— (2005e). *Ausbaugegner verlangen Abbruch der Erörterung. Antrag für Montag angekündigt / "Versuch, die Region auseinander zu dividieren"*

*/ Zaun um Baustelle der A-380-Werft steht bereits, 10. September 2005.* Frankfurter Rundschau, Stephan Börnecke.

— (2005f). "BUND: Lüge bei Gutachten. Kritik an Fluggastprognose, 6. Dezember 2005." Frankfurter Rundschau, Stephan Börnecke.

— (2005g). *EU-Kritik an Airbus-Werft. Aber keine Bedenken gegen Flughafen-Ausbau / Koch in Brüssel, 2. Juli 2005.* Frankfurter Rundschau, Jörg Reckmann.

— (2005h). *Experten rügen Lärmgutachten. Bei der Erörterung des Frankfurter Flughafenausbaus schafft ein Auftritt von Fraport-Chef Bender Irritationen, 19. November 2005.* Frankfurter Rundschau, Stephan Börnecke.

— (2005i). *Frankfurter Ausbaugegner kommen schon früh zu Wort. Erörterung zur Erweiterung des Flughafens beginnt am 12. September / 60 Bürger werden persönlich vorsprechen, 5. August 2005.* Frankfurter Rundschau, Ben Reichardt & Wolgang Schubert.

— (2005j). *Fraport hält Kritik am Parkhaus für unberechtigt, 15. November 2005.* Frankfurter Rundschau, mae.

— (2005k). *Initiative kritisiert fehlende Ausbaugegner, 15. September 2005.* Frankfurter Rundschau, ran.

— (2005l). *Land gegen Abbruch des Genehmigungsverfahrens, 26. September 2005.* Frankfurter Rundschau, Stephan Börnecke.

— (2005m). *Misstrauisch bei der Zahl der Flüge. In der Offenbacher Anhörung über den Ausbau des Frankfurter Flughafens geht es hoch her, aber mit Worten, 13. September 2005.* Frankfurter Rundschau, Wolfgang Schubert & Stephan Börnecke.

— (2005n). *Nervosität im Bannwald. In Offenbach beginnt die Anhörung zum Ausbau des Frankfurter Flughafens, geichzeitig fallen die ersten Bäume, 12. September 2005.* Frankfurter Rundschau, Stephan Börnecke.

— (2005o). *Rodung für A-380-Werft beginnt. 350 Ausbaugegner demonstrieren im Hessen-Forst / Offenbachs OB fordert Abbruch des Erörterungstermins, 13. September 2005.* Frankfurter Rundschau, Stephan Börnecke & Wolfgang Schubert & Andrea Rost.

— (2006). *Anwälte verlangen Aufschub im Erörterungstermin, 7. Januar 2006.* Frankfurter Rundschau, Stephan Börnecke.

— (2007). *Grüne sehen Anlass zur Klage, 21. Dezember 2007*. Frankfurter Rundschau, Jürgen Schultheis.

— (2008a). *Auch Darmstadt erwägt Klage gegen Ausbau, 24. Januar 2008*. Frankfurter Rundschau, dpa & lhe.

— (2008b). *Lufthansa will nachts fliegen, 8. Februar 2008*. Frankfurter Rundschau, dpa & rtr.

— (2008c). *Stadt klagt gegen Beschluss, 17. Januar 2008*. Frankfurter Rundschau, Jürgen Schultheis.

— (2008d). *TUI, Condor und Post klagen gegen Ausbaupläne, 7. Februar 2008*. Frankfurter Rundschau, dpa & lhe.

— (2011). *Vor 30 Jahren fiel das Hüttendorf, 2. November 2011*. Frankfurter Rundschau, Oliver Heil.

— (2013a). *"Wir kämpfen auf jeden Fall weiter", 28. Juni 2013*. Frankfurter Rundschau, Andrea Rost.

— (2013b). *Flörsheim zieht vor Gericht, 13. Juni 2013*. Frankfurter Rundschau, dpa.

— (2013c). *Flugzeug wirbelt Dachziegel bei Landeanflug herunter, 3. März 2013*. Frankfurter Rundschau, dpa/lha.

— (2013d). *Weiter Wirbel um Dachschäden, 10. Mai 2013*. Frankfurter Rundschau, Andrea Rost.

— (2014a). *Vermutlich waren es Wirbel, 19. Mai 2014*. Frankfurter Rundschau, fr.

— (2014b). *Wirbelschleppen: Al-Wazir will Sicherungszonen ausweiten, 19. Mai 2014*. Frankfurter Rundschau, dpa/lhe.

— (2015). *Terminal 3 geht 2022 in Betrieb, 15. April 2015*. Frankfurter Rundschau, Jutta Rippegather.

FRA (2014). *Daten & Fakten. Entwicklungsgeschichte*. Frankfurt Airport (FRA), Website: http://www.frankfurt-airport.de – rev. December 28th, 2014.

Frankfurt (1971). *Ein Beitrag zur Diskussion: Regionalstadt Frankfurt*. Frankfurt am Main: Wurm.

— (2015a). *Beteiligungsportfolio der Stadt Frankfurt am Main, Stand 31.12.2013*. City of Frankfurt am Main, Website: http://www.frankfurt.de – rev. January 5th, 2015.

— (2015b). *Leben in Frankfurt. Umwelt. Nachbar Flughafen.* City of Frankfurt am Main, Website: http://www.frankfurt.de – rev. January 5th, 2015.

— (2015c). *Leben in Frankfurt. Umwelt. Nachbar Flughafen. Flughafenausbau.* City of Frankfurt am Main, Website: http://www.frankfurt.de – rev. 5th January 2015.

Frantz, Christiane and Kerstin Martens (2006). *Nichtregierungsorganisationen (NGOs).* Wiesbaden: VS Verlag für Sozialwissenschaften.

Fraport (2001). *Ausbauprogramm Flughafen Frankfurt/Main. Unterlagen zum Raumordnungsverfahren, Ordner 1 bis 16.* Frankfurt am Main: Fraport AG. Available at http://www.dfld.de – rev. June 26th, 2015. Frankfurt am Main: Fraport AG.

— (2002). *Pressemitteilung: Fraport AG begrüßt ROV-Abschluss, 11. Juni 2002.* Fraport AG.

— (2003a). *A380-Werft Flughafen Frankfurt Main. Unterlagen zum Planfeststellungsverfahren, Ordner 1 bis 7.* Frankfurt am Main: Fraport AG. Available at http://www.dfld.de – rev. June 26th, 2015.

— (2003b). *Pressemitteilung: Flughafen-Ausbau: Voraussichtlich letzter Scoping-Tag in Sossenheim, 11. April 2003.* Fraport AG.

— (2004a). *A380-Werft Flughafen Frankfurt Main. Unterlagen zum Planfeststellungsverfahren. Planänderung (nur geänderte und neue Unterlagen), Ordner 1 bis 7.* Frankfurt am Main: Fraport AG. Available at http://www.dfld.de – rev. June 26th, 2015. Frankfurt am Main: Fraport AG.

— (2004b). *Ausbau Flughafen Frankfurt Main. Unterlagen zum Planfeststellungsverfahren, Version 1, Ordner 1 bis 60.* Available at http://www.dfld.de – rev. June 29th, 2015. Frankfurt am Main: Fraport AG.

— (2004c). *Pressemitteilung: Erörterungstermin zur A380-Werft am Frankfurter Flughafen beendet, 5. März 2004 (33/04).* Fraport AG, jh.

— (2004d). "Pressemitteilung: Fraport AG und BUND im Gespräch, 23. November 2004 (149/04)." Fraport AG, bu.

— (2004e). *Pressemitteilung: Fraport: Umweltfreundlichere Lösung für A380-Werft. Schölch: Optimierte Planung verringert Eingriffe in den Wald, 15. Juli 2004 (91/04).* Fraport AG, dw.

— (2004f). *Pressemitteilung: Vorwürfe von Landrat Siehr unberechtigt, 14. November 2005 (153/05)*. Fraport AG.

— (2007a). *Ausbau Flughafen Frankfurt Main. Unterlagen zum Planfeststellungsverfahren, Version 2, Ordner 1 bis 65*. Frankfurt am Main: Fraport AG. Available at http://www.dfld.de – rev. June 29th, 2015. Frankfurt am Main: Fraport AG.

— (2007b). *Bender: "Historischer Tag für Frankfurter Flughafen". Fraport-Chef: Meilenstein auf dem Weg zum Flughafenausbau, 18. Dezember 2007 (139/07)*. Fraport AG.

— (2011). *CargoCity Frankfurt – Task Force Entwicklungen und Projekte*. Frankfurt am Main: Fraport AG.

— (2013). *Vorsorgeprogramm: Sicherung von Dacheindeckungen. Informationsbroschüre für Anwohner*. Frankfurt am Main: Fraport AG Frankfurt Airport Services Worldwide.

— (2014). *Zahlen, Daten, Fakten 2014 zum Flughafen Frankfurt*. Frankfurt am Main: Fraport AG Frankfurt Airport Services Worldwide.

— (2015a). "Flughäfen." Fraport AG, Website: http://www.fraport.de – rev. January 5th, 2015.

Freeman, Richard (2008). "Learning in Public Policy." In: *The Oxford Handbook of Public Policy*. Ed. by Robert E. Goodin, Michael Moran, and Martin Rein. Oxford: Oxford University Press, pp. 376–388.

Freund, Bodo (2003). "The Frankfurt Rhine-Main Region." In: *Metropolitan Governance and Spatial Planning. Comparative Case Studies of European City-Regions*. Ed. by Willem Salet, Andy Thornley, and Anton Kreukels. London & New York: Spon Press, pp. 125–144.

Froschauer, Ulrike and Manfred Lueger (2009). "ExpertInnengespräche in der interpretativen Organisationsforschung." In: *Experteninterviews. Theorien, Methoden, Anwendungsfelder*. Ed. by Alexander Bogner, Beate Littig, and Wolfgang Menz. 3rd ed. Wiesbaden: VS Verlag für Sozialwissenschaften, pp. 239–258.

Früh, Werner (2011). *Inhaltsanalyse. Theorie und Praxis*. 7th ed. Konstanz: UTB.

Fukuyama, Francis (2000). *The Great Disruption: Human Nature and the Reconstitution of Social Order*. New York etc.: Touchstone.

— (2006). *The End of History and the Last Man*. New York etc.: Free Press Paperback.

Fürst, Dietrich (2007). "Regional Governance." In: *Handbuch Governance. Theoretische Grundlagen und empirische Anwendungsfelder*. Ed. by Arthur Benz, Susanne Lütz, and Uwe Schimank. Wiesbaden: VS Verlag für Sozialwissenschaften, pp. 352–365.

— (2010). "Regional Governance." In: *Governance – Regieren in komplexen Regelsystemen. Eine Einführung*. Ed. by Arthur Benz and Nicolai Dose. 2nd ed. Wiesbaden: VS Verlag für Sozialwissenschaften, pp. 49–68.

Fürst, Dietrich and Jörg Knieling (2002). *Regional Governance. New Modes of Self-Government in the European Community. EuroConference 19.-21. April 2001 in Hannover*. Hannover: Verlag der ARL.

Ganser, Karl (2005). "Die Region die Stadt der Zukunft – Perspektiven für die Regionalisierung." In: *Zukunft von Stadt und Region. Perspektiven der Regionalisierung,* , volume II of *Beiträge zum Forschungsverbund*. Ed. by Deutsches Institut für Urbanistik. Vol. II. Beiträge zum Forschungsver-bund "Stadt 2030" Wiesbaden: VS Verlag für Sozialwissenschaften, pp. 19–26.

Gehrlein, Ulrich (2009). *Perspektiven und Potentiale metropolnaher Regionen*. Vol. 6. Studien zur politischen Gesellschaft. Frankfurt am Main: Institut für Ländliche Strukturforschung (IfLS).

Geis, Anna (2002). "Neue Konflikte durch kooperative Politikformen. Das Mediationsverfahren und das Regionale Dialogforum zur zukünftigen Entwicklung des Frankfurter Flughafens." In: *Kommunale Entscheidungsprozesse im Wandel. Theoretische und empirische Analysen*. Ed. by Jörg Bogumil. Opladen: Leske + Budrich, pp. 267–284.

— (2005). *Regieren mit Mediation. Das Beteiligungsverfahren zur zukünftigen Entwicklung des Frankfurter Flughafens*. Vol. 6. Studien zur politischen Gesellschaft. Wiesbaden: VS Verlag für Sozialwissenschaften.

Geißel, Brigitte (2007). "Zur (Un-)Möglichkeit von Local Governance mit Zivilgesellschaft: Konzepte und empirische Befunde." In: *Local Governance – mehr Transparenz und Bürgernähe*. Ed. by Lilian Schwalb and Heike Walk. Wiesbaden: VS Verlag für Sozialwissenschaften, pp. 23–38.

— (2008). "Zur Evalution demokratischer Innovationen – die lokale Ebene." In: *Lokale Politikforschung heute*. Ed. by Hubert Heinelt and Angelika Vetter. Wiesbaden: VS Verlag für Sozialwissenschaften, pp. 227–248.

— (2009). "Kritische Bürger – Gefahr oder Ressource für die Leistungsfähigkeit eines politischen Systems?" In: *Gesellschaft – Wirtschaft – Politik* 2009.III, pp. 387–396.

— (2011a). "Kritische Bürger und demokratische Innovationen - Aktuelle Herausforderungen lokaler Beteiligung." In: *Raum-Planung, Fachzeitschrift für räumliche Planung und Forschung* 2011.CLVI/CLVII.

— (2011b). "Partizipative Innovationen auf lokaler Ebene – Potentiale oder Gefahren für inklusive, responsive, "accountable" Repräsentation?" In: *Krise und Reform politischer Repräsentation*. Ed. by Markus Linden and Winfried Thaa. Baden-Baden: Nomos, pp. 195–216.

— (2012a). "Democratic Innovations. Theoretical and Empirical Challanges of Evaluation." In: *Evaluating Democratic Innovations. Curing the Democratic Malaise?* Ed. by Brigitte Geißel and Kenneth Newton. London & New York: Routledge, pp. 209–214.

— (2012b). "Impacts of democratic innovation in Europe." In: *Evaluating Democratic Innovations. Curing the democratic malaise?* Ed. by Brigitte Geißel and Kenneth Newton. London & New York: Routledge, pp. 163–183.

— (2013). "Introduction: On the Evaluation of Participatory Innovations – A Preliminary Framework." In: *Participatory Democratic Innovations in Europe. Improving the Quality of Democracy?* Ed. by Brigitte Geißel and Marko Joas. Berlin & Toronto: Barbara Budrich Publishers & Opladen, pp. 8–31.

Geißel, Brigitte and Marko Joas (2013). *Participatory Democratic Innovations in Europe. Improving the Quality of Democracy?* Opladen etc.: Barbara Budrich Publishers.

Geißel, Brigitte and Kenneth Newton (2012). *Evaluating Democratic Innovations. Curing the Democratic Malaise?* London & New York: Routledge.

Geißel, Brigitte, Roland Roth, et al. (2014a). "Partizipation und Demokratie im Wandel – Wie verändert sich unsere Demokratie durch neue Kombinationen repräsentativer, deliberativer und direktdemokratischer Elemente?" In: *Partizipation im Wandel. Unsere Demokratie zwischen Wählern, Mitmachern und Entscheidern*. Ed. by Bertelsmann Stiftung and Staatsministerium Baden-Wüttemberg. Gütersloh: Verlag Bertelsmann Stiftung, pp. 11–39.

— (2014b). "Partizipation und Demokratie im Wandel: Wie unsere Demokratie durch neue Verfahren und Kombinationen repräsentativer, deliberativer und direkt Bteiligung gestärkt wird." In: *Partizipation im Wandel. Unsere Demokratie zwischen Wählern, Mitmachern und Entscheidern*. Ed. by Bertelsmann Stiftung and Staatsministerium Baden-Wüttemberg. Gütersloh: Verlag Bertelsmann Stiftung, pp. 489–503.

Gering, Hein K. (1981). *Brief an den Vorsitzenden des Haushaltsausschusses des Hessischen Landtag, 28. Januar 1981.*

Giddens, Anthony (2012). *The Consequences of Modernity*. 20th ed. Cambridge & Malden: Polity Press.

Gilardi, Fabizio (2010). "Who Learns from What in Policy Diffusion Processes?" In: *American Journal of Political Science* 54.III, pp. 650–666.

Gilardi, Fabrizio and Claudio M. Radaelli (2012). "Governance and Learning." In: *Oxford Handbook of Governance*. Ed. by David Levi-Faur. Oxford & New York: Oxford University Press, pp. 155–168.

GKF (1998). *Mediation – Eine Zukunftsregion im offenen Dialog. Vereinbarungen über das weitere Vorfahren mit der Diskussion um den Flughafen Frankfurt*. Wiesbaden: Gesprächskreis Flughafen (GKF).

Glaser, Barney G. and Anselm L. Strauss (2005). *Grounded Theory. Strategien qualitativer Forschung*. 2nd ed. Bern: Hans Huber.

Gläser, Jochen and Grit Laudel (2009). *Experteninterviews und qualitative Inhaltsanalyse als Instrumente rekonstruierender Untersuchungen*. 3rd ed. Wiesbaden: VS Verlag für Sozialwissenschaften.

Glassmann, Ulrich and Helmut Voelzkow (2003). "The Governance of Local Economies in Germany." In: *Local Production Systems in Europe. Rise or Demise?* Ed. by Colin Crouch et al. Reprint. Oxford etc.: Oxford University Press, pp. 79–116.

Glückler, Johannes (2000). "Netzwerke, Lernen und evolutionäre Regionalentwicklung." In: *Zeitschrift für Wirtschaftsgeographie* 44.III/IV, pp. 167–182.

Gohl, Christopher and Hans-Peter Meister (2012). *Politische Mediation bei Infrastrukturprojekten. Das Beispiel des Ausbaus des Flughafens Frankfurt 1998 - 2008*. Vol. 70. Region – Nation – Europa. Münster etc.: LIT Verlag.

Gourevitch, Peter Alexis and James J. Shinn (2005). *Political Power and Corporate Control. The New Global Politics of Corporate Governance*. Princton & Oxford: Princeton University Press.

Grabher, Gernot (1994a). *Lob der Verschwendung. Redundanz in der Regionalen-twicklung: Ein sozioökonomisches Plädoyer*. Berlin: edition sigma.

— (1994b). "Rediscovering the Social in the Economic Interfirm Relations." In: *The Embedded Firm. On the Socioeconomics of Industrial Networks*. Ed. by Gernot Grabher. Reprint. London: Routledge, pp. 1–31.

— (1994c). *The Embedded Firm. On the Socioeconomics of Industrial Networks*. Reprint. London: Routledge.

Grande, Edgar (2009). "Perspektiven der Governance-Forschung: Grundzüge des Forschungsprogramms des Münchner Centrums für Governance-Forschung." In: *Perspektiven der Governance-Forschung*. Ed. by Edgar Grande and Stefan May. Baden-Baden: Nomos, pp. 77–89.

Granovetter, Mark S. (1973). "The Strength of Weak Ties." In: *The American Journal of Sociology* 78.VI, pp. 1360–1380.

— (1983). "The Strength of Weak Ties: A Network Theory Revisited." In: *Sociological Theory* 1, pp. 201–233.

— (1985). "Economic Action and Social Structure: The Problem of Embeddedness." In: *The American Journal of Sociology* 91.III, pp. 481–510.

— (1988). "The Sociological and Economic Approaches to Labor Market Analysis. A Social Structural View." In: *Industries, Firms, and Jobs. Sociological and Economic Approaches*. Ed. by Georges Farkas and Paula England. New York: Plenum Press, pp. 187–216.

— (1990a). "The Contenders: Mark Granovovetter, Interview from 12 May 1988, Port Jefferson, New York." In: *Economics and Sociology. Redefining Their Boundaries: Conversations With Economists and Sociologists*. Ed. by Richard Swedberg. Prinston: Prinston University Press, pp. 96–114.

— (1990b). "The Myth of Social Network Analysis as a Special Method in Social Science." In: *Connections* 13.I-II, pp. 13–16.

— (1990c). "The old and new economic sociology. A History and an Agenda." In: *Beyond the marketplace. Rethinking economy and society*. Ed. by Roger Friedland and A. F. Robertson. New York: De Gruyter, pp. 89–112.

— (1992). "Economic Institutions as Social Constructions: A Framework for Analysis." In: *Acta Sociologica* 35.I, pp. 3–11.

— (1995a). "The Economic Sociology of Firms and Entrepreneurs." In: *The Economic Sociology of Immigration*. Ed. by Alejandro Portes. New York: Russell Sage Foundation, pp. 128–165.

— (1995b). *Getting a Job. A Study of Contacts and Careers*. 2nd ed. Chicago & London: The University of Chicago Press.

— (2002). "A Theoretical Agenda for Economic Sociology." In: *The New Economic Sociology. Developments in an Emerging Field*. Ed. by Mauro Guillén et al. New York: Russell Sage Foundation, pp. 35–60.

— (2005). "The Impact of Social Structure on Economic Outcomes." In: *The Journal of Economic Perspectives* 19.I, pp. 33–50.

Grin, John and Anne Loeber (2007). "Theories of Policy Learning: Agency, Strucutre and Change." In: *Handbook od Public Policy Analysis. Theory, Politcs, and Methods*. Ed. by Frank Fischer and Sidney Mara S. Miller, Gerald J. Public Administration and Public Policy 125. Boca Raton: CRC Press, pp. 201–219.

GRÜNE (1998). *Pressemitteilung 35/1998*. BÜNDNIS 90/DIE GRÜNEN, Hessen.

— (2004). *Pressemitteilung, 19. Juli 2004*. BÜNDNIS 90/DIE GRÜNEN, Hessen.

— (2005). *Pressemitteilung: Revision dringend erforderlich, 28. Juni 2005*. BÜNDNIS 90/DIE GRÜNEN, Hessen.

— (2006). *Pressemitteilung: Rechtswidriges Verfahren beenden, 12. Januar 2005*. BÜNDNIS 90/DIE GRÜNEN, Hessen.

— (2007). *Pressemitteilung: Planfeststellungsbeschluss zum Flughafen. Wortbruch vollendet, 18. Dezember 2007*. BÜNDNIS 90/DIE GRÜNEN, Hessen.

Gschwend, Thomas and Frank Schimmelfennig (2007). "Forschungsdesign in der Politikwissenschaft: Ein Dialog zwischen Theorie und Daten." In: *Forschungsdesign in der Politikwissenschaft. Probleme – Strategien – Anwendungen*. Ed. by Thomas Gschwend and Frank

Schimmelfennig. Frankfurt am Main & New York: Campus Verlag, pp. 13–35.

Habermas, Jürgen (1981). *Theorie des kommunikativen Handelns, Band 2*. Frankfurt am Main: Suhrkamp Verlag.

— (2001). *Faktizität und Geltung: Beiträge zur Diskurstheorie des Rechts und des demokratischen Rechtsstaats*. 2nd ed. Frankfurt am Main: Suhrkamp Verlag.

Haft, Fritjof (2009). “Verhandlung und Mediation.” In: *Handbuch Mediation*. Ed. by Fritjof Haft and Katharina von Schlieffen. 2nd ed. München: C.H. Beck, pp. 69–79.

Haft, Fritjof and Katharina von Schlieffen (2009). *Handbuch Mediation*. 2nd ed. München: C.H. Beck.

Hahn, Isabelle (2012). *Governance-Mix in Wirtschaftsclustern: Eine Untersuchung von Governance-Formen am Beispiel verschiedener Clusterinitiativen*. Frankfurt am Main: Hochschulschrift: Goethe-Universität Frankfurt.

Haller, Gret (2002). *Die Grenzen der Solidarität. Europa und die USA im Umgang mit Staat, Nation und Religion*. 2nd ed. Berlin: Aufbau-Verlag.

Handwerkstag, Hessischer (1981). “Brief des Hessischen Handwerkstags an den Vorsitzenden des Haushaltsausschusses im Hessischen Landtag Herrn Dr. Lang, 21. Januar 1981.”

Hänsch, Klaus, Frank Niethammer, and Kurt Oeser (2000). *Bericht. Mediation Flughafen Frankfurt/Main*. Frankfurt am Main: Mediationsgruppe Flughafen Frankfurt/Main.

Harrison, Bennett (1992). “Industrial Districts: Old Wine in New Bottles?” In: *Regional Studies* 26.V, pp. 469–483.

Haselberg, Kl. von (1981). *Zu Protokoll gegebener Schriftsatz nebst Anlage zum medizinischen Gutachen zum Ausbau des Rhein-Main-Flughafens von Prof. Dr. med. D. phil. Gerd Jansen, 5. Februar 1981*.

Hasse, Raimund and Georg Krücken (2005). *Neo-Institutionalismus. Mit einem Vorwort von John Meyer*. 2nd ed. Bielefeld: Transcript Verlag.

Haus, Michael (2010). “Von government zu governance? Bürgergesellschaft und Engagementpolitik im Kontext neuer Formen des Regierens.” In: *Engagementpolitik. Die Entwicklung der Zivilgesellschaft als politische Aufgabe*. Ed. by Thomas Olk,

Ansgar Klein, and Birger Hardnuß. Wiesbaden: VS Verlag für Sozialwissenschaften, pp. 210–232.

Häußermann, Hartmut (1977). *Die Politik der Bürokratie. Einführung in die Soziologie der staatlichen Verwaltung*. Frankfurt am Main & New York: Campus Verlag.

Heidenreich, Martin (2012). "State of the Art. The Social Embeddedness of Multinational Companies: A Literatur Review." In: *Socio-Economic Review* 10.III, pp. 549–579.

Héritier, Adrienne (2002). "Introduction." In: *Common Goods. Reinventing European and International Governance*. Ed. by Adrienne Héritier. Governance in Europe. Lanham etc.: Rowman & Littlefield Publishers, pp. 1–12.

Héritier, Adrienne and Dirk Lehmkuhl (2008). "The Shadow of Hierarchy and New Modes of Governance." In: *Journal of Public Policy* 28.I, pp. 1–17.

Héritier, Adrienne and Martin Rhodes (2011). *New Modes of Governance in Europe: Governing in the Shadow of Hierarchy*. Basingstoke: Palgrave Macmillan.

Hermann, Mario, Tobias Pentek, and Boris Otto (2015). "Design Principles for Industrie 4.0 Scenarios: A Literature Review." In: *Working Papers* 2015.I, pp. 3–16.

HessL (2000a). *Plenarprotokoll 15/44. 15. Wahlperiode, 44. Sitzung, Wiesbaden, den 21. Juni 2000*. Wiesbaden: Hessischer Landtag (HessL).

— (2000b). *Protokoll. Anhörung der kommunalen Gebietskörperschaften zum Frankfurter Flughafen am 5. Juni 2000, Teil 1: Anhörung*. Wiesbaden: Hessischer Landtag (HessL).

— (2000c). *Protokoll. Anhörung zum Frankfurter Flughafen 1. Tag, 10. Mai 2000, Teil 1*. Wiesbaden: Hessischer Landtag (HessL).

— (2000d). *Protokoll. Anhörung zum Frankfurter Flughafen 1. Tag, 10. Mai 2000, Teil 2*. Wiesbaden: Hessischer Landtag (HessL).

— (2000e). *Protokoll. Anhörung zum Frankfurter Flughafen 2. Tag, 11. Mai 2000, Teil 1*. Wiesbaden: Hessischer Landtag (HessL).

— (2000f). *Protokoll. Anhörung zum Frankfurter Flughafen 2. Tag, 11. Mai 2000, Teil 2*. Wiesbaden: Hessischer Landtag (HessL).

— (2000g). *Protokoll. Anhörung zum Frankfurter Flughafen 3. Tag, 12. Mai 2000, Teil 1*. Wiesbaden: Hessischer Landtag (HessL).

— (2000h). *Protokoll. Anhörung zum Frankfurter Flughafen 3. Tag, 12. Mai 2000, Teil 2*. Wiesbaden: Hessischer Landtag (HessL).

— (2013). *Kleine Anfrage des Abg. Frank Kaufmann (BÜNDNIS 90/DIE GRÜNEN) vom 13.02.2013 betreffend Umsetzung des Urteils des Bundesverwaltungsgerichts zum Flughafenausbau in Frankfurt vom 04.04.2012 und Antwort des Ministers für Wirtschaft, Verkehr und Landesentwicklung*, Vol. 18/6994 of *Drucksache*. Wiesbaden: Hessischer Landtag (HessL).

HessSt (2008). *Forum Flughafen und Region. Struktur des zukünftigen Dialogs zwischen Luftverkehrsakteuren und der Region (Kabinettsbeschluss vom 13. Juni 2008)*. Wiesbaden: Hessische Staatskanzlei (HessSt).

HessVGH (2006). *Pressemitteilung: Erweiterung des Flughafens Frankfurt/Main – Ausbaugegner setzen Einsicht in die Verfahrensakten gerichtlich durch, 5. Januar 2006 (1/2006)*. Hessischer Verwaltungsgerichtshof Kassel (VGH).

— (2009). *Presseinformation: Ausbau des Flughafens Frankfurt Main. Musterklagen gegen den Bau der neuen Landebahn überwiegend abgewiesen. Neue Entscheidung über Nachtflüge erforderlich, 21. August 2009 (24/2009)*. Hessischer Verwaltungsgerichtshof Kassel (VGH).

— (2013). "Presseinformation: Eilanträge auf Betriebseinschränkungen für Nordwest-Landebahn abgelehnt, 30. Juli 2013 (19/2013)." Hessischer Verwaltungsgerichtshof Kassel (VGH).

— (2015). *Startseite. Aktuell. Wochenplan für die 18. Kalenderwoche*. Hessischer Verwaltungsgerichtshof Kassel (VGH), Website: https://vgh-kassel-justiz.hessen.de – rev. April 28th, 2015.

Hirsch, Joachim (1974). *Staatsapparat und Reproduktion des Kapitals*. Frankfurt am Main: Suhrkamp Verlag.

HLGL (ny). "Zeitgeschichte in Hessen - Daten, Fakten, Hintergründe." Hessisches Landesamt für geschichtliche Landeskunde, Website: https://www.lagis-hessen.de – rev. May, 17th 2014.

HMIS (2015). *Gesetz über die Metropolregion Frankfurt/Rhein-Main*. Hessisches Ministerium des Inneres und für Sport (HMIS), Website: http://verwaltung.hessen.de – rev. January 5th, 2015.

HMLWLFN (1993). "Erklärung von Waldflächen in den Gemarkungen Sachsenhausen, Oberrad, Fechenheim, Schwanheim, Griesheim, Nied, Bergen-Enkheim und Wald, Stadt Frankfurt am Main, in der Gemarkung Offenbach, Stadt Offenbach am Main, in den Gemarkungen Zeppelinheim und Neu-Isenburg, Landkreis Offenbach, sowie in der Gemarkung Bischofsheim, Main-Kinzig-Kreis, zu Bannwald vom 6.7.1993." In: *Staatsanzeiger für das Land Hessen* 1993.XXIX, pp. 1784–1799.

— (1994). *Neue Chancen. 1. Rhein-Main-Konferenz der Länder Hessen, Bayern und Rheinland-Pfalz, am 4. März 1994, im Frankfurt-Airport-Conference-Center Flughafen Frankfurt/Main.* Wiesbaden: Hessisches Ministerium für Landesentwicklung, Wohnen, Landwirtschaft, Forsten und Naturschutz (HMLWLFN).

HMWEVL (2014). *Planergänzungsbeschluss zur Ergänzung der unter A XI 2.3 des Planfeststellungsbeschlusses zum Ausbau des Verkehrsflughafens Frankfurt Main vom 18. Dezember 2007 und mit Planergänzungsbeschluss vom 10. Mai 2013 verfügten Nebenbestimmungen zu Wirbelschleppen*, Vol. 66p 01.03.04/31. Wiesbaden: Hessisches Ministerium für Wirtschaft, Energie, Verkehr und Landesentwicklung (HMWEVL).

, HMWEVL (2015a). *Hessische Cluster- & Netzwerk-Initiativen.* Hessisches Ministerium für Wirtschaft, Energie, Verkehr und Landesentwicklung (HMWEVL), Website: http://www.hessen-cluster.de – rev. January 8th, 2015.

— (2015b). "Verkehr." Hessisches Ministerium für Wirtschaft, Energie, Verkehr und Landesentwicklung (HMWEVL), Website: https://wirtschaft.hessen.de – rev. January 5th, 2015.

— (2015c). *Verkehr. Luftverkehr. Ausbau Flughafen Frankfurt.* Hessisches Ministerium für Wirtschaft, Energie, Verkehr und Landesentwicklung (HMWEVL), Website: https://wirtschaft.hessen.de – rev. January 5th, 2015.

HMWT (1971). "Raumordnungsverfahren nach dem Hessischen Landesplanungsgesetz für den Ausbau des Flughafens Frankfurt/Main." In: *Staatsanzeiger für das Land Hessen* 1971.XVIII, pp. 752–759.

HMWV (1966). *Eine dritte Startbahen für den Flughafen Frankfurt am Main.* Wiesbaden: Hessisches Ministerium für Wirtschaft und Verkehr. Wiesbaden: Hessisches für wirtschaft und verkehr edition.

HMWVL (2000). *Landesentwicklungsplan Hessen 2000, 13. Dezember 2000.* Wiesbaden: Hessisches Ministerium für Wirtschaft, Verkehr und Landesentwicklung (HMWVL).

— (2004). *Planfeststellungsbeschluss zur Errichtung einer A380-Werft am Verkehrsflughafen Frankfurt Main.* Vol. PF-66 m 04.03.06.01. Wiesbaden: Hessisches Ministerium für Wirtschaft, Verkehr und Landesentwicklung (HMWVL).

— (2007). *Planfeststellungsbeschluss zum Ausbau des Verkehrsflughafens Frankfurt Main*, volume PF-66 p -V-. Wiesbaden: Hessisches Ministerium für Wirtschaft, Verkehr und Landesentwicklung (HMWVL).

— (2013a). *Planergänzungsbeschluss zur Ergänzung der unter A XI 2.3 des Planfeststellungsbeschlusses zum Ausbau des Verkehrsflughafens Frankfurt Main vom 18. Dezember 2007 zwecks Umsetzung des Urteils des Bundesverwaltungsgerichts vom 4. April 2012 (BVerwG 4 C 6.10).* Vol. 66p 01.03.04/27. Wiesbaden: Hessisches Ministerium für Wirtschaft, Verkehr und Landesentwicklung (HMWVL).

— (2013b). *Planergänzungsbeschluss zur Ergänzung des Planfeststellungsbeschlusses zum Ausbau des Verkehrsflughafens Frankfurt Main vom 18. Dezember 2007 verfügten Nebenbestimmung zu Wirbelschleppen.* Vol. 66p 01.03.04/29. Wiesbaden: Hessisches Ministerium für Wirtschaft, Verkehr und Landesentwicklung (HMWVL).

— (2013c). *Planfeststellungsbeschluss zur Änderung des Planfeststellungsbeschlusses zum Ausbau des Verkehrsflughafens Frankfurt Main vom 18. Dezember 2007. Än-derung Ausbaubereich Südost (21. Änderungsplanfeststellungsbeschluss).* Vol. 66p 01.03.04/26. Wiesbaden: Hessisches Ministerium für Wirtschaft, Verkehr und Landesentwicklung (HMWVL).

— (2013d). *Pressemitteilung: Rentsch ordnet Vorbeugung an, 14. Mai 2013.* Hessisches Ministerium für Wirtschaft, Verkehr und Landesentwicklung (HMWVL).

Hochfeld, Christian, et al. (2000). *Dokumentation zum Mediationsverfahren Flughafen Frankfurt am Main. Leitfaden durch den Diskussionsprozess und die Ergebnisse.* Wiesbaden: Hessische Staatskanzlei.

Höhnberg, Ulrich (2005). "Raumordnungsverfahren." In: *Handwörterbuch der Raumordnung.* Ed. by Akademie für Raumforschung und Landesplanung. 4th ed. Hannover: Verlag der Akademie für Raumforschung und Landesplanung, pp. 884–891.

Hollingsworth, J. Rogers and Leon N. Lindberg (1985). "The Governance of the American Economy: The Role of Markets, Clans, Hierarchies, and Associative Behaviour." In: *Private Interest Government. Beyond Market and State.* Ed. by Wolfgang Streeck and Philippe C. Schmitter. London: Sage Publications, pp. 221–254.

Hollingsworth, J. Rogers, Philippe C. Schmitter, et al. (1994). *Governing Capitalist Economies. Performance and Conrol of Economic Sectors.* New York & Oxford: Oxford University Press.

HOLM (2015a). *Idee: Das House of Logistics & Mobility (HOLM).* House of Logistics & Mobility (HOLM) GmbH, Website: http://www.frankfurt-holm.de – rev. January 7th, 2015.

— (2015b). *Mitgliederliste HOLM e.V., Stand: 26.02.2015.* House of Logistics & Mobility (HOLM) GmbH, Website: http://www.frankfurt-holm.de – rev. July 3rd, 2015.

— (2015c). *Partner.* House of Logistics & Mobility (HOLM) GmbH, Website: http://www.frankfurt-holm.de – rev. January 7th, 2015.

— (2015d). *Satzung des HOLM (House of Logistics and Mobility) e.V. mit Sitz in Bessie-Coleman-Straße 7 60549 Frankfurt am Main Frankfurt am Main 01.04.2015.* House of Logistics & Mobility (HOLM) GmbH, Website: http://www.frankfurt-holm.de – rev. July 3rd, 2015.

Holtkamp, Lars, Jörg Bogumil, et al. (2006). *Kooperative Demokratie.* Vol. 9. Studien der Demokratieforschung. Frankfurt am Main & New York: Campus Verlag.

Holz-Rau, Christian and Joachim Scheiner (2015). "Mobilitätsbiografien und Mobilitätssozialisation: Neue Zugänge zu einem alten Thema." In: *Räumliche Mobilität und Lebenslauf. Studien zu Mobilitätsbiografien und Mobilitätssozialisation.* Ed. by Joachim Scheiner and Christian Holz-Rau. Wiesbaden: Springer VS, Studien zur Mobilitäts- und Verkehrsforschung, pp. 3–22.

Homeyer, Ingmar von (2010). "Emerging Experimentalism in EU Environmental Governance", In: *Experimentalist Governance in the European Union. Towards a New Architecture.* Ed. by Charles F. Sabel and Jonathan Zeitlin. Oxford etc.: Oxford University Press, pp. 121–150.

Hooton, Ted (1980). *Comparison of capacity: 2 runways versus 3 runways at Frankfurt Airport.*

— (1981). *Letter to the Vorsitzende des Haushaltsausschusses of the Hessische Landtag, 21. January 1981.*

Hösch, Ulrich (2009). "Der Entwurf eines neuen Flughafenkonzepts." In: *Aktuelle Probleme des Luftverkehrs-, Planfeststellungs- und umweltrechts 2009. Vorträge auf den Elften Speyerer Planungsrechtstagen und dem Seyerer Luftverkehrsrechtstag vom 4. bis 6. März 2009 an der Deutschen Hochschule für Verwaltungswissenschaften Speyer*. Ed. by Jan Ziekow. Berlin: Duncker & Humblot, pp. 73–101.

Hoyler, Michael, Tim Freytag, et al. (2008a). "Advantageous Fragmentation? Reimagining Metropolitan Governance and Spatial Planning in Rhine-Main." In: *Built Environment* 32.II, pp. 124–136.

— (2008b). "Connecting Rhine-Main: The Production of Multi-Scalar Polycentricities through Knowledge-Intensive Business Services." In: *Regional Studies* 42.VIII, pp. 1095–1111.

HR1 (1980). *Transcript of the radio broadcast: Politisches Forum, Sendung im Hessischen Rundfunk, 1. Programm am 13.11.1980. Autor: Dietrich Jörn Weder; Sprecher: Hans Helmut Sievert und Walter Hillsbecher.*

Hunecke, Marcel (2015). *Mobilitätsverhalten verstehen und verändern. Psychologische Beiträge zur interdisziplinären Mobilitätsforschung*. Studien zur Mobilitäts- und Verkehrsforschung. Wiesbaden: Springer VS.

Huschke, Wolfgang J. (2008). *Die Rosinenbomber. Die berliner Luftbrücke 1948/49, ihre technischen Voraussetzungen und deren erfolgreiche Umsetzung*. Berlin: Berliner Wissenschafts-Verlag.

Hustedt, Thurid, Arndt Wonka, et al. (2014). *Verwaltungsstrukturen in der Europäischen Union: Kommission, Komitologie, Agenturen und Verwaltungsnetzwerke*. Wiesbaden: Springer VS.

IFOK (ny). *Liste der im Rahmen des Mediationsverfahrens offen gebliebenen Fragestellungen.* forum flughafen & region (FFR), Website: http://www.forum-flughafen-region.de – rev. July 3rd, 2015.

IHK (1981). *Stellungnahmen Arbeitsgemeinschaft hessischer Industrie- und Handelskammern, 22. Januar 1981.*

Ingram, Helen and Anne L. Schneider (2006). "Policy Analysis for Democracy." In: *The Oxford Handbook of Contextual Political*

*Analysis*. Ed. by Michael Moran, Martin Rein, and Robert E. Goodin. Oxford & New York etc.: Oxford University Press, pp. 169–184.

ISG (2014a). *Beständeübersicht der Abteilung "Sammlungen". Politische Parteien, Vereine und Verbände*. Institut für Stadtgeschichte Frankfurt am Main (ISG), Website: http://www.stadtgeschichte-ffm.de – rev. December 29th, 2014.

— (2014b). *Chronik der Stadt Frankfurt am Main. Stadtchronik*. Institut für Stadtgeschichte Frankfurt am Main (ISG), Website: http://www.stadtgeschichte-ffm.de – rev. December 28th, 2014.

Jachtenfuchs, Markus and Beate Kohler-Koch (2004). "Governance in der Europäischen Union." In: *Governance – Regieren in komplexen Regelsystemen. Eine Einführung*. Ed. by Arthur Benz and Nicolai Dose. Wiesbaden: VS Verlag für Sozialwissenschaften, pp. 77–101.

Jahn, Egbert (1984). "Gewalt in der Auseinandersetzung um die Startbahn 18 West." In: *Leviathan* 12.III, pp. 308–335.

James, Oliver and Martin Lodge (2003). "The Limitations of 'Policy Transfer' and 'Lesson Drawing' for Public Policy Research." In: *Political Studies Review* 1.II, pp. 179–193.

Janka, Bettina (2011). *Corporate Governance in Deutschland und Spanien*, Vol. 41. Schriften der Deutsch-Spanischen Juristenvereinigung. Frankfurt am Main: Peter Lang.

Jann, Werner and Kai Wegrich (2009). "Phasenmodelle und Politikprozesse: Der Policy Cycle." In: *Lehrbuch der Politikfeldanalyse 2.0*. Ed. by Klaus Schubert and Nils C. Bandelow. 2nd ed. München: Oldenbourg Verlag, pp. 75–113.

Jassop, Bob (2002). "Governance and Metagovernance: On Reflexivity, Requisite Variety, and Requisite Irony." In: *Participatory Governance in Multi-Level Context. Concepts and Experience*. Ed. by Hubert Heinelt et al. Opladen: Leske + Budrich, pp. 33–57.

Jenkins, William Ieuan (1978). *Policy Analysis: A political and Organizational Perspective*. London: Robertson.

Jensen, Michael C. and William H. Meckling (1976). "Theory of the Firm: Manageriav Behavior, Agency Costs and Ownership Structure." In: *Journal of Financial Economics* 3.IV, pp. 305–360.

Jessop, Bob (1998). "The Rise of Governance and the Risk of Failure: The Case of Economic Development." In: *International Social Science Journal* 50.CLV, pp. 29–45.

Joerges, Christian and Jürgen Neyer (1988). "Transforming Strategic Interaction Into Deliberative Problem-Solving: European Comitology in the Foodstuffs Sector." In: *Journal of European Public Policy* 4.IV, pp. 608–625.

John, Peter (2001). *Local Governance in Western Europe*. Sage Politics Texts. London etc.: Sage Publications.

Johnsen, Hartmut (1996). *Der Startbahn-West-Konflikt. Ein politisches Lehrstück?. Zeitzeugen ziehen Zwischenbilanz*. Frankfurt am Main: Societäts-Verlag.

Jones, Victor (1942). *Metropolitan Government*. Chicago: The University of Chicago Press.

KAF (1981). *Brief an den Hessischen Landtag – Anhörungsausschuß, 28. Januar 1981*. Kommunale Arbeitsgemeinschaft Flughafenerweiterung (KAF).

Kalberg, Stephan (2000). "Strukturierte Mißverständnisse. Unterschiede der politischen Kultur in Amerika und Deutschland." In: *Europa oder Amerika? Zur Zukunft des Westens*. Ed. by Karl H. Bohrer and Kurt Scheel. MERKUR Son-derheft. München: Klett-Cotta, pp. 948–957.

Kappelhoff, Peter (1992). "Die Auflösung des Sozialen." In: *Analyse & Kritik* 14.II, pp. 221–238.

Kasarada, John D. (2010). *Global Airport Cities*. Twickenham: Insight Media.

Kay, Inkook (2013). *Regulierung als Erscheinungsform der Gewährleistungsverwaltung. Eine rechtsdogmatische Untersuchung zur Einordnung der Regulierung in das Staats- und Verwaltungsrecht*. Frankfurt am Main: Peter Lang.

Kelsterbach (1998a). *Beschluss Nr. 10/5: Resolution gegen den geplanten Ausbau des Flughafens Frankfurt*. Kelsterbach: Stadtverwaltung Kelsterbach.

— (1998b). *Schreiben des Bürgermeisters Engish an Herrn Dr. Ruhr, Flughafenerweiterung; hier: Bau einer zusätzlichen Landebahn zur Kapazitätensteigerung der Flugbewegungen auf dem Flughafen Frankfurt (AZ: I/1-go-Wd 10.11.1998)*. Stadtarchiv Kelsterbach.

Kenis, Patrick and Jörg Raab (2008). "Politiknetzwerke als Governanceform. Versuche einer Bestandsaufnahme und Neuausrichtung der Diskussion." In: *Governance in einer sich wandelenden Welt*. Ed. by Gunnar Folke Schuppert and Michael Zürn. Wiesbaden: VS Verlag für Sozialwissenschaften. PVS – Politische Vierteljahresschrift, Sonderheft 41/2008, pp. 132–148.

Kenis, Patrick and Volker Schneider (1991). "Policy Networks and Policy Analysis: Scrutinizing a New Analytical Tool." In: *Policy Networks. Empirical Evidence and Theoretical Considerations*. Ed. by Bernd Marin and Renate Mayntz. Frankfurt & Boulder: Campus Verlag & Westview Press, pp. 25–59.

Kerwer (2004). "Banking on Privat Actors. Financial Market Regulation and the Limits of Transnational Governance." In: *European and International Regulation after the Nation State*. Ed. by Adrienne Héritier, Michael Stolleis, and Fritz W. Scharpf. Common Goods: Law, Politics and Economics, Vol. 12. Baden-Baden: Nomos, pp. 205–223.

Ketels, Christian (2015). *Cities – A (not so) New Arena for Competitiveness Upgrading*. The Global Practitioners Network for Competitiveness, Clusters and Innovation (TCI), Website: http://www.tci-network.org – rev. May 27th, 2015.

Kieling, Jörg (2003). "Kooperative Regionalplanung und Regional Governance: Praxisbeispiele, Theoriebezüge und Perspektiven." In: *Informationen zur Raumentwicklung* 2003.VIII/IX, pp. 463–478.

Kiese, Matthias (2008). "Stand und Perspektiven der regionalen Clusterforschung." In: *Cluster und Regionalentwicklung. Theorie, Beratung und praktische Umsetzung*. Ed. by Matthias Kiese and Ludwig Schätzl. 2nd ed. Dortmund: Rohn, pp. 9–50.

Kingsbury, Benedict, Nico Krisch, and Richard B. Stewart (2005). "The Emergence of Global Administrative Law." In: *Law and Contemporary Problems* 68.III/IV, pp. 15–61.

Kleinfeld, Ralf, Harald Plamper, et al. (2006a). *Regional Governance. Steuerung, Koordination und Kommunikation in regionalen Netzwerken als neue Formen des Regierens. Band 1*. Göttingen: V&R unipress.

— (2006b). *Regional Governance. Steuerung, Koordination und Kommunikation in regionalen Netzwerken als neue Formen des Regierens. Band 2*. Göttingen: V&R unipress.

Knieling, Jörg (2009). "Metropolregionen: Zukunft oder Zeitgeist?" In: *Metropolregionen. Innovation, Wettbewerb, Handlungsfähigkeit. Metropolregionen und Raumentwicklung Teil 3*. Ed. by Jörg Kieling. Hannover: Akademie für Raumforschung und Landesplanung, pp. 1–8.

Knippenberger, Ute (2012). *Regionale Governance des Funktionswandels von Flughäfen. Eine Analyse am Beispiel der "Airport City" Frankfurt am Main*. Wiesbaden: VS Verlag für Sozialwissenschaften.

Koch, Philippe (2013). "Overestimating the Shift from Government to Governance: Evidence from Swiss Metropolitan Areas." In: *Governance: An International Journal of Policy, Administration, and Institutions* 26.III, pp. 397–423.

König, Wolfgang (2001). "Vorwort." In: *Metropolitane Region in der Vernetzung. Der Fall Frankfurt Rhein/Main*. Ed. by Josef Esser and Eike W. Schamp. Frankfurt am Main & New York: Campus Verlag, page 7.

Königsstein, Medizinmeteorologische Beratungsstelle (1961). *Medizinmeteorologisches Gutachen über die Bedeutung des Waldgürtels südlich von Frankfurt, 5. Juli 1961*.

Kooiman, Jan (1993). "Social-Political Governance: Introduction." In: *Modern Governance. New Government – Society Interactions*. Ed. by Jan Kooiman. London etc.: Sage Publications, pp. 1–8.

— (1994). *Modern Governance: New Government – Society Interactions*. Reprint. London etc.: Sage Publications.

— (2010). *Governing as Governance*. London etc.: Sage Publications.

Krippner, Greta R. (2001). "The Elusive Market: Embeddedness and the Paradigm of Economic Sociology." In: *Theory and Society* 30.VI, pp. 775–810.

Krippner, Greta R. and Anthony S. Alvarez (2007). "Embeddedness and the Interlectual Projects of Economic Sociology." In: *The Annual Review of Sociology* 33, pp. 219–240.

Krippner, Greta R., Mark S. Granovetter, et al. (2004). "Polanyi Symposium: a conversation on embeddedness." In: *Socio-Economic Review* 2.I, pp. 109–135.

Krüger-Röth, Doris (2008). "Metropolregion Frankfurt/Rhein-Main." In: *Metropolregionen in Deutschland. 11 Beispiele für Regional Governance*. Ed. by Jürgen Ludwig et al. Baden-Baden: Nomos, pp. 54–67.

Krugman, Paul (1991). *Geography and Trade*. Gaston Eyskens Lectures Series. Leuven etc.: Leuven University Press & MIT Press.

Kübler, Daniel (2012). "Introduction: Metropolitanisation and Metropolitan Governance." In: *European Political Science* 11.III, pp. 402–408.

Kuckartz, Udo (2009). *Einführung in die computergestützte Analyse qualitativer Daten*. 3rd ed. Wiesbaden: VS Verlag für Sozialwissenschaften.

Kujath, Hans Joachim (2005). "Einleitung. Metropolregionen - Die Bedeutung von Metropolregionen für die Wissensökonomie." In: *Knoten im Netz. Zur neuen Rolle der Metropolregionen in der Dienstleistungswirtschaft und Wissensökonomie*. Ed. by Hans Joachim Kujath. Münster: LIT Verlag, pp. 9–20.

Lackowska, Marta (2011a). "Frankfurt/Rhine Main: Governance without coordination?" In: *Metropolitan Governance. Different Paths in Contrasting Contexts: Germany and Israel*. Ed. by Hubert Heinelt, Eran Razin, and Karsten Zimmer-mann. Vol. 9. Interdisciplinary Urabn Research. Frankfurt am Main: Campus Verlag, pp. 79–117.

— (2011b). "Metropolitan Governance: The Academic Debate." In: *Metropolitan Governance. Different Paths in Constrasting Contexts: Germany and Israel*. Ed. by Hubert Heinelt, Eran Razin, and Karsten Zimmermann. Vol. 9. Interdisci-plinary Urban Research. Frankfurt am Main: Campus Verlag, pp. 21–31.

Ladeur, Karl-Heinz (1983). "Politische Ökonomie verwaltungsgerichtlicher Planungskontrolle." In: *Gegentendenzen zur Verrechtlichung*. volume 9 of *Jahrbuch für Rechtssoziologie und Rechtstheorie*. . Ed. by Voigt Rüdiger. Vol. 9. Jahrbuch für Rechtssoziologie und Rechtstheorie. Opladen: Westdeutscher Ver-lag, pp. 102–115.

Lahmann, Erdwin and Karl-Ernst Prescher (1979). *Luftverunreinigungen in der Umgebung vo Flughäfen*. Schriftenreihe des Vereins für Wasser-, Boden- und Lufthygiene. Stuttgart: Gustav Fischer.

Lamnek, Siegfried (2010). *Qualitative Sozialforschung*. 5th ed. Basal: Beltz Verlag.

Langhagen-Rohrbach, Christian (2005). "Aktuelle Regionalisierungsprozesse in der Region Rhein-Main." In: *Raumforschung und Raumordnung* 62.I, pp. 58–66.

Lasswell, Harold Dwight (1956). *The Decision Process: Seven Categories of Functional Analysis*. College Park: University of Maryland Press.

Lauschmann, Elisabeth (1976). *Grundlagen einer Theorie der Regionalpolitik*. 3rd ed. Vol. 2. of Taschenbücher zur Raumplanung. Hannover: Hermann Schroedel Verlag.

Le Galès, Patrick (1995). "Du gouvernement des villes à la gouvernance urbaine." In: *Revue française de science politique* 45.I, pp. 57–95.

— (1997). "Gouvernement et gouvernance des régions: faiblesses structurelles et nouvelles mobilisations." In: *Les paradoxes de régions en Europe*. Ed. by Patrick Le Galès and Christian Lequesne. Paris: La Découverte, pp. 237–263.

Le Galès, Patrick and Carlo Trigilia (2004). "Conclusions." In: *Neokorporatismus und Gesundheitswesen*. Ed. by Colin Crouch et al. Reprint. Baden-Baden: Nomos, pp. 331–343.

Le Galès, Patrick and Helmut Voelzkow (2003). "Introduction: The Governance of Local Economies." In: *Local Production Systems in Europe. Rise or Demise?* Ed. by Colin Crouch et al. Reprint. Oxford etc.: Oxford University Press, pp. 1–24.

Lefèvre, Christian (1998). "Metropolitan Government and Governance in Western Countries: A Critical Review." In: *International Journal of Urban and Regional Research* 22.I, pp. 9–25.

Lehmbruch, Gerhard (1988). "Der Neokorporatismus der Bundesrepublik im internationalen Vergleich und die "Konzertierte Aktion im Gesundheitswesen"." In: *Neokorporatismus und Gesundheitswesen*. Ed. by Gérard Gäfgen. Baden-Baden: Nomos, pp. 3–18.

Levi-Faur, David (2012). "From "Big Government" to "Big Governance"." In: *Oxford Handbook of Governacne*. Ed. by David Levi-Faur. Oxford & New York: Oxford University Press, pp. 3–18.

LFU (2012). "News: Task Force für Luftfrachtstandort Frankfurt gegründet." In: *Logistik für Unternehmen. Das Fachmagazin der internen und externen Logistik* 2012.

Lindblom, Charles E. (1965). *The Intelligence of Democracy. Decision Making Through Mutual Adjustment*. New York & London: The Free Press Collier-Macmillan.

Lindner, Peter, Stefan Ouma, et al. (2014). *Industriestudie. Frankfurt am Main 2013*. Frankfurt am Main: Peter Lang.

Loer, Thomas (2007). *Die Region. Eine Begriffsbestimmung am Fall des Ruhrgebietes*. Qualitative Soziologie. Stuttgart: Lucius & Lucius.

LRM (2015). *Ziele von Logistik RheinMain*. Logistik RheinMain (LRM), Website: http://www.logistik-rhein-main.de – rev. January 7th, 2015.

Ludwig, Jürgen, Klaus Mandel, et al. (2008). *Metropolregionen in Deutschland. 11 Beispiele für Regional Governance*. Baden-Baden: Nomos.

Luhmann, Niklas (1971). *Politische Planung. Aufsätze zur Soziologie von Politik und Verwaltung*. Opladen: Westdeutscher Verlag.

Lütz, Susanne (2006). "Einleitung: Governance in der politischen Ökonomie." In: *Governance in der politischen Ökonomie. Struktur und Wandel des modernen Kapitalismus*. Ed. by Susanne Lütz. Wiesbaden: VS Verlag für Sozialwissenschaften, pp. 13–56.

— (2010). "Governance in der politischen Ökonomie I: Makro- und Mesoperspektiven." In: *Governance – Regieren in komplexen Regelsystemen*. Ed. by Arthur & Nicolai Dose Benz. Wiesbaden: VS Verlag für Sozialwissenschaften, pp. 137–154.

Marcussen, Martin and Jacob Torfing (2007). *Democratic Network Governance in Europe*. Houndsmills & New York: Palgrave Macmillan.

Marin, Bernd and Renate Mayntz (1991). "Policy Networks and Policy Analysis: Scrutinizing a New Analytical Tool." In: *Policy Networks. Empirical Evidence and Theoretical Considerations*. Ed. by Bernd Marin and Renate Mayntz. Frankfurt am Main & Boulder: Campus Verlag & Westview Press, pp. 11–23.

Marshall, Alfred (1997). *Industry and Trade: A Study of Industrial Technique and Business Organization, and of their Influences on the Conditions of Various Classes and Nations*. Bristol: Overstone Press.

Martin, Lawrence L. and Jeannie Hock Schiff (2011). "City-County Consolidations: Promise Versus Performance." In: *State and Local Government Review* 43.II, pp. 167–177.

May, Judith P (1978). *The Policy Cycle*. California: Sage Publications.

May, Peter J. (1992). "Policy Learning and Failure." In: *Journal of Public Policy* 12.IV, pp. 331–354.

Mayntz, Renate (1977). "Sociology, Value Freedom, and the Problems of Political Counseling." In: *Using Social Research in Public Policy-Making*. Ed. by Carol H. Weiss. Lexington: Lexington Books, pp. 55–65.

— (1979). "Regulative Politik in der Krise?" In: *Sozialer Wandel in Westeuropa. Verhandlungen des 19. Deutschen Soziologentages, 17.-19. April 1979 im Internationalen Congress Centrum (ICC) Berlin*. Ed. by Joachim Matthes. Frankfurt am Main & New York: Campus Verlag, pp. 55–81.

— (1982). "Problemverarbeitung durch politisch-administrative Systeme: Zum Stand der Forschung." In: *Politikwissenschaft und Verwaltungswissenschaft*. Ed. by Joachim Jens Hesse. Opladen: Westdeutscher Verlag, pp. 74–89.

— (1983). "Zur Einleitung: Probleme der Theoriebildung in der Implementationsforschung." In: *Implementation politischer Programme II. Ansätze zur Theoriebildung*. Ed. by Renate Mayntz. Opladen: Westdeutscher Verlag, pp. 7–24.

— (1987). "Politische Steuerung und gesellschaftliche Steuerungsprobleme. Anmerkungen zu einem theoretischen Paradigma." In: *Jahrbuch zur Staats- und Verwaltungswissenschaft, Band 1*. Ed. by Thomas Ellwein et al. Baden-Baden: Nomos, pp. 89–110.

— (1992). *Verbände zwischen Mitgliederinteresse und Gemeinwohl*. Gütersloh: Verlag Bertelmann Stiftung.

— (1996). "Politische Steuerung: Aufstieg, Niedergang und Transformation einer Theorie." In: *Politische Theorien inder der Ära der Transformation*. Ed. by Klaus von Beyme and Claus Offe. Opladen: Westdeutscher Verlag, Politische Vierteljahresschrift, Sonderheft 26/1995. Opladen: Westdeutscher Verlag, pp. 148–168.

— (2003). "Governance im modernen Staat." In: *Governance. Eine Einführung. Dreifachkurseinheit*. Ed. by Fachbereich Kultur- und Sozialwissenschaften Fer-nUniversität in Hagen. Hagen: FernUniversität in Hagen, pp. 71–83.

— (2006). "Governance Theorie als fortentwickelte Steuerungstheorie." In: *Governance-Forschung. Vergewisserung über Stand und Entwicklungslinien*. Ed. by Gunnar Folke Schuppert. 2nd ed. Baden-Baden: Nomos, pp. 11–20.

— (2007). "The Architecture of Multi-level Governance of Economic Sectors." In: *MPIfG Discussion Paper* 2007.13, pp. 1–28.

— (2009a). "Governancetheorie: Erkenntnisse und offene Fragen." In: *Perspektiven der Governance-Forschung*. Ed. by Edgar Grande and Stefan May. Baden-Baden: Nomos, pp. 9–19.

— (2009b). *Über Governance. Institutionen und Prozesse politischer Regelung*. Schriften aus dem Max-Planck-Institut für Gesellschaftsforschung. Frankfurt am Main & New York: Campus Verlag.

Mayntz, Renate and Fritz W. Scharpf (1973). *Planungsorganisation. Die Diskussion um die Reform von Regierung und Verwaltung des Bundes*. München: R. Piper & Co. Verlag.

— (1995). "Der Ansatz des akteurszentrierten Instituionalismus." In: *Gesellschaftliche Selbstregelung und politische Steuerung*, , volume 23 of *Schriften des Max-Planck-Instituts für Gesellschaftsforschung Köln*. Frankfurt am Main & New York: Campus Verlag, pp. 39–72.

Mayntz, Renate and Volker Schneider (1995). "Die Entwicklung technischer Infrastruktursysteme zwischen Steuerung und Selbstorganisation." In: *Gesellschaftliche Selbstregelung und politische Steuerung*. Ed. by Renate Mayntz and Fritz W. Scharpf. Frankfurt am Main & New York: Campus Verlag, Schriften des Max-Planck-Instituts für Gesellschaftsforschung Köln, Band 23, pp. 73–100.

Mediationsgruppe (2000a). *Das Mediationsverfahren Flughafen Frankfurt: Gutachten, Zusammenfassungen, Endbericht und Präsentation der Ergebnisse (CD-ROM)*. Frankfurt am Main: Mediationsgruppe Flughafen Frankfurt/Main.

— (2000b). "Dialogbriefe." In: *Das Mediationsverfahren Flughafen Frankfurt: Gutachten, Zusammenfassungen, Endbericht und Präsentation der Ergebnisse (CD-ROM)*. Ed. by Mediationsgruppe. Frankfurt am Main: Mediationsgruppe Flughafen Frankfurt/Main.

— (2000c). "Geschäftsordnung der Mediationsgruppe Flughafen Frankfurt vom 16. Juli 1998 (Fassung gemäß Beschluss vom 01.10.1999)." In: *Das Mediationsverfahren Flughafen Frankfurt: Gutachten, Zusammenfassungen, Endbericht und Präsentation der Ergebnisse (CD-ROM)*. Ed. by Mediationsgruppe. Frankfurt am Main: Mediationsgruppe Flughafen Frankfurt/Main.

— (2000d). "Projektteams." In: *Das Mediationsverfahren Flughafen Frankfurt: Gutachten, Zusammenfassungen, Endbericht und Präsentation der Ergebnisse (CD-ROM)*. Ed. by Mediationsgruppe. Frankfurt am Main: Mediationsgruppe Flughafen Frankfurt/Main.

— (2000e). "Sitzungsprotokolle." In: *Das Mediationsverfahren Flughafen Frankfurt: Gutachten, Zusammenfassungen, Endbericht und Präsentation der Ergebnisse (CD-ROM)*. Ed. by Mediationsgruppe. Frankfurt am Main: Mediationsgruppe Flughafen Frankfurt/Main.

Meier, Andreas (2012). *eDemocracy & eGovernment. Stages of a Democratic Knowledge Society*. Berlin & Heidelberg: Springer.

Menzel, Ulrich (1998). *Globalisierung versus Fragmentierung*. Frankfurt am Main: Suhrkamp Verlag.

— (2010). "Entwicklungstheorie." In: *Entwicklungstheorie. Theorien - Probleme - Strategien*. Ed. by Reinhard Stockmann, Ulrich Menzel, and Franz Nuscheler. München: Oldenbourg Wissenschaftsverlag, pp. 11–159.

Merk, Olaf (2014). "Metropolitan Governance of Transport and Land Use in Chicago." In: *OECD Regional Development Working Papers* 2014.VIII, pp. 1–20.

Merkur (2012a). *Bürgerentscheid: Dritte Startbahn ist vom Tisch, 18. Juni 2012*. Münchner Merkur.

— (2012b). "München gegen 3. Startbahn: Die Stimmen, 18. Juni 2012." Münchner Merkur, Johannes Welte & David Costanzo.

Merten, Klaus (2009). *Inhaltsanalyse. Einführung in Theorie, Methode und Praxis*. 2nd ed. Wiesbaden: VS Verlag für Sozialwissenschaften.

Merton, Robert K. (2004). *Auf den Schultern von Riesen*. Frankfurt am Main: Suhrkamp.

Merton, Robert K. and Elinor Barber (2004). *The Travels and Adventures of Serendipity: A Study in Sociological Semantics and the Sociology of Science*. Princeton: Princeton University Press.

Michels, Ank (2011). "Innovations in Democratic Governance." In: *International Review of Administrative Science* 77.II, pp. 275–293.

Milner, Helen V. and Robert O. Keohane (1996). "Internationalization and Domestic Politics: An Introduction." In: *Internationalization and Domestic Politics*. Ed. by Robert O. Keohane and Helen V. Milner. Cambridge etc.: Cambridge University Press, pp. 3–24.

Mock, J and P Gonsowski (1981). *Stellungnahme: Grundwasser und Wasserbilanz im Bereich der geplanten Startbahn West Fughafen Frankfurt*. 2nd ed. Darmstadt: Institut für Wasserbau, Technische Hochschule Darmstadt.

Monar, Jörg (2010). "Experimentalist Governance in Justice and Home Affairs." In: *Experimentalist Governance in the European Union. Towards a New Architecture*. Ed. by Charles F. Sabel and Jonathan Zeitlin. Oxford etc.: Oxford University Press, pp. 237–260.

Motlanthe, Kgalema (2010). "Social Origins." In: *The Historical Roots of the ANC. Understanding the ANC today*. Ed. by Ben Turkor. Auckland Park: Jacana, pp. 13–20.

MS (2005). *Studie mit zwei Probanden. Jühe nennt Fraport Gutachten "wissenschaftlich eine Katastrophe", 11. November 2005*. Main-Spitze, Susanne Wildmeister.

— (2013). *Wirbelschleppen: "Für Flörsheim" will gegen Planergänzungsbeschluss klagen, 3. Juli 2013*. Main-Spitze, Markus Jäger.

Müller, Karlhans and Milan Skaryd (1972). *Airport der Zukunft. Airport of the Future*. 2nd ed. Frankfurt am Main: Hans J. Rummel & Co KG.

NABU (1998). "Resolution: Keine Erweiterung des Flughafens Frankfurt, Wetzlar 18. Dezember 1998." Naturschutzbund Deutschland (NABU), Hessen.

Nance, Mark T. and Patrick Cottrell (2014). "A turn Toward Experimentalism? Rethinking Security and Governance in the Twenty-First Century." In: *Review of International Studies* 40.II, pp. 277–301.

Naßmacher, Hiltrud (2010). *Politikwissenschaft*. 6th ed. München: Oldenbourg Verlag.

Neale, Walter C. (1990). "Karl Polanyi and American Institutionalism: A Strage Case of Convergance." In: *The Life and Work of Karl Polanyi. A Celebration*. Ed. by Kari Polanyi-Levitt. Montreal & New York: Black Rose Books, pp. 145–251.

Nelles, Jen and Frédéric Durand (2014). "Political Rescaling and Metropolitan Governance in Cross-Border Regions: Comparing the Cross-Border Metropolitan Areas of Lille and Luxembourg." In: *European Urban and Regional Studies* 21.I, pp. 277–301.

Neu-Isenburg (1997). *Pressemitteilung, 10. Dezember 1997*.

Newman, Abraham L. (2010). "Innovating European Data Privacy Regulation: Unintended Pathways to Experimentalist Governance." In: *Experimentalist Governance in the European Union. Towards a New Architecture*. Ed. by Charles F. Sabel and Jonathan Zeitlin. Oxford etc.: Oxford University Press, pp. 29–42.

Newton, Kenneth (2012). "Metropolitan Governance." In: *European Political Science* 11.III, pp. 409–419.

Nibbeling, Joachim (2001). *Die Privatisierung des Haftvollzugs*. Vol. 3057. Eu-ropäische Hochschulschriften. Rechtswissenschaft. Frankfurt am Main etc.: Peter Lang.

Nölke, Andreas, Henk Overbeek, and Bastiaan van Apeldoorn (2007). "Marketization, Transnationalization, Commodification, and the Shifts in Corporate Governance Regulation. A Conclusion." In: *The Transnational Politics of Corporate Governance Regulation*. Ed. by Henk Overbeek, Bastiaan van Apeldoorn, and Andreas Nölke. Routledge/RIPE Studies in Global Political Economy. London & New York: Routledge, pp. 203–219.

Nuscheler, Franz (2010). "Weltprobleme." In: *Entwicklungstheorie. Theorien - Probleme - Strategien*. Ed. by Reinhard Stockmann, Ulrich Menzel, and Franz Nuscheler. München: Oldenbourg Wissenschaftsverlag, pp. 161–349.

O'Donnell, Rory and Charles F. Sabel (2001). "Democratic Experimentalism: What to do About Wicked Problems After Whitehall." In: *Devolution and Globalisation. Implications for Local Decision-Makers. Governance*. Ed. by OECD. Paris: Organisation for Economic Co-operation and Development (OECD), pp. 67–90.

Oeser, Kurt (1981). *Zu Protokoll gegebene Stellungsnahme anlässlich der Anhörung im Hessischen Landtag vom 2. bis 5. Februar 1981 zum Ausbau der Startbahn 18 West des Frankfurter Flughafens, 2. Februar 1981*.

Offe, Claus (1987). "Die Staatstheorie auf der Suche nach ihrem Gegenstand. Beobachtungen zur aktuellen Diskussion." In: *Jahrbuch zur Staats- und Verwaltungswissenschaft, Band 1*. Ed. by Thomas Ellwein et al. Baden-Baden: Nomos, pp. 309–320.

— (2006). *Strukturprobleme des kapitalistischen Staates*. Neuausgabe. Frankfurt am Main & New York: Campus Verlag.

Offe, Klaus (2008). “Governance – “Empty signifier” oder sozialwissenschaftliches Forschungsprogramm?” In: *Governance in einer sich wandelenden Welt*. Ed. by Gunnar Folke Schuppert and Michael Zürn. PVS – Politische Vierteljahresschrift, Sonderheft 41/2008. Wiesbaden: VS Verlag für Sozialwissenschaften, pp. 61–76.

Ohmae, Kenichi (1995). *The End of the Nation State. The Rise of Regional Economies*. New York etc.: The Free Press.

Ostrom, Vincent, Charles M. Tiebout, and Robert Warren (1961). “The Organization of Government in Metropolitan Areas: A Theoretical Inquiry.” In: *The American Political Science Review* 55.IV, pp. 831–842.

Overdevest, Christine and Jonathan Zeitlin (2014). “Assembling an Experimentalist Regime: Transnational Governance Interactions in the Forest Sector.” In: *Regulation & Governance* 8.II, pp. 22–48.

Padgett, Carol (2012). *Corporate Governance. Theory and Practice*. Basingsstoke & New York: Palgrave Macmillan.

Panebianco, Stefano (2013). *Standortfaktor Regional Governance auf dem Prüfstand. Theoretische Überlegungen und empirische Analysen zur Bedeutung regionaler Steuerungssysteme für die wirtschaftliche Entwicklung von Regionen*. Hamburg: Verlag Dr. Kovač.

Parsons, Talcott (1976). *Zur Theorie sozialer Systeme*. Studienbücher zur Sozialwissenschaft. Opladen: Westdeutscher Verlag.

Parsons, Wayne (1995). *Public Policy. An Introduction to the Theory and Practice of Policy Analysis*. Cheltenham & Northampton: Edward Elgar.

Perkmann, Markus and Ngai-Ling Sum (2002). “Globalization, Regionalization and Cross-Border Regions: Scales, Discourses and Governance.” In: *Globalization, Regionalization, and Cross-Border Regions*. Ed. by Markus Perkmann and Ngai-Ling Sum. Houndsmills & New York: Palgrave Macmillan, pp. 3–21.

Perroux, François (1968). “Les Espace Économiques.” In: *L'Espace et les Pôles De Croissance*. Ed. by Jacques-R. Boudeville. Paris: Presses Universitaires de France, pp. 5–22.

Peters, B. Guy (2012). “Governance as Political Theory.” In: *Oxford Handbook of Governance*. Ed. by David Levi-Faur. Oxford & New York: Oxford University Press, pp. 19–32.

Pfeiffer, Werner and Enno Weiß (1992). *Lean Management: Grundlagen der Führung und Organisation industrieller Unternehmen*. Berlin: Erich Schmidt Verlag.

Phares, Don (2009). *Governing Metropolitan Regions in the 21st Century*. New York: M.E. Sharpe.

Piagessi, Danilo, Kristian Sund, et al. (2011). *Global Strategy and Practice of E-Governance. Examples from Around the World*. Hershey: IGI Global.

Piecha, Oliver M. (2005). "Der Traum von Groß-Frankfurt. Visionen zu Rhein-Main aus den Zwanziger Jahren." In: *Forschung Frankfurt* 2005.I, pp. 41–44.

Pierre, Jon (2000a). "Conclusion: Governance beyond State Strenght." In: *Debating Governance*. Ed. by Jon Pierre. Oxford & New York: Oxford University Press, pp. 241–246.

— (2000b). "Introduction: Understanding Governance." In: *Debating Governance*. Ed. by Jon Pierre. Oxford & New York: Oxford University Press, pp. 1–10.

Pierre, Jon and B. Guy Peters (2000). *Governance, Politics, and the State*. New York: St. Martin's Press.

Piore, Michael J. and Charles F. Sabel (1984). *The Second Industrial Divide. Possibilities for Prosperity*. New York: BasicBooks.

Bi-Plenum (1981). *Fragenkatalog der Bürgerinitiativen für das Landtagshearing zur Startbahn West (entsprechend Beschluß des Bi Plenums vom 15.1.81)*.

Plummer, Ken (1983). *Documents of Life. An Introduction to the Problems and Literature of a Humanistic Method*. London etc.: Georg Allen & Unwin.

— (2005). *Documents of Life 2. An Invitation to a Critical Humanism*. Reprint. London etc.: Sage Publications.

Polanyi, Karl (2009). *The Great Transformation. The Political and Economic Origins of our Time*. ebrary. Boston: Beacon Press.

Pollack, Mark A. and Gregory C. Shaffer (2010). "Biotechnology Policy: Between National Fears and Globl Disciplines." In: *Policy-making in the European Union*. Ed. by Helen Wallace, Nark A. Pollack, and Alasdair R. Young. 6th ed. The New Uropean Series. Oxford etc.: Oxford University Press, pp. 331–354.

Pollak, Johannes and Peter Slominski (2009). "Experimentalist but not Accountable Governance? The Role of Frontex in Managing the EU's External Borders." In: *West European Politics* 32.V, pp. 904–924.

Popper, Karl R. (1979). *Die beiden Grundprobleme der Erkenntnistheorie*. Tübingen: J.C.B. Mohr (Paul Siebeck).

Porter, Michael Eugen (1998). "Clusters and the New Economics of Competition." In: *Harvard Business Review* 76.VI, pp. 77–90.

— (2008a). *On Competition*. 2nd ed. Boston: Harvard Business School Publishing.

— (2008b). "Regions and the New Economics of Competition." In: *Global City-Regions. Trends, Theory, Policy*. Ed. by Allen John Scott. Oxford: Oxford University Press, pp. 139–157.

— (2014). *Wettbewerbsvorteile. Spitzenleistungen erreichen und behaupten*. 8th ed. Frankfurt am Main & New York: Campus Verlag.

PoS (2013a). *Final SEPA Determination of Non-Significance (DNS) of Proposed Action. Seattle-Tacoma International Airport (Sea-Tac Airport). Part 150 Noise Compatibility Planning Study*. Port of Seattle (PoS), http://www.airportsites.net – rev. 21st February 2015.

— (2013b). *Part 150 Noise Compatibility Study Update. Final*. Seattle: Port of Seattle (PoS).

— (2013c). *Sound Insulation Statistics*. Port of Seattle (PoS), https://www.portseattle.org/programs/sound-insulation – rev. February 22nd, 2015.

Posner, Elliot (2010). "The Lamfalussy Process: Polyarchic Origins of Networked Financial Rule-Making in the EU." In: *Experimentalist Governance in the European Union. Towards a New Architecture*. Ed. by Charles F. Sabel and Jonathan Zeitlin. Oxford etc.: Oxford University Press, pp. 43–60.

Pötschke, Manuela (2010). "Datengewinnung und Aufbereitung." In: *Handbuch der sozialwissenschaftlichen Datenanalyse*. Ed. by Christof Wolf and Henning Best. 2nd ed. Wiesbaden: VS Verlag für Sozialwissenschaften, pp. 41–64.

Potter, David (1992). "The Autonomy of the Third World States within the Global Economy." In: *Global Politics. Globalization and the Nation State*. Ed. by Anthony G. McGrew and Paul G. Lewis et. al. Wiesbaden: VS Verlag für Sozialwissenschaften, pp. 216–232.

Poulantzas, Nicos (1977). *L'Etat, le Pouvoir, le Socialisme*. Paris: Presses Universitaires de France.

Powell, Walter W. (1990). "Neither Market nor Hierarchy: Network Forms of Organization." In: *Research in Organizational Behavior* 12, pp. 295–336.

Powell, Walter W. and Paul J. DiMaggio (1983). "The Iron Cage Revisited: Institutional Isomorphism and Collective Rationality in Organizational Fields." In: *American Sociological Review* 48.II, pp. 147–160.

Prebisch, Raúl (1950). *The Economic Development of Latin America and its Principal Problems*. 3rd ed. Lake Success: United Nations Department of Economic Affairs.

Prognos (2010). *Gutachten zu den funktionalen Verflechtungen in der Metropolregion FrankfurtRheinMain (FRM). Endergebnisse, Gutachten im Auftrag der Handwerkskammer Rhein-Main (HWK), Industrie- und Handelskammer Frankfurt am Main (IHK) und der Vereinigung der hessischen Unternehmerverbände e.V.*. Reprint. Stuttgar & Berlin: Prognos AG.

Putnam, Robert (1993). *Making Democracy Work. Civic traditions in modern Italy*. Princeton: Princeton University Press.

Pütz, Marco (2004). *Regional Governance: Theoretisch-konzeptionelle Grundlagen und eine Analyse nachhaltiger Siedlungsentwicklung in der Metropolregion München*. München: oekom verlag.

Radaelli, Claudio M. (2009). "Measuring Policy Learning: Regulatory Impact Assessment in Europe." In: *Journal of European Public Policy* 16.VIII, pp. 1145–1164.

Raschke, Falk W. (2009). *Regionale Wettbewerbsvorteile. Identifikation, Analyse und Management von Clustern am Beispiel der Logistik im Rhein-Main-Gebiet*. Wiesbaden: Gabler.

Rausch, Karl-Friedrich (1999). "Der Frankfurter Flughafen als Drehscheibe des internationalen Luftverkehrs." In: *Sechtes Kolloquium Luftverkehr an der Technischen Universität Darmstadt. August-Euler-Luftfahrtpreis-Verleihung. "Der Ausbau des Frankfurter Flughafens – eine Notwendigkeit für die Region"*. Ed. by TUD. Darmstadt: Technische Universität Darmstadt (TUD), pp. 52–131.

RB (1979). *Wirtschaftsfaktor Flughafen Frankfurt/Main. Welche verkehrs- und regionalwirtschaftlichen Konsequenzen ergeben sich unter besonderer Berücksichtigung der Auswirkungen auf Arbeitsplatzangebot und Arbeitsplatzsicherheit aus dem im Planfeststellungsbeschlußals notwendig erkannten und vorgesehenen Bau der Startbahn 18 West?* München: Roland Berger & Partner GmbH.

— (1980). *Wirtschaftsfaktor Flughafen Frankfurt/Main – Vertiefungsstudie – dargestellt am Ein- und Ausgabeverhalten privater Haushalte, der Abhängigkeit von Unternehmen, dem Steueraufkommen von Gemeinden und den Multiplikatorwirkungen.* München: Roland Berger & Partner GmbH.

RDF (2000). "Codex. Allgemeine Verfahrensregeln für das Regionale Dialogforum." forum flughafen & region (FFR), Website: http://www.forum-flughafen-region.de – rev. July 3rd, 2015.

— (2001). *Jahresbericht RDF Juni 2000/Juni 2001.* Frankfurt am Main: Regionales Dialogforum Flughafen Frankfurt/Main (RDF).

– (2002a). *dialogbrief, April/Mai 2002.* Regionales Dialogforum Flughafen Frank-furt/Main (RDF), Website: http://www.forum-flughafen-region.de – rev. Febru-ary 4th, 2015.

– (2002b). *Jahresbericht RDF Juni 2001/Juni 2002.* Frankfurt am Main: Regionales Dialogforum Flughafen Frankfurt/Main (RDF).

– (2003). *Jahresbericht RDF Juli 2002/Juni 2003.* Frankfurt am Main: Regionales Dialogforum Flughafen Frankfurt/Main (RDF).

– (2004). *Jahresbericht RDF Juli 2003/Juni 2004.* Frankfurt am Main: Regionales Dialogforum Flughafen Frankfurt/Main (RDF).

– (2005). *Jahresbericht RDF Juli 2004/Juni 2005.* Frankfurt am Main: Regionales Dialogforum Flughafen Frankfurt/Main (RDF).

– (2005b). *RDF Papier zur Bewertung der Ergebnisse des Gutachtens "Praxisori-entiertes Umsetzungskonzept zur Verlagerung der Flugbewegungen in den Zeiten des geplanten Nachtflugverbots" und zum weiteren Vorgehen.* Frankfurt am Main: Regionales Dialogforum Flughafen Frankfurt/Main (RDF).

— (2006a). *Jahresbericht RDF Juli 2005/Juni 2006.* Frankfurt am Main: Re-gionales Dialogforum Flughafen Frankfurt/Main (RDF).

— (2006b). *Positionen des RDF zum Thema Kooperation von Flughäfen zur Verkehrsver-lagerung*. Positionen. Mainz: Regionales Dialogforum Flughafen Frankfurt/Main (RDF).

– (2006c). *Positionen des RDF zum Thema Volkswirtschaftlicher Nutzen des Flughafens Frankfurt/Main*. Positionen. Mainz: Regionales Dialogforum Flughafen Frank-furt/Main (RDF).

— (2007a). *Anti-Lärm-Pakt. Zusammenfasssung*. Frankfurt am Main: Regionales Dialogforum Flughafen Frankfurt/Main (RDF).

— (2007b). *Positionen des RDF zum Thema Flugbetriebsbedingte Luftschadstoffe*. Positionen. Frankfurt am Main: Regionales Dialogforum Flughafen Frankfurt/Main (RDF).

— (2007c). *Positionen des RDF zum Thema Fluglärmmonitoring*. Positionen. Frankfurt am Main: Regionales Dialogforum Flughafen Frankfurt/Main (RDF).

— (2007d). *Positionen des RDF zum Thema Kapazitätssteigernde Maßnahmen an Flughäfen*. Positionen. Königsstein im Taunus: Regionales Dialogforum Flughafen Frankfurt/Main (RDF).

— (2007e). *Positionen des RDF zum Thema lärmabhängige Entgelte*. Positionen. Frankfurt am Main: Regionales Dialogforum Flughafen Frankfurt/Main (RDF).

— (2008a). *dialogbrief, April 2008*. Regionales Dialogforum Flughafen Frankfurt/Main (RDF), Website: http://www.forum-flughafen-region.de – rev. February 4th, 2015.

— (2008b). *Jahresbericht RDF Juli 2006/Juni 2008*. Frankfurt am Main: Regionales Dialogforum Flughafen Frankfurt/Main (RDF).

— (2008c). *Positionen des RDF zum Thema Immobilienmanagement im Umfeld des Flughafen Frankfurt*. Positionen. Frankfurt am Main: Regionales Dialogforum Flughafen Frankfurt/Main (RDF).

– (2008d). *Positionen des RDF zum Thema Tieffrequenter Lärm*. Positionen. Bad Schwalbach: Regionales Dialogforum Flughafen Frankfurt/Main (RDF).

Regionalpark (2014). "Regionalparkportal. Über uns. Idee und Ziele." Regionalpark RheinMain, Website: http://www.regionalpark-rheinmain.de – rev. December 30th, 2014.

Regionalverband (2012). *Regionales Monitoring 2012. Daten und Fakten – Metropolregion FrankfurtRheinMain*. Frankfurt am Main: Regionalverband FrankfurtRheinMain.

— (2013). *The Regional Authority FrankfurtRheinMain. Structure, tasks and services*. Frankfurt am Main: Regionalverband FrankfurtRheinMain.

— (2015). *Cluster-Netzwerke*. Regionalverband FrankfurtRheinMain, Website: http://www.metropolregion-frankfurtrheinmain.de – rev. January 8th, 2015.

Rhodes, Rod A.W. (2001). *Understanding Governance. Policy Networks, Reflexivity and Accountability*. Reprint. Buckingham & Philadelphia: Open University Press.

Risse, Thomas (2008). "Regieren in "Räumen begrenzter Staatlichkeit": Zur *Reisefähigkeit* des Governance-Konzepts." In: *Governance in einer sich wandelenden Welt*. Ed. by Gunnar Folke Schuppert and Michael Zürn. Wiesbaden: VS Verlag für Sozialwissenschaften, PVS – Politische Vierteljahresschrift, Sonderheft 41/2008, pp. 149–170.

Ritzer, George (2007). *The Blackwell Companion to Globalization*. Malden etc.: Blackwell Publishing.

Robertson, Roland (1995). "Glocalization: Time-Space and Homogeneity-Hetrogenety." In: *Global Modernities*. Ed. by Mike Featherstone, Scott Lash, and Roland Robertson. London etc.: Sage Publications, pp. 25–44.

— (2012). "Globalisation or glocalisation?" In: *Journal of International Communication* 18.II, pp. 191–208.

Roller, Gerhard (1999). *Aufhebbarkeit von Bannwalderklärungen. Gutachten im Rahmen der Qualitätssicherung, erstellt im Auftrag des projektbüros der Mediationsgruppe Flughafen Frankfurt/Main*. Frankfurt am Main: Mediationsgruppe Flughafen Frankfurt/Main.

Roller, Gerhard and Jennifer Steinwachs (2001). *Die Aufhebbarkeit von Bannwalderklärungen – Eine Untersuchung aus aktuellem Anlaß*, volume 01-4 of *Sofia-Diskussionsbeiträge zur Institutionenanalyse*. Darmstadt: Sofia.

Rosenau, James N. (1992). "Citizenship in a Changing Global Order." In: *Governance without Government: Order and Change in World Politics*.

Ed. by James N. Rosenau and Ernst-Otto Czempiel. Cambridge: Cambridge University Press, pp. 272–294.

Rosenau, James N. and Ernst-Otto Czempiel (1992). *Governance without Government: Order and Change in World Politics*. Cambridge: Cambridge University Press.

Rössger, Edgar (1965). *Maßnahmen zur Erhöhung der Start- und Landekapazitäten des Flughafens Frankfurt-Rhein Main*. Berlin: Institut für Flugführung und Luftverkehr der Technischen Universität Berlin.

Rössler, Patrick (2010). *Inhaltsanalyse*. 2nd ed. Konstanz: UTB.

Rothstein, Bo (2012). "Good Governance." In: *Oxford Handbook of Governance*. Ed. by David Levi-Faur. Oxford & New York: Oxford University Press, pp. 143–154.

RP (2004). *Frankfurt: Vergeblicher Widerstand gegen Startbahn-West*. Rheinische Post.

RPDA (1999). *Regionalplan Südhessen 2000, 10. Dezember 1999*. Darmstadt: Regierungspräsidium Darmstadt (RPDA).

— (2001). "Raumordnungsverfahren nach dem Hessischen Landesplanungsgesetz für den Ausbau des Flughafens Frankfurt/Main." In: *Staatsanzeiger für das Land Hessen* 2001.XLIII, page 3693.

— (2002a). "Erklärung zur Änderung der Erklärung von Waldflächen in den Gemarkungen Sachsenhausen, Oberrad, Fechenheim, Schwanheim, Griesheim, Nied, Bergen-Enkheim und Wald, Stadt Frankfurt am Main, in den Gemarkungen Zeppelinheim und Neu-Isenburg, Landkreis Offenbach, sowie in der Gemarkung Bischofsheim, Main-Kinzig-Kreis, zu Bannwald vom 24.7.2002." In: *Staatsanzeiger für das Land Hessen* 2002.XXXII, pp. 3053–3062.

— (2002b). *Raumordnungsverfahren Flughafen Frankfurt Main. Landesplanerische Beurteilung*. Darmstadt: Regierungspräsidium Darmstadt (RPDA).

— (2002c). *Stenografisches Protokoll. Projektgruppe Flughafen. Erörterungstermin in dem Raumordnungsverfahren für das Vorhaben der Fraport AG zur Erweiterung des Flughafens Frankfurt/Main, Frankfurt am Main, 10. April 2002*. Darmstadt: Regierungspräsidium Darmstadt (RPDA).

— (2002d). *Stenografisches Protokoll. Projektgruppe Flughafen. Erörterungstermin in dem Raumordnungsverfahren für das Vorhaben der Fraport AG zur Erweiterung des Flughafens Frankfurt/Main, Frankfurt am Main, 11. April 2002.* Darmstadt: Regierungspräsidium Darmstadt (RPDA).

— (2002e). *Stenografisches Protokoll. Projektgruppe Flughafen. Erörterungstermin in dem Raumordnungsverfahren für das Vorhaben der Fraport AG zur Erweiterung des Flughafens Frankfurt/Main, Frankfurt am Main, 12. April 2002.* Darmstadt: Regierungspräsidium Darmstadt (RPDA).

— (2002f). *Stenografisches Protokoll. Projektgruppe Flughafen. Erörterungstermin in dem Raumordnungsverfahren für das Vorhaben der Fraport AG zur Erweiterung des Flughafens Frankfurt/Main, Frankfurt am Main, 13. April 2002.* Darmstadt: Regierungspräsidium Darmstadt (RPDA).

— (2002g). *Stenografisches Protokoll. Projektgruppe Flughafen. Erörterungstermin in dem Raumordnungsverfahren für das Vorhaben der Fraport AG zur Erweiterung des Flughafens Frankfurt/Main, Frankfurt am Main, 15. April 2002.* Darmstadt: Regierungspräsidium Darmstadt (RPDA).

— (2002h). *Stenografisches Protokoll. Projektgruppe Flughafen. Erörterungstermin in dem Raumordnungsverfahren für das Vorhaben der Fraport AG zur Erweiterung des Flughafens Frankfurt/Main, Frankfurt am Main, 16. April 2002.* Darmstadt: Regierungspräsidium Darmstadt (RPDA).

— (2002i). *Stenografisches Protokoll. Projektgruppe Flughafen. Erörterungstermin in dem Raumordnungsverfahren für das Vorhaben der Fraport AG zur Erweiterung des Flughafens Frankfurt/Main, Frankfurt am Main, 17. April 2002.* Darmstadt: Regierungspräsidium Darmstadt (RPDA).

— (2002j). *Stenografisches Protokoll. Projektgruppe Flughafen. Erörterungstermin in dem Raumordnungsverfahren für das Vorhaben der Fraport AG zur Erweiterung des Flughafens Frankfurt/Main, Frankfurt am Main, 18. April 2002.* Darmstadt: Regierungspräsidium Darmstadt (RPDA).

— (2002k). *Stenografisches Protokoll. Projektgruppe Flughafen. Erörterungstermin in dem Raumordnungsverfahren für das Vorhaben der Fraport AG zur Erweiterung des Flughafens Frankfurt/Main,*

*Frankfurt am Main, 19. April 2002*. Darmstadt: Regierungspräsidium Darmstadt (RPDA).

— (2002l). *Stenografisches Protokoll. Projektgruppe Flughafen. Erörterungstermin in dem Raumordnungsverfahren für das Vorhaben der Fraport AG zur Erweiterung des Flughafens Frankfurt/Main, Frankfurt am Main, 22. April 2002*. Darmstadt: Regierungspräsidium Darmstadt (RPDA).

— (2002m). *Stenografisches Protokoll. Projektgruppe Flughafen. Erörterungstermin in dem Raumordnungsverfahren für das Vorhaben der Fraport AG zur Erweiterung des Flughafens Frankfurt/Main, Frankfurt am Main, 23. April 2002*. Darmstadt: Regierungspräsidium Darmstadt (RPDA).

— (2002n). *Stenografisches Protokoll. Projektgruppe Flughafen. Erörterungstermin in dem Raumordnungsverfahren für das Vorhaben der Fraport AG zur Erweiterung des Flughafens Frankfurt/Main, Frankfurt am Main, 24. April 2002*. Darmstadt: Regierungspräsidium Darmstadt (RPDA).

— (2002o). *Stenografisches Protokoll. Projektgruppe Flughafen. Erörterungstermin in dem Raumordnungsverfahren für das Vorhaben der Fraport AG zur Erweiterung des Flughafens Frankfurt/Main, Frankfurt am Main, 8. April 2002*. Darmstadt: Regierungspräsidium Darmstadt (RPDA).

— (2002p). *Stenografisches Protokoll. Projektgruppe Flughafen. Erörterungstermin in dem Raumordnungsverfahren für das Vorhaben der Fraport AG zur Erweiterung des Flughafens Frankfurt/Main, Frankfurt am Main, 9. April 2002*. Darmstadt: Regierungspräsidium Darmstadt (RPDA).

— (2003a). *Stenografisches Protokoll: Erörterungstermin in dem Raumordnungsverfahren für das Vorhaben der Fraport AG zur Erweiterung des Flughafens Frankfurt/Main (Landebahn Nordwest), Frankfurt am Main, 10. April 2003*. Darmstadt: Regierungspräsidium Darmstadt (RPDA).

— (2003b). *Stenografisches Protokoll: Erörterungstermin in dem Raumordnungsverfahren für das Vorhaben der Fraport AG zur Erweiterung des Flughafens Frankfurt/Main (Landebahn Nordwest), Frankfurt am Main, 11. April 2003*. Darmstadt: Regierungspräsidium Darmstadt (RPDA).

— (2003c). *Stenografisches Protokoll: Erörterungstermin in dem Raumordnungsverfahren für das Vorhaben der Fraport AG zur Erweiterung des Flughafens Frankfurt/Main (Landebahn Nordwest), Frankfurt am Main, 8. April 2003*. Darmstadt: Regierungspräsidium Darmstadt (RPDA).

— (2003d). *Stenografisches Protokoll: Erörterungstermin in dem Raumordnungsverfahren für das Vorhaben der Fraport AG zur Erweiterung des Flughafens Frankfurt/Main (Landebahn Nordwest), Frankfurt am Main, 9. April 2003*. Darmstadt: Regierungspräsidium Darmstadt (RPDA).

— (2003e). *Stenografisches Protokoll: Erörterungstermin in dem Raumordnungsverfahren für das Vorhaben der Fraport AG zur Erweiterung des Flughafens Frankfurt/Main, Frankfurt am Main, 7. April 2003*. Darmstadt: Regierungspräsidium Darmstadt (RPDA).

— (2004a). *Protokoll: Erörterungstermin zum Planfeststellungsverfahren betreffend Erweiterungsmaßnahmen für die Wartung von Airbus-A380-Flugzeugen – A380-Werft – am Flughafen Frankfurt/Main, 10. Februar 2004*. Darmstadt: Regierungspräsidium Darmstadt.

— (2004b). *Protokoll: Erörterungstermin zum Planfeststellungsverfahren betreffend Erweiterungsmaßnahmen für die Wartung von Airbus-A380-Flugzeugen – A380-Werft – am Flughafen Frankfurt/Main, 12. Februar 2004*. Darmstadt: Regierungspräsidium Darmstadt.

— (2004c). *Protokoll: Erörterungstermin zum Planfeststellungsverfahren betreffend Erweiterungsmaßnahmen für die Wartung von Airbus-A380-Flugzeugen – A380-Werft – am Flughafen Frankfurt/Main, 13. Februar 2004*. Darmstadt: Regierungspräsidium Darmstadt.

— (2004d). *Protokoll: Erörterungstermin zum Planfeststellungsverfahren betreffend Erweiterungsmaßnahmen für die Wartung von Airbus-A380-Flugzeugen – A380-Werft – am Flughafen Frankfurt/Main, 15. Januar 2004*. Darmstadt: Regierungspräsidium Darmstadt.

— (2004e). *Protokoll: Erörterungstermin zum Planfeststellungsverfahren betreffend Erweiterungsmaßnahmen für die Wartung von Airbus-A380-Flugzeugen – A380-Werft – am Flughafen Frankfurt/Main, 16. Februar 2004*. Darmstadt: Regierungspräsidium Darmstadt.

— (2004f). *Protokoll: Erörterungstermin zum Planfeststellungsverfahren betreffend Erweiterungsmaßnahmen für die Wartung von Airbus-A380-Flugzeugen – A380-Werft – am Flughafen Frankfurt/Main, 16. Januar 2004*. Darmstadt: Regierungspräsidium Darmstadt.

— (2004g). *Protokoll: Erörterungstermin zum Planfeststellungsverfahren betreffend Erweiterungsmaßnahmen für die Wartung von Airbus-A380-Flugzeugen – A380-Werft – am Flughafen Frankfurt/Main, 17. Februar 2004*. Darmstadt: Regierungspräsidium Darmstadt.

— (2004h). *Protokoll: Erörterungstermin zum Planfeststellungsverfahren betreffend Erweiterungsmaßnahmen für die Wartung von Airbus-A380-Flugzeugen – A380-Werft – am Flughafen Frankfurt/Main, 19. Februar 2004*. Darmstadt: Regierungspräsidium Darmstadt.

— (2004i). *Protokoll: Erörterungstermin zum Planfeststellungsverfahren betreffend Erweiterungsmaßnahmen für die Wartung von Airbus-A380-Flugzeugen – A380-Werft – am Flughafen Frankfurt/Main, 19. Januar 2004*. Darmstadt: Regierungspräsidium Darmstadt.

— (2004j). *Protokoll: Erörterungstermin zum Planfeststellungsverfahren betreffend Erweiterungsmaßnahmen für die Wartung von Airbus-A380-Flugzeugen – A380-Werft – am Flughafen Frankfurt/Main, 2. Februar 2004*. Darmstadt: Regierungspräsidium Darmstadt.

— (2004k). *Protokoll: Erörterungstermin zum Planfeststellungsverfahren betreffend Erweiterungsmaßnahmen für die Wartung von Airbus-A380-Flugzeugen – A380-Werft – am Flughafen Frankfurt/Main, 20. Februar 2004*. Darmstadt: Regierungspräsidium Darmstadt.

— (2004l). *Protokoll: Erörterungstermin zum Planfeststellungsverfahren betreffend Erweiterungsmaßnahmen für die Wartung von Airbus-A380-Flugzeugen – A380-Werft – am Flughafen Frankfurt/Main, 20. Januar 2004*. Darmstadt: Regierungspräsidium Darmstadt.

— (2004m). *Protokoll: Erörterungstermin zum Planfeststellungsverfahren betreffend Erweiterungsmaßnahmen für die Wartung von Airbus-A380-Flugzeugen – A380-Werft – am Flughafen Frankfurt/Main, 22. Januar 2004*. Darmstadt: Regierungspräsidium Darmstadt.

— (2004n). *Protokoll: Erörterungstermin zum Planfeststellungsverfahren betreffend Erweiterungsmaßnahmen für die Wartung von Airbus-A380-Flugzeugen – A380-Werft – am Flughafen Frankfurt/Main, 23. Februar 2004*. Darmstadt: Regierungspräsidium Darmstadt.

— (2004o). *Protokoll: Erörterungstermin zum Planfeststellungsverfahren betreffend Erweiterungsmaßnahmen für die Wartung von Airbus-A380-Flugzeugen – A380-Werft – am Flughafen Frankfurt/Main, 23. Januar 2004*. Darmstadt: Regierungspräsidium Darmstadt.

— (2004p). *Protokoll: Erörterungstermin zum Planfeststellungsverfahren betreffend Erweiterungsmaßnahmen für die Wartung von Airbus-A380-Flugzeugen – A380-Werft – am Flughafen Frankfurt/Main, 24. Februar 2004*. Darmstadt: Regierungspräsidium Darmstadt.

— (2004q). *Protokoll: Erörterungstermin zum Planfeststellungsverfahren betreffend Erweiterungsmaßnahmen für die Wartung von Airbus-A380-Flugzeugen – A380-Werft – am Flughafen Frankfurt/Main, 26. Februar 2004*. Darmstadt: Regierungspräsidium Darmstadt.

— (2004r). *Protokoll: Erörterungstermin zum Planfeststellungsverfahren betreffend Erweiterungsmaßnahmen für die Wartung von Airbus-A380-Flugzeugen – A380-Werft – am Flughafen Frankfurt/Main, 26. Januar 2004*. Darmstadt: Regierungspräsidium Darmstadt.

— (2004s). *Protokoll: Erörterungstermin zum Planfeststellungsverfahren betreffend Erweiterungsmaßnahmen für die Wartung von Airbus-A380-Flugzeugen – A380-Werft – am Flughafen Frankfurt/Main, 27. Januar 2004*. Darmstadt: Regierungspräsidium Darmstadt.

— (2004t). *Protokoll: Erörterungstermin zum Planfeststellungsverfahren betreffend Erweiterungsmaßnahmen für die Wartung von Airbus-A380-Flugzeugen – A380-Werft – am Flughafen Frankfurt/Main, 29. Januar 2004*. Darmstadt: Regierungspräsidium Darmstadt.

— (2004u). *Protokoll: Erörterungstermin zum Planfeststellungsverfahren betreffend Erweiterungsmaßnahmen für die Wartung von Airbus-A380-Flugzeugen – A380-Werft – am Flughafen Frankfurt/Main, 3. Februar 2004*. Darmstadt: Regierungspräsidium Darmstadt.

— (2004v). *Protokoll: Erörterungstermin zum Planfeststellungsverfahren betreffend Erweiterungsmaßnahmen für die Wartung von Airbus-A380-Flugzeugen – A380-Werft – am Flughafen Frankfurt/Main, 30. Januar 2004*. Darmstadt: Regierungspräsidium Darmstadt.

— (2004w). *Protokoll: Erörterungstermin zum Planfeststellungsverfahren betreffend Erweiterungsmaßnahmen für die Wartung von Airbus-A380-Flugzeugen – A380-Werft – am Flughafen Frankfurt/Main, 5. Februar 2004*. Darmstadt: Regierungspräsidium Darmstadt.

— (2004x). *Protokoll: Erörterungstermin zum Planfeststellungsverfahren betreffend Erweiterungsmaßnahmen für die Wartung von Airbus-A380-Flugzeugen – A380-Werft – am Flughafen Frankfurt/Main, 6. Februar 2004.* Darmstadt: Regierungspräsidium Darmstadt.

— (2004y). *Protokoll: Erörterungstermin zum Planfeststellungsverfahren betreffend Erweiterungsmaßnahmen für die Wartung von Airbus-A380-Flugzeugen – A380-Werft – am Flughafen Frankfurt/Main, 9. Februar 2004.* Darmstadt: Regierungspräsidium Darmstadt.

— (2005a). *Protokoll: Erörterungstermin zum Planfeststellungsverfahren für den geplanten Ausbau des Flughafen Frankfurt Main, 04. Oktober 2005, Offenbach am Main.* Darmstadt: Regierungspräsidium Darmstadt (RPDA).

— (2005b). *Protokoll: Erörterungstermin zum Planfeststellungsverfahren für den geplanten Ausbau des Flughafen Frankfurt Main, 10. Oktober 2005, Offenbach am Main.* Darmstadt: Regierungspräsidium Darmstadt (RPDA).

— (2005c). *Protokoll: Erörterungstermin zum Planfeststellungsverfahren für den geplanten Ausbau des Flughafen Frankfurt Main, 11. Oktober 2005, Offenbach am Main.* Darmstadt: Regierungspräsidium Darmstadt (RPDA).

— (2005d). *Protokoll: Erörterungstermin zum Planfeststellungsverfahren für den geplanten Ausbau des Flughafen Frankfurt Main, 12. September 2005, Offenbach am Main.* Darmstadt: Regierungspräsidium Darmstadt (RPDA).

— (2005e). *Protokoll: Erörterungstermin zum Planfeststellungsverfahren für den geplanten Ausbau des Flughafen Frankfurt Main, 13. Oktober 2005, Offenbach am Main.* Darmstadt: Regierungspräsidium Darmstadt (RPDA).

— (2005f). *Protokoll: Erörterungstermin zum Planfeststellungsverfahren für den geplanten Ausbau des Flughafen Frankfurt Main, 13. September 2005, Offenbach am Main.* Darmstadt: Regierungspräsidium Darmstadt (RPDA).

— (2005g). *Protokoll: Erörterungstermin zum Planfeststellungsverfahren für den geplanten Ausbau des Flughafen Frankfurt Main, 14. Oktober 2005, Offenbach am Main.* Darmstadt: Regierungspräsidium Darmstadt (RPDA).

— (2005h). *Protokoll: Erörterungstermin zum Planfeststellungsverfahren für den geplanten Ausbau des Flughafen Frankfurt Main, 15. September 2005, Offenbach am Main.* Darmstadt: Regierungspräsidium Darmstadt (RPDA).

— (2005i). *Protokoll: Erörterungstermin zum Planfeststellungsverfahren für den geplanten Ausbau des Flughafen Frankfurt Main, 16. September 2005, Offenbach am Main.* Darmstadt: Regierungspräsidium Darmstadt (RPDA).

— (2005j). *Protokoll: Erörterungstermin zum Planfeststellungsverfahren für den geplanten Ausbau des Flughafen Frankfurt Main, 17. Oktober 2005, Offenbach am Main.* Darmstadt: Regierungspräsidium Darmstadt (RPDA).

— (2005k). *Protokoll: Erörterungstermin zum Planfeststellungsverfahren für den geplanten Ausbau des Flughafen Frankfurt Main, 18. Oktober 2005, Offenbach am Main.* Darmstadt: Regierungspräsidium Darmstadt (RPDA).

— (2005l). *Protokoll: Erörterungstermin zum Planfeststellungsverfahren für den geplanten Ausbau des Flughafen Frankfurt Main, 19. September 2005, Offenbach am Main.* Darmstadt: Regierungspräsidium Darmstadt (RPDA).

— (2005m). *Protokoll: Erörterungstermin zum Planfeststellungsverfahren für den geplanten Ausbau des Flughafen Frankfurt Main, 20. Oktober 2005, Offenbach am Main.* Darmstadt: Regierungspräsidium Darmstadt (RPDA).

— (2005n). *Protokoll: Erörterungstermin zum Planfeststellungsverfahren für den geplanten Ausbau des Flughafen Frankfurt Main, 20. September 2005, Offenbach am Main.* Darmstadt: Regierungspräsidium Darmstadt (RPDA).

— (2005o). *Protokoll: Erörterungstermin zum Planfeststellungsverfahren für den geplanten Ausbau des Flughafen Frankfurt Main, 21. Oktober 2005, Offenbach am Main.* Darmstadt: Regierungspräsidium Darmstadt (RPDA).

— (2005p). *Protokoll: Erörterungstermin zum Planfeststellungsverfahren für den geplanten Ausbau des Flughafen Frankfurt Main, 22. September 2005, Offenbach am Main.* Darmstadt: Regierungspräsidium Darmstadt (RPDA).

— (2005q). *Protokoll: Erörterungstermin zum Planfeststellungsverfahren für den geplanten Ausbau des Flughafen Frankfurt Main, 23. September 2005, Offenbach am Main.* Darmstadt: Regierungspräsidium Darmstadt (RPDA).

— (2005r). *Protokoll: Erörterungstermin zum Planfeststellungsverfahren für den geplanten Ausbau des Flughafen Frankfurt Main, 24. Oktober 2005, Offenbach am Main.* Darmstadt: Regierungspräsidium Darmstadt (RPDA).

— (2005s). *Protokoll: Erörterungstermin zum Planfeststellungsverfahren für den geplanten Ausbau des Flughafen Frankfurt Main, 25. Oktober 2005, Offenbach am Main.* Darmstadt: Regierungspräsidium Darmstadt (RPDA).

— (2005t). *Protokoll: Erörterungstermin zum Planfeststellungsverfahren für den geplanten Ausbau des Flughafen Frankfurt Main, 26. September 2005, Offenbach am Main.* Darmstadt: Regierungspräsidium Darmstadt (RPDA).

— (2005u). *Protokoll: Erörterungstermin zum Planfeststellungsverfahren für den geplanten Ausbau des Flughafen Frankfurt Main, 27. September 2005, Offenbach am Main.* Darmstadt: Regierungspräsidium Darmstadt (RPDA).

— (2005v). *Protokoll: Erörterungstermin zum Planfeststellungsverfahren für den geplanten Ausbau des Flughafen Frankfurt Main, 29. September 2005, Offenbach am Main.* Darmstadt: Regierungspräsidium Darmstadt (RPDA).

— (2005w). *Protokoll: Erörterungstermin zum Planfeststellungsverfahren für den geplanten Ausbau des Flughafen Frankfurt Main, 30. September 2005, Offenbach am Main.* Darmstadt: Regierungspräsidium Darmstadt (RPDA).

— (2005x). *Protokoll: Erörterungstermin zum Planfeststellungsverfahren für den geplanten Ausbau des Flughafen Frankfurt Main, 6. Oktober 2005, Offenbach am Main.* Darmstadt: Regierungspräsidium Darmstadt (RPDA).

— (2005y). *Protokoll: Erörterungstermin zum Planfeststellungsverfahren für den geplanten Ausbau des Flughafen Frankfurt Main, 7. Oktober 2005, Offenbach am Main.* Darmstadt: Regierungspräsidium Darmstadt (RPDA).

— (2006a). *Protokoll: Erörterungstermin zum Planfeststellungsverfahren für den geplanten Ausbau des Flughafen Frankfurt Main, 02. Februar 2006, Offenbach am Main*. Darmstadt: Regierungspräsidium Darmstadt (RPDA).

— (2006b). *Protokoll: Erörterungstermin zum Planfeststellungsverfahren für den geplanten Ausbau des Flughafen Frankfurt Main, 03. Februar 2006, Offenbach am Main*. Darmstadt: Regierungspräsidium Darmstadt (RPDA).

— (2006c). *Protokoll: Erörterungstermin zum Planfeststellungsverfahren für den geplanten Ausbau des Flughafen Frankfurt Main, 06. Februar 2006, Offenbach am Main*. Darmstadt: Regierungspräsidium Darmstadt (RPDA).

— (2006d). *Protokoll: Erörterungstermin zum Planfeststellungsverfahren für den geplanten Ausbau des Flughafen Frankfurt Main, 07. Februar 2006, Offenbach am Main*. Darmstadt: Regierungspräsidium Darmstadt (RPDA).

— (2006e). *Protokoll: Erörterungstermin zum Planfeststellungsverfahren für den geplanten Ausbau des Flughafen Frankfurt Main, 07. Februar 2006, Offenbach am Main*. Darmstadt: Regierungspräsidium Darmstadt (RPDA).

RPDA (2006f). *Protokoll: Erörterungstermin zum Planfeststellungsverfahren für den geplanten Ausbau des Flughafen Frankfurt Main, 09. Februar 2006, Offenbach am Main*. Darmstadt: Regierungspräsidium Darmstadt (RPDA).

— (2006g). *Protokoll: Erörterungstermin zum Planfeststellungsverfahren für den geplanten Ausbau des Flughafen Frankfurt Main, 10. Februar 2006, Offenbach am Main*. Darmstadt: Regierungspräsidium Darmstadt (RPDA).

— (2006h). *Protokoll: Erörterungstermin zum Planfeststellungsverfahren für den geplanten Ausbau des Flughafen Frankfurt Main, 10. Januar 2006, Offenbach am Main*. Darmstadt: Regierungspräsidium Darmstadt (RPDA).

— (2006i). *Protokoll: Erörterungstermin zum Planfeststellungsverfahren für den geplanten Ausbau des Flughafen Frankfurt Main, 12. Januar 2006, Offenbach am Main*. Darmstadt: Regierungspräsidium Darmstadt (RPDA).

— (2006j). *Protokoll: Erörterungstermin zum Planfeststellungsverfahren für den geplanten Ausbau des Flughafen Frankfurt Main, 13. Januar 2006, Offenbach am Main*. Darmstadt: Regierungspräsidium Darmstadt (RPDA).

— (2006k). *Protokoll: Erörterungstermin zum Planfeststellungsverfahren für den geplanten Ausbau des Flughafen Frankfurt Main, 14. Februar 2006, Offenbach am Main*. Darmstadt: Regierungspräsidium Darmstadt (RPDA).

— (2006l). *Protokoll: Erörterungstermin zum Planfeststellungsverfahren für den geplanten Ausbau des Flughafen Frankfurt Main, 16. Januar 2006, Offenbach am Main*. Darmstadt: Regierungspräsidium Darmstadt (RPDA).

— (2006m). *Protokoll: Erörterungstermin zum Planfeststellungsverfahren für den geplanten Ausbau des Flughafen Frankfurt Main, 17. Januar 2006, Offenbach am Main*. Darmstadt: Regierungspräsidium Darmstadt (RPDA).

— (2006n). *Protokoll: Erörterungstermin zum Planfeststellungsverfahren für den geplanten Ausbau des Flughafen Frankfurt Main, 19. Januar 2006, Offenbach am Main*. Darmstadt: Regierungspräsidium Darmstadt (RPDA).

— (2006o). *Protokoll: Erörterungstermin zum Planfeststellungsverfahren für den geplanten Ausbau des Flughafen Frankfurt Main, 20. Januar 2006, Offenbach am Main*. Darmstadt: Regierungspräsidium Darmstadt (RPDA).

— (2006p). *Protokoll: Erörterungstermin zum Planfeststellungsverfahren für den geplanten Ausbau des Flughafen Frankfurt Main, 23. Januar 2006, Offenbach am Main*. Darmstadt: Regierungspräsidium Darmstadt (RPDA).

— (2006q). *Protokoll: Erörterungstermin zum Planfeststellungsverfahren für den geplanten Ausbau des Flughafen Frankfurt Main, 24. Januar 2006, Offenbach am Main*. Darmstadt: Regierungspräsidium Darmstadt (RPDA).

— (2006r). *Protokoll: Erörterungstermin zum Planfeststellungsverfahren für den geplanten Ausbau des Flughafen Frankfurt Main, 26. Januar 2006, Offenbach am Main*. Darmstadt: Regierungspräsidium Darmstadt (RPDA).

— (2006s). *Protokoll: Erörterungstermin zum Planfeststellungsverfahren für den geplanten Ausbau des Flughafen Frankfurt Main, 27. Januar 2006, Offenbach am Main*. Darmstadt: Regierungspräsidium Darmstadt (RPDA).

— (2006t). *Protokoll: Erörterungstermin zum Planfeststellungsverfahren für den geplanten Ausbau des Flughafen Frankfurt Main, 30. Januar 2006, Offenbach am Main*. Darmstadt: Regierungspräsidium Darmstadt (RPDA).

— (2006u). *Protokoll: Erörterungstermin zum Planfeststellungsverfahren für den geplanten Ausbau des Flughafen Frankfurt Main, 31. Januar 2006, Offenbach am Main*. Darmstadt: Regierungspräsidium Darmstadt (RPDA).

— (2006v). *Protokoll: Erörterungstermin zum Planfeststellungsverfahren für den geplanten Ausbau des Flughafen Frankfurt Main, 5. Januar 2006, Offenbach am Main*. Darmstadt: Regierungspräsidium Darmstadt (RPDA).

— (2006w). *Protokoll: Erörterungstermin zum Planfeststellungsverfahren für den geplanten Ausbau des Flughafen Frankfurt Main, 6. Januar 2006, Offenbach am Main*. Darmstadt: Regierungspräsidium Darmstadt (RPDA).

— (2006x). *Protokoll: Erörterungstermin zum Planfeststellungsverfahren für den geplanten Ausbau des Flughafen Frankfurt Main, 9. Januar 2006, Offenbach am Main*. Darmstadt: Regierungspräsidium Darmstadt (RPDA).

– (2015). *Planung & Verkehr. Verkehr. Luftverkehr. Flughafenausbau.* Regierungsprä-sidium Darmstadt (RPDA), Website: http://www.rp-darmstadt.hessen.de – rev. February 14th, 2015.

– (2005bg). *Informationen für die Teilnehmerinnen und Teilnehmer am Erörterung-stermin im Planfeststellungsverfahren für den geplanten Ausbau des Flughafens Frankfurt Main ab 12. September 2005, 10.00 Uhr in der Stadthalle Offenbach, Waldstraße 312, 63071 Offenbach am Main*. Darmstadt: Regierungspräsidium Darmstadt (RPDA).

– (2006ba). *Mitteilung: Einzelheiten zur Weiterführung des Erörterungstermins aufgrund von durch Einsichtnahme in die Verfahrensakten der Anhörungsbehörde gewonnenen neuen Erkenntnissen, 19. Januar 2006*. Regierungspräsidium Darmstadt (RPDA).

– (2006ax). *Planfeststellungsverfahren für den Ausbau de Flughafens Frankfurt Main. Stellungnahme der Anhörungsbehörde vom 29. September 2006*. Darmstadt: Regierungspräsidium Darmstadt (RPDA).

– (2005be). *Pressemitteilung: Ab Freitag öffentliche Bekanntmachung des Flughafen-Erörterungstermins, 19. August 2005*. Regierungspräsidium Darmstadt (RPDA).

– (2004a[a]). *Pressemitteilung: Befangenheitsanträge abgelehnt – Verfahrensfragen im Vordergrund, 19. Januar 2004*. Regierungspräsidium Darmstadt.

– (2005bf). *Pressemitteilung: EÖT Flughafenausbau: Abbruch- und Aussetzungsanträge abgelehnt, 21. September 2005*. Regierungspräsidium Darmstadt (RPDA).

– (2004a[b]). *Pressemitteilung: ET A 380 Wartungshalle: Verfahrensfragen und Antragsgegenstand wurden beraten, 20. Januar 2004*. Regierungspräsidium Darmstadt.

– (2006az). *Pressemitteilung: ET Flughafenausbau: Nacherörterung beginnt, 22. März 2006*. Regierungspräsidium Darmstadt (RPDA).

– (2006ay). *Pressemitteilung: Fortsetzung des Erörterungstermins. Zur Erörterung von durch Einsichtnahme in die Verfahrensakten der Anhörungsbehörde gewonnenen neuen Erkenntnissen, 14. März 2006*. Regierungspräsidium Darmstadt (RPDA).

– (2004a[c]). *Protokoll: Erörterungstermin zum Planfeststellungsverfahren betreffend Erweiterungsmaßnahmen für die Wartung von Airbus-A380-Flugzeugen – A380-Werft – am Flughafen Frankfurt/Main, 1. März 2004*. Darmstadt: Regierungspräsidium Darmstadt.

– (2004a[d]). *Protokoll: Erörterungstermin zum Planfeststellungsverfahren betreffend Erweiterungsmaßnahmen für die Wartung von Airbus-A380-Flugzeugen – A380-Werft – am Flughafen Frankfurt/Main, 2. März 2004*. Darmstadt: Regierungspräsidium Darmstadt.

– (2004aa). *Protokoll: Erörterungstermin zum Planfeststellungsverfahren betreffend Erweiterungsmaßnahmen für die Wartung von Airbus-A380-Flugzeugen – A380-Werft – am Flughafen Frankfurt/Main, 27. Februar 2004*. Darmstadt: Regierungspräsidium Darmstadt.

– (2004a[e]). *Protokoll: Erörterungstermin zum Planfeststellungsverfahren betreffend Erweiterungsmaßnahmen für die Wartung von Airbus-A380-Flugzeugen – A380-Werft – am Flughafen Frankfurt/Main, 4. März 2004*. Darmstadt: Regierungspräsidium Darmstadt.

– (2004a[f]). *Protokoll: Erörterungstermin zum Planfeststellungsverfahren betreffend Erweiterungsmaßnahmen für die Wartung von Airbus-A380-Flugzeugen – A380-Werft – am Flughafen Frankfurt/Main, 5. März 2004*. Darmstadt: Regierungspräsidium Darmstadt.

– (2005ae). *Protokoll: Erörterungstermin zum Planfeststellungsverfahren für den geplanten Ausbau des Flughafen Frankfurt Main, 1. November 2005, Offenbach am Main*. Darmstadt: Regierungspräsidium Darmstadt (RPDA).

– (2006an). *Protokoll: Erörterungstermin zum Planfeststellungsverfahren für den geplanten Ausbau des Flughafen Frankfurt Main, 10. März 2006, Offenbach am Main*. Darmstadt: Regierungspräsidium Darmstadt (RPDA).

– (2005ai). *Protokoll: Erörterungstermin zum Planfeststellungsverfahren für den geplanten Ausbau des Flughafen Frankfurt Main, 10. November 2005, Offenbach am Main*. Darmstadt: Regierungspräsidium Darmstadt (RPDA).

– (2005aj). *Protokoll: Erörterungstermin zum Planfeststellungsverfahren für den geplanten Ausbau des Flughafen Frankfurt Main, 11. November 2005, Offenbach am Main*. Darmstadt: Regierungspräsidium Darmstadt (RPDA).

RPDA (2005az). *Protokoll: Erörterungstermin zum Planfeststellungsverfahren für den geplanten Ausbau des Flughafen Frankfurt Main, 12. Dezember 2005, Offenbach am Main*. Darmstadt: Regierungspräsidium Darmstadt (RPDA).

– (2005ba). *Protokoll: Erörterungstermin zum Planfeststellungsverfahren für den geplanten Ausbau des Flughafen Frankfurt Main, 13. Dezember 2005, Offenbach am Main*. Darmstadt: Regierungspräsidium Darmstadt (RPDA).

– (2006ao). *Protokoll: Erörterungstermin zum Planfeststellungsverfahren für den geplanten Ausbau des Flughafen Frankfurt Main, 13. März 2006, Offenbach am Main*. Darmstadt: Regierungspräsidium Darmstadt (RPDA).

– (2006ap). *Protokoll: Erörterungstermin zum Planfeststellungsverfahren für den geplanten Ausbau des Flughafen Frankfurt Main, 14. März 2006, Offenbach am Main.* Darmstadt: Regierungspräsidium Darmstadt (RPDA).

– (2005ak). *Protokoll: Erörterungstermin zum Planfeststellungsverfahren für den geplanten Ausbau des Flughafen Frankfurt Main, 14. November 2005, Offenbach am Main.* Darmstadt: Regierungspräsidium Darmstadt (RPDA).

– (2005bb). *Protokoll: Erörterungstermin zum Planfeststellungsverfahren für den geplanten Ausbau des Flughafen Frankfurt Main, 15. Dezember 2005, Offenbach am Main.* Darmstadt: Regierungspräsidium Darmstadt (RPDA).

– (2005al). *Protokoll: Erörterungstermin zum Planfeststellungsverfahren für den geplanten Ausbau des Flughafen Frankfurt Main, 15. November 2005, Offenbach am Main.* Darmstadt: Regierungspräsidium Darmstadt (RPDA).

– (2005bc). *Protokoll: Erörterungstermin zum Planfeststellungsverfahren für den geplanten Ausbau des Flughafen Frankfurt Main, 16. Dezember 2005, Offenbach am Main.* Darmstadt: Regierungspräsidium Darmstadt (RPDA).

– (2006aa). *Protokoll: Erörterungstermin zum Planfeststellungsverfahren für den geplanten Ausbau des Flughafen Frankfurt Main, 16. Februar 2006, Offenbach am Main.* Darmstadt: Regierungspräsidium Darmstadt (RPDA).

– (2006aq). *Protokoll: Erörterungstermin zum Planfeststellungsverfahren für den geplanten Ausbau des Flughafen Frankfurt Main, 16. März 2006, Offenbach am Main.* Darmstadt: Regierungspräsidium Darmstadt (RPDA).

– (2006ab). *Protokoll: Erörterungstermin zum Planfeststellungsverfahren für den geplanten Ausbau des Flughafen Frankfurt Main, 17. Februar 2006, Offenbach am Main.* Darmstadt: Regierungspräsidium Darmstadt (RPDA).

– (2006ar). *Protokoll: Erörterungstermin zum Planfeststellungsverfahren für den geplanten Ausbau des Flughafen Frankfurt Main, 17. März 2006, Offenbach am Main.* Darmstadt: Regierungspräsidium Darmstadt (RPDA).

– (2005am). *Protokoll: Erörterungstermin zum Planfeststellungsverfahren für den geplanten Ausbau des Flughafen Frankfurt Main, 17. November 2005, Offenbach am Main.* Darmstadt: Regierungspräsidium Darmstadt (RPDA).

– (2005an). *Protokoll: Erörterungstermin zum Planfeststellungsverfahren für den geplanten Ausbau des Flughafen Frankfurt Main, 18. November 2005, Offenbach am Main.* Darmstadt: Regierungspräsidium Darmstadt (RPDA).

– (2005bd). *Protokoll: Erörterungstermin zum Planfeststellungsverfahren für den geplanten Ausbau des Flughafen Frankfurt Main, 19. Dezember 2005, Offenbach am Main.* Darmstadt: Regierungspräsidium Darmstadt (RPDA).

– (2005au). *Protokoll: Erörterungstermin zum Planfeststellungsverfahren für den geplanten Ausbau des Flughafen Frankfurt Main, 2. Dezember 2005, Offenbach am Main.* Darmstadt: Regierungspräsidium Darmstadt (RPDA).

– (2006ai). *Protokoll: Erörterungstermin zum Planfeststellungsverfahren für den geplanten Ausbau des Flughafen Frankfurt Main, 2. März 2006, Offenbach am Main.* Darmstadt: Regierungspräsidium Darmstadt (RPDA).

– (2006ac). *Protokoll: Erörterungstermin zum Planfeststellungsverfahren für den geplanten Ausbau des Flughafen Frankfurt Main, 20. Februar 2006, Offenbach am Main.* Darmstadt: Regierungspräsidium Darmstadt (RPDA).

– (2006as). *Protokoll: Erörterungstermin zum Planfeststellungsverfahren für den geplanten Ausbau des Flughafen Frankfurt Main, 20. März 2006, Offenbach am Main.* Darmstadt: Regierungspräsidium Darmstadt (RPDA).

– (2006ad). *Protokoll: Erörterungstermin zum Planfeststellungsverfahren für den geplanten Ausbau des Flughafen Frankfurt Main, 21. Februar 2006, Offenbach am Main.* Darmstadt: Regierungspräsidium Darmstadt (RPDA).

– (2006at). *Protokoll: Erörterungstermin zum Planfeststellungsverfahren für den geplanten Ausbau des Flughafen Frankfurt Main, 21. März 2006, Offenbach am Main.* Darmstadt: Regierungspräsidium Darmstadt (RPDA).

– (2005ao). *Protokoll: Erörterungstermin zum Planfeststellungsverfahren für den geplanten Ausbau des Flughafen Frankfurt Main, 21. November 2005, Offenbach am Main*. Darmstadt: Regierungspräsidium Darmstadt (RPDA).

– (2005ap). *Protokoll: Erörterungstermin zum Planfeststellungsverfahren für den geplanten Ausbau des Flughafen Frankfurt Main, 22. November 2005, Offenbach am Main*. Darmstadt: Regierungspräsidium Darmstadt (RPDA).

– (2006ae). *Protokoll: Erörterungstermin zum Planfeststellungsverfahren für den geplanten Ausbau des Flughafen Frankfurt Main, 23. Februar 2006, Offenbach am Main*. Darmstadt: Regierungspräsidium Darmstadt (RPDA).

– (2006au). *Protokoll: Erörterungstermin zum Planfeststellungsverfahren für den geplanten Ausbau des Flughafen Frankfurt Main, 23. März 2006, Offenbach am Main*. Darmstadt: Regierungspräsidium Darmstadt (RPDA).

– (2006af). *Protokoll: Erörterungstermin zum Planfeststellungsverfahren für den geplanten Ausbau des Flughafen Frankfurt Main, 24. Februar 2006, Offenbach am Main*. Darmstadt: Regierungspräsidium Darmstadt (RPDA).

– (2006av). *Protokoll: Erörterungstermin zum Planfeststellungsverfahren für den geplanten Ausbau des Flughafen Frankfurt Main, 24. März 2006, Offenbach am Main*. Darmstadt: Regierungspräsidium Darmstadt (RPDA).

– (2005aq). *Protokoll: Erörterungstermin zum Planfeststellungsverfahren für den geplanten Ausbau des Flughafen Frankfurt Main, 24. November 2005, Offenbach am Main*. Darmstadt: Regierungspräsidium Darmstadt (RPDA).

– (2005ar). *Protokoll: Erörterungstermin zum Planfeststellungsverfahren für den geplanten Ausbau des Flughafen Frankfurt Main, 25. November 2005, Offenbach am Main*. Darmstadt: Regierungspräsidium Darmstadt (RPDA).

– (2006ag). *Protokoll: Erörterungstermin zum Planfeststellungsverfahren für den geplanten Ausbau des Flughafen Frankfurt Main, 27. Februar 2006, Offenbach am Main*. Darmstadt: Regierungspräsidium Darmstadt (RPDA).

– (2006aw). *Protokoll: Erörterungstermin zum Planfeststellungsverfahren für den geplanten Ausbau des Flughafen Frankfurt Main, 27. März 2006, Offenbach am Main.* Darmstadt: Regierungspräsidium Darmstadt (RPDA).

– (2005aa). *Protokoll: Erörterungstermin zum Planfeststellungsverfahren für den geplanten Ausbau des Flughafen Frankfurt Main, 27. Oktober 2005, Offenbach am Main.* Darmstadt: Regierungspräsidium Darmstadt (RPDA).

– (2005ab). *Protokoll: Erörterungstermin zum Planfeststellungsverfahren für den geplanten Ausbau des Flughafen Frankfurt Main, 27. Oktober 2005, Offenbach am Main.* Darmstadt: Regierungspräsidium Darmstadt (RPDA).

– (2006ah). *Protokoll: Erörterungstermin zum Planfeststellungsverfahren für den geplanten Ausbau des Flughafen Frankfurt Main, 28. Februar 2006, Offenbach am Main.* Darmstadt: Regierungspräsidium Darmstadt (RPDA).

– (2005as). *Protokoll: Erörterungstermin zum Planfeststellungsverfahren für den geplanten Ausbau des Flughafen Frankfurt Main, 28. November 2005, Offenbach am Main.* Darmstadt: Regierungspräsidium Darmstadt (RPDA).

– (2005ac). *Protokoll: Erörterungstermin zum Planfeststellungsverfahren für den geplanten Ausbau des Flughafen Frankfurt Main, 28. Oktober 2005, Offenbach am Main.* Darmstadt: Regierungspräsidium Darmstadt (RPDA).

– (2005at). *Protokoll: Erörterungstermin zum Planfeststellungsverfahren für den geplanten Ausbau des Flughafen Frankfurt Main, 29. November 2005, Offenbach am Main.* Darmstadt: Regierungspräsidium Darmstadt (RPDA).

– (2006aj). *Protokoll: Erörterungstermin zum Planfeststellungsverfahren für den geplanten Ausbau des Flughafen Frankfurt Main, 3. März 2006, Offenbach am Main.* Darmstadt: Regierungspräsidium Darmstadt (RPDA).

– (2005ad). *Protokoll: Erörterungstermin zum Planfeststellungsverfahren für den geplanten Ausbau des Flughafen Frankfurt Main, 31. Oktober 2005, Offenbach am Main.* Darmstadt: Regierungspräsidium Darmstadt (RPDA).

– (2005af). *Protokoll: Erörterungstermin zum Planfeststellungsverfahren für den geplanten Ausbau des Flughafen Frankfurt Main, 4. November 2005, Offenbach am Main.* Darmstadt: Regierungspräsidium Darmstadt (RPDA).

– (2005av). *Protokoll: Erörterungstermin zum Planfeststellungsverfahren für den geplanten Ausbau des Flughafen Frankfurt Main, 5. Dezember 2005, Offen-bach am Main.* Darmstadt: Regierungspräsidium Darmstadt (RPDA).

– (2005aw). *Protokoll: Erörterungstermin zum Planfeststellungsverfahren für den geplanten Ausbau des Flughafen Frankfurt Main, 6. Dezember 2005, Offenbach am Main.* Darmstadt: Regierungspräsidium Darmstadt (RPDA).

– (2006ak). *Protokoll: Erörterungstermin zum Planfeststellungsverfahren für den geplanten Ausbau des Flughafen Frankfurt Main, 6. März 2006, Offenbach am Main.* Darmstadt: Regierungspräsidium Darmstadt (RPDA).

– (2006al). *Protokoll: Erörterungstermin zum Planfeststellungsverfahren für den geplanten Ausbau des Flughafen Frankfurt Main, 7. März 2006, Offenbach am Main.* Darmstadt: Regierungspräsidium Darmstadt (RPDA).

– (2005ag). *Protokoll: Erörterungstermin zum Planfeststellungsverfahren für den geplanten Ausbau des Flughafen Frankfurt Main, 7. November 2005, Offenbach am Main.* Darmstadt: Regierungspräsidium Darmstadt (RPDA).

– (2005ax). *Protokoll: Erörterungstermin zum Planfeststellungsverfahren für den geplanten Ausbau des Flughafen Frankfurt Main, 8. Dezember 2005, Offenbach am Main.* Darmstadt: Regierungspräsidium Darmstadt (RPDA).

– (2005ah). *Protokoll: Erörterungstermin zum Planfeststellungsverfahren für den geplanten Ausbau des Flughafen Frankfurt Main, 8. November 2005, Offenbach am Main.* Darmstadt: Regierungspräsidium Darmstadt (RPDA).

– (2005ay). *Protokoll: Erörterungstermin zum Planfeststellungsverfahren für den geplanten Ausbau des Flughafen Frankfurt Main, 9. Dezember 2005, Offenbach am Main.* Darmstadt: Regierungspräsidium Darmstadt (RPDA).

– (2006am). *Protokoll: Erörterungstermin zum Planfeststellungsverfahren für den geplanten Ausbau des Flughafen Frankfurt Main, 9. März 2006, Offenbach am Main.* Darmstadt: Regierungspräsidium Darmstadt (RPDA).

RPDA and Regionalverband (2011). *Regionalplan Südhessen/Regionaler Flächennutzungsplan 2010, 17. Oktober 2011.* Darmstadt: Regierungspräsidium Darmstadt (RPDA).

Rucht, Dieter and Stefan Sauer (1984). "Fallstudie: Startbahn West." In: *Flughafenprojekte als Politikum. Die Konflikte in Stuttgart, München und Frankfurt.* Ed. by Dieter Rucht. Frankfurt am Main & New York: Campus Verlag, pp. 195–272.

Rüdiger, Robert and Norbert Konegen (2006). *Globalisierung und Lokalisierung – Zur Neubestimmung des Kommunalen in Deutschland.* Münster: Waxmann.

Rudzio, Wolfgang (2015). *Das politische System der Bundesrepublik Deutschland.* 9th ed. Wiesbaden: Springer VS.

Rüssesheim, Dekanat (1981). *Brief des Dekanats Rüssesheim an die Abgeordneten des Hessischen Landtages, 10. Januar 1981.*

RW (2002). *Pressemitteilung: Einen "raumverträglichen" Ausbau des Frankfurter Flughafens kann es nicht geben. ROBIN WOOD kritisiert Ergebnis des Raumordnungsverfahrens, 11. Juni 2002.* Robin Wood.

Sabatier, Paul A. (2007). "The Need for Better Theories." In: *Theories of the Policy Process.* Ed. by Paul A. Sabatier. 2nd ed. Boulder: Westview Press, pp. 3–17.

Sabel, Charles F. (1994). "Learning by Monitoring: The Institutions of Economic Development." In: *The Handbook of Economic Sociology.* Ed. by Neil J. Smelser and Swedberg Richard. Princeton & New York: Princeton University Press & Russel Sage Foundation, pp. 137–165.

– (1996). "Experimental Regionalism and the Dilemmas of Regional Economic Policy." In: *Paper presented on the conference on Socio-Economic Systems of Japan, the United States, the United Kingdom, Germany, and France, Institute of Fiscal and Monetary Policy, Tokyo (Japan), 16th February 1996.*

– (2004a). "Beyond Principal-Agent Governance: Experimentalist Organizations, Learning and Accountability." In: *De Staat van de Democratie.*

*Democratie voorbij de Staat*. Ed. by Ewald R. Engelen and Monika Sie Dhian Ho. Amsterdam: Amsterdam University Press, pp. 173–195.

– (2004b). "Mondo in bottiglia o finestra sul mondo? Domande aperte sui distretti industriali nello spirito di Sebastiano Brusco." In: *Stato e Mercato* 25.I, pp. 143–158.

Sabel, Charles F. (2007). "A Real-Time Revolution in Routines." In: *The Firm as a Collaborative Community. Reconstructing Trust in the Knowledge Economy*. Ed. by Charles Heckscher and Paul S. Adler. Reprint. New York etc.: Oxford University Press, pp. 106–156.

– (2012). "Dewey, Democracy, and Democratic Experimentalism." In: *Contempo-rary Pragmatism* 9.II, pp. 35–55.

Sabel, Charles F. and William H. Simon (2004). "Destabilization Rights: How Public Law Litigation Succeeds." In: *Harvard Law Review* 117.IV, pp. 1015–1101.

– (2008). "Minimalism and Experimentalism in the Administrative State." In: *Georgetown Law Journal* 100.I, pp. 53–93.

Sabel, Charles F. and Jonathan Zeitlin (2010a). *Experimentalist Governance in the European Union. Towards a New Architecture*. Oxford etc.: Oxford University Press.

– (2010b). "Learning from the Difference: The New Architecture of Experimentalist Governance in the EU." In: *Experimentalist Governance in the European Union. Towards a New Architecture*. Ed. by Charles F. Sabel and Jonathan Zeitlin. Oxford etc.: Oxford University Press, pp. 1–28.

– (2012a). "Experimentalism in the EU: Common Ground and Persistent Differ-ences." In: *Regulation & Governance* 6.III, pp. 410–426.

– (2012b). "Experimentalist Governance." In: *Oxford Handbook of Governance*. Ed. by David Levi-Faur. Oxford & New York: Oxford University Press, pp. 169–183.

Salet, Willem, Andy Thornley, and Anton Kreukels (2003). "Institutional and Spatial Coordination in European Metropolitan Regions." In: *Metropolitan Gover-nance and Spatial Planning*. Ed. by Willem Salet, Andy Thornley, and Anton Kreukels. London & New York: Spon Press, pp. 3–19.

Sassen, Saskia (2002). "The State and Globalization." In: *The Emergence of Pri-vate Authority in Global Governance*. Ed. by Rodney Bruce Hall and Thomas J. Biersteker. Cambridge Studies in International Relations 85. Cambridge etc.: Cambridge University Press, pp. 91–112.

– (2005). "The Global City: Introducing a Concept." In: *Brown Journal of World Affairs* 12.II, pp. 27–43.

– (2008a). *Cities in a World Economy*. 3rd ed. Thousand Oaks: Pine Forge Press.

– (2008b). "Global Cities and Global City-Regions: A Comparison." In: *Global City-Regions. Trends, Theory, Policy*. Ed. by Allen John Scott. Oxford: Oxford University Press, pp. 78–95.

Saxenian, AnnaLee (1996). *Regional Advantage. Culture and Competition in Silicon Valley and Route 128*. Cambridge: Harvard University Press.

Schäfer, Lars (2011). *Mediation im öffentlichen Bereich. Eine Untersuchung ihrer rechtlichen Möglichkeiten und Grenzen am Beispiel des Mediationsverfahrens Flughafen Frankfurt/Main; mit Entwurf eine Mediationsgesetzes*. Vol. 15. Außergerichtliche Konfliktbeilegung. Hamburg: Verlag Dr. Kovač.

Schäfer-Gümbel, Thorsten (ny). *Neue Balance in der Region. Moratorium für das Terminal 3 ist Voraussetzung für Flughafen-Dialog*. Thorsten Schäfer-Gümbel, Website: http://www.schaeferguembel.de – rev. February 19th, 2015.

Schaffer, Martin and Christoph Scheck (2006). *Regionale Kooperationen im Rhein-Main-Gebiet. Anforderungen und Handlungsempfehlungen für eine zukunftsfähige Weiterentwicklung*. Vol. 18. Materialien zur Regionalentwicklung und Raumord-nung. Kaiserslautern: Technische Universität Kaiserslautern.

Scharpf, Fritz W. (1973). *Planung als politischer Prozess. Aufsätze zur Theorie der planenden Demokratie*. Frankfurt am Main: Suhrkamp Verlag.

– (1997). *Games Real Actors Play. Actor-Centered Institutionalism in Policy Research*. Theoretical Lenses on Public Policy. Boulder: Westview Press.

Scheiner, Joachim and Christian Holz-Rau (2015). *Räumliche Mobilität und Lebenslauf. Studien zu Mobilitätsbiografien und Mobilitätssozialisation*. Studien zur Mobilitätsund Verkehrsforschung. Wiesbaden: Springer VS.

Schlieffen, Katharina von (2010). “Professionalität, Profession und Mediation.” In: *Professionalisierung und Mediation*. Ed. by Katharina von Schlieffen. München: C.H. Beck, pp. 1–13.

Schmidt, Susanne K. (2010). ““Governance of Industries” – die Transformation staatsnaher Wirtschaftssektoren im Zuge von Liberalisierung und Europäisierung”. In: *Governance – Regieren in komplexen Regelsystemen*. Ed. by Arthur & Nicolai Dose Benz. Wiesbaden: VS Verlag für Sozialwissenschaften, pp. 167–218.

Schmidt-Schaun, Gerhard (1981). *Brief an Dr. Erwin Lang, MdL, Hessischer Landtag, 24. Januar 1981*.

Schnell, Rainer, Paul B. Hill, and Elke Esser (2013). *Methoden der empirischen Sozialforschung*. 10th ed. München: Oldenbourg Verlag.

Scholl, Hans J. (2010). *E-Government. information, technology and transformation*. Armonk: M.E. Sharpe.

Scholles, Frank (2008). “Das System der Projektzulassung in Deutschland.” In: *Handbuch: Theorie und Methoden der Raum- und Umweltplanung*. Ed. by Frank Scholles and Dietrich Fürst. 3rd ed. Dortmund: Rohn, pp. 100–132.

Schulze, Hans-Dietrich (2006). *Metropolitanes & Mediterranes. Beiträge aus der Humangeographie*. Vol. 107. Berliner Geographische Arbeiten. Berlin: Geographis-ches Institut der Humboldt-Universität zu Berlin.

Schulze, Verena (2013). *Safety and Security Governance. Kommunale Politik der Inneren Sicherheit aus der Perspektive des Governance-Ansatzes*. Frankfurt am Main: Verlag für Polizeiwissenschaften.

Schumpeter, Joseph Alois (2003). *Capitalism, Socialism and Democracy*. London & New York: Routledge.

Schuppert, Gunnar Folke (2005). *Der Gewährleistungsstaat – Ein Leitbild auf dem Prüfstand*. Baden-Baden: Nomos.

– (2007). “Was ist und wozu Governance¿‘ In: *Die Verwaltung. Zeitschrift für Verwaltungsrecht und Verwaltungswissenschaften* 40.IV, pp. 463–511.

– (2008). “Governance – auf der Suche nach Konturen eines “anerkannt uneindeuti-gen Begriffs”.” In: *Governance in einer sich wandelenden Welt*. Ed. by Gunnar Folke Schuppert and Michael Zürn. PVS – Politische Vierteljahresschrift, Son-derheft 41/2008. Wiesbaden: VS Verlag für Sozialwissenschaften, pp. 13–40.

– (2011). *Governance und Rechtsetzung: Grundfragen einer modernen Regelungswis-senschaft*. Baden-Baden: Nomos.

Schutter, Olivier de (2010). "The Role of Evaluation in Experimentalist Governance: Learning by Monitoring in the Establishment of the Area of Freedom, Security and Justice." In: *Experimentalist Governance in the European Union. Towards a New Architecture*. Ed. by Charles F. Sabel and Jonathan Zeitlin. Oxford etc.: Oxford University Press, pp. 261–296.

Schwedes, Oliver (2014). *Öffentliche Mobilität. Perspektiven für eine nachhaltige Verkehrsentwicklung*. 2nd ed. Wiesbaden: Springer VS.

Scott, Allen John (1988). *New Industrial Spaces. Flexible Production Organization and Regional Development in North America and Western Europe*. Vol. 3. Studies in Society and Space. London: Pion.

– (2008). "Introduction." In: *Global City-Regions. Trends, Theory, Policy*. Ed. by Allen John Scott. Oxford: Oxford University Press, pp. 1–8.

Scott, Allen John et al. (2008). "Global City-Region." In: *Global City-Regions. Trends, Theory, Policy*. Ed. by Allen John Scott. Oxford: Oxford University Press, pp. 11–30.

Sebald, Martin and Alexander Straßner (2004). *Verbände in der Bundesrepublik Deutschland. Eine Einführung*. Wiesbaden: VS Verlag für Sozialwissenschaften.

Seeger, Sarah (2008). "Die Institutionen- und Machtarchitektur der Europäischen Union mit dem Vertrag von Lissabon." In: *Lissabon in der Analyse. Der Refor-mvertrag der Europäischen Union*. Ed. by Werner Weidenfeld. Münchner Beiträge zur Europäischen Einigung, 20. Baden-Baden: Nomos, pp. 63–98.

Sellers, Jefferey and Vincent Hoffmann-Martinot (2005). "Introduction: Metropoliza-tion and Political Change." In: *Metropolization and Political Change*. Ed. by Vin-cent Hoffmann-Martinot and Jefferey Sellers. Wiesbaden: VS Verlag für Sozial-wissenschften, pp. 11–14.

Setälä, Maija (2006). "On the Problems of Responsibility and Accountability in Referendums." In: *European Journal of Political Research* 45.IV, pp. 699–721.

Shonfield, Andrew (1968). *Geplanter Kapitalismus*. Köln & Berlin: Kiepenheuer & Witsch.

Simon, Hermann (2012). *Hidden Champions. Aufbruch nach Globalia. Die Erfol-gsstrategien unbekannter Weltmarktführer*. Frankfurt am Main & New York: Campus Verlag.

Singer, Hans Peter (1949). *United Nations: Postwar Price Relations in Trade between Underdeveloped and Industrialized Countries*. Lake Success: United Nations Department of Economic Affairs.

Sinz, Manfred (2005). "Region." In: *Handwörterbuch der Raumordnung*. Ed. by ARL. 4th ed. Hannover: Verlag der Akademie für Raumforschung und Landesplanung (ARL), pp. 919–923.

Slack, Enid and Rupak Chattopadhyay (2013). "Introduction." In: *Governance and Finance of Metropolitan Areas in Federal Systems*. Ed. by Enid Slack and Rupak Chattopadhyay. Ontario: Oxford University Press, pp. 1–9.

Smith, Graham (2005). *Beyond the Ballot. 57 Democratic Innovations from Around the World. A Report for the POWER Inquiry*. London: The POWER Inquiry.

Smout, Marie-Claude (1998). "The Proper Use of Governance in International Relations." In: *International Social Science Journal* 50, pp. 81–89.

Sørensen, Eva and Jacob Torfing (2008). *Theories of Democratic Network Governance*. Paperback. Houndsmills & New York: Palgrave Macmillan.

Sorg, Jens M. (2007). *Rechts- und Staatsverständnis (USA - Islam - Europa). Bedeutung von Staat, Nation und Religion für die Identität in verschiedenen Kulturen*. Frankfurt am Main: Juristisches Seminar: Grundlagen des Rechts, Summer Term 2007, Rechts- und Staatsverständnis (USA - Islam - Europa), Prof. Dr. Ulfrid Neumann & Dr. Dr. h.c. Margareta Haller.

– (2009). *Zum Potential von Sozialkapital für die Entwicklungsanalyse. Bedeutungen und Möglichkeiten von Sozialkapital für die Entwicklungszusammenarbeit*. Frankfurt am Main: Hochschulschrift: Goethe-Universität Frankfurt.

Sorg, Jens M. et al. (2014). *Wertsicherung 2014. Eine Analyse des Fondsmarkts für Wertsicherungsstrategien*. Frankfurt am Main: GREEN finance consulting e.V.

Spannowsky, Willy, Peter Runkel, and Konrad Goppel (2010). *Raumordnungsgesetz (ROG): Kommentar*. München: C.H. Beck.

SPD (1998). *Pressemitteilung, 4. März 1998*. Sozialdemokratische Partei Deutschlands, Hessen.

– (2006). *Pressemitteilung: Walter: VGH-Entscheidung – Weitere Verzögerung des Flughafen-Ausbaus droht, 12. Januar 2006*. Sozialdemokratische Partei Deutschlands, Hessen.

SPD and GRÜNE (1995). *Koalitionsvereinbarung für die 14. Wahlperiode des Hessischen Landtages zwischen BÜNDNIS90/DIE GRÜNEN & SPD*. Wiesbaden: Sozialdemokratische Partei Deutschlands, Hessen & BÜNDNIS90/DIE GRÜNEN, Hessen.

SPIEGEL (1982). *Django mit Latte, 6. November 1982*. Der Spiegel.

– (2008). *Die Wucht des Widerstandes, 12. April 2008*. Der Spiegel, Steffen Gerth.

SSCI (2014a). *Analysis of SSCI (2014-09-22) - Experimentalist Governance*. Social Sciences Citation Index, Website: http://apps.webofknowledge.com – rev. September 22, 2014.

– (2014b). *Analysis of SSCI (2014-09-22) - Governance*. Social Sciences Citation Index, Website: http://apps.webofknowledge.com – rev. September 22, 2014.

Statista (2015). *Top 10 Fluglinien weltweit nach der Anzahl der Passagiere im Jahr 2013 (in Millionen)*. Statista, Website: http://de.statista.com – rev. January 5, 2015.

Stelkens, Paul, Heinz Joachim Bonk, and Michael Sachs (2014). *Verwaltungsverfahrensgesetz. Kommentar*. 8th ed. München: C.H. Beck.

Stephans, G. Ross and Nelson Wikstrom (2000). *Metropolitan Government and Governance. Theoretical Perspectives, Empirical Analysis, and the Future*. Oxford & New York: Oxford University Press.

Stewart Fenner, Jr. (2011). "The Primacy of Delaware and the Embeddedness of the Firm." In: *The Embedded Firm. Corporate Governance, Labor, and Finance Capitalism*. Ed. by Cynthia A. Williams and Peer Zumbansen. Cambridge & New York: Cambridge University Press, pp. 104–118.

Stockmann, Reinhard (2010). "Einleitung: Entwicklung." In: *Entwicklungstheorie. Theorien - Probleme - Strategien*. Ed. by Reinhard Stockmann, Ulrich Menzel, and Franz Nuscheler. München: Oldenbourg Wissenschaftsverlag, pp. 1–9.

Stokes, Michael C. (1997). *Plato: Apology*. Warminster: Aris & Phillipsa.

Storper, Michael (1995). "The Resurgence of Regional Economies, Ten Years Later: The Region as a Nexus of Untraded Interdependencies." In: *European Urban and Regional Studies* 2.III, pp. 191–221.

– (1997a). "Territories, Flows, and Hierarchies in the Global Economy." In: *Spaces of Globalization: Reasserting the Power of the Local*. Ed. by Kevin R. Cox. New York: Guilfor Press, pp. 19–44.

– (1997b). *The Regional World: Territorial Development in a Global Economy*. Paperback Reissue. New York & London: The Guilford Press.

– (2009). "Roepke Lecture in Economic Geography: Regional Context and Global Trade." In: *Economic Geography* 85.I, pp. 1–21.

Storper, Michael and Allen John Scott (2003). "Regions, Globalization, Development." In: *Regional Studies* 37.VI & VII, pp. 579–221.

Strange, Susan (1996). *The Retreat of the State. The Diffusion of Power in the World Economy*. Vol. 49. Cambridge Studies in International Relations. Cambridge: Cambridge University Press.

Streeck, Wolfgang and Philippe C. Schmitter (1985). "Community, Market, State – and Associations? The Prospective Contribution of Interest Governance to Social Order." In: *Private Interest Government. Beyond Market and State*. Ed. by Wolfgang Streeck and Philippe C. Schmitter. London: Sage Publications, pp. 1–29.

Strempel, Dieter (2002). "Rechtspolitische Aspekte der Mediation." In: *Handbuch Mediation*. Ed. by Fritjof Haft and Katharina von Schlieffen. München: C.H. Beck, pp. 104–136.

Stroker, Gerry (1998). "Governance as Theory: Five Propositions." In: *International Social Science Journal* 50.CLV, pp. 17–28.

Studenski, Paul (1930). *The Government of Metropolitan Areas in the United States*. New York: National Municipal League.

Svetiev, Yana (2010). "Network Competition Governance in the EU: Delegation, Decentralization, or Experimentalist Architecture¿' In: *Experimentalist Gover-nance in the European Union. Towards a New*

*Architecture*. Ed. by Charles F. Sabel and Jonathan Zeitlin. Oxford etc.: Oxford University Press, pp. 79–120.

SWR (2013). *Die Startbahn West - Deutschlands umstrittenster Flughafen-Ausbau, April 2013*. Südwestrundfunk.

Swyngedouw, Erik (1992). "The Mammon Quest. 'Glocalisation', Interspatial Competition and the Monetary Order: The Construction of New Scale." In: *Cities and Regions in the New Europe: The Global-Local Interplay and Spatial Development Strategies*. Ed. by Mick Dunford and Grigoris Kafkalas. London: Belhaven Press, pp. 39–67.

– (1997). "Neither Global nor Local. "Glocalization" and the Politics of Scale." In: *Spaces of Globalization: reasserting the power of the local*. Ed. by Kevin R. Cox. New York: Guilfor Press, pp. 137–166.

Talpin, Julien (2012). "When Democratic Innovations Let the People Decide." In: *Evaluating Democratic Innovations*. Ed. by Brigitte Geißel and Kenneth Newton. London & New York: Routledge, pp. 184–206.

taz (1991). *Neuer Prozeßgegen Startbahngegner "Aschu" Schubart*. die tageszeitung, Heide Platen.

– (2005). *Im Gewirr der Rechtsgutachten, 4. Oktober 2005*. die tageszeitung, Heide Platen.

Theisen, Heinz (1995). "Zukunftsvorsorge als Staatsaufgabe." In: *Staatswissenschaften und Staatspraxis* 6.I, pp. 111–125.

Thießen, Friedrich (2012). *Grenzen der Demokratie. Die gesellschaftliche Auseinan-dersetzung bei Großprojekten*. 2nd ed. Wiesbaden: Springer SV.

Tiebout, Charles M. (1956). "A Pure Theory of Local Expenditures." In: *Journal of Political Economy* 64.V, pp. 416–424.

Tilly, Charles (2006). "Why and How History Matters." In: *The Oxford Handbook of Contextual Political Analysis*. Ed. by Robert E. Goodin and Charles Tilly. Oxford & New York etc.: Oxford University Press, pp. 417–437.

Tömmel, Ingeborg and Amy Verdun (2009). *Innovative Governance in the European Union: The Politics of Multilevel Policymaking*. Boulder: Lynner Rinner.

Treibel, Werner (1992). *Geschichte der deutschen Verkehrsflughäfen. Eine Dokumentation von 1909 bis 1989*. Bonn: Bernard & Graefe Verlag.

Troost, Hans J. (2001). "Neue Vernetzungsstrategien in der metropolitanen Region Rhein-Main: Das Beispiel des Mediationsverfahrens Flughafen Frankfurt." In: *Metropolitane Region in der Vernetzung. Der Fall Frankfurt Rhein/Main*. Ed. by Josef Esser and Eike W. Schamp. Frankfurt am Main & New York: Campus Verlag, pp. 245–279.

Tsebelis, George (2002). *Veto Players: How Political Institutions Work*. Princeton: Princeton University Press.

Tulmets, Elsa (2010). "Experimentalist Governance in EU External Relations: Enlargement and European Neighbourhood Policy." In: *Experimentalist Governance in the European Union. Towards a New Architecture*. Ed. by Charles F. Sabel and Jonathan Zeitlin. Oxford etc.: Oxford University Press, pp. 297–324.

UN (2006). "Definition of Basic Concepts and Terminologies in Governance and Public Administration." In: *Committee of Experts on Public Administration, Fifth session, New York, 27-31 March 2006* E/C.16/2006/4.Agenda item 5, pp. 1–15.

Verdun, Amy (2012). "Experimentalist Governance in the European Union: A Commentary." In: *Regulation & Governance* 6.III, pp. 385–393.

VhU (2005). *Pressemitteilung: VhU zum Erörterungstermin Flughafen am 12. Septem-ber, 9. September 2005*. Vereinigung der hessischen Unternehmerverbände e.V. (VhU).

– (2007). *Pressemitteilung: VhU zum Planfeststellungsbeschluss Flughafenausbau, 18. Dezember 2007*. Vereinigung der hessischen Unternehmerverbände e.V. (VhU).

Volden, Craig, Michael M. Ting, and Daniel P. Carpenter (2008). "A Formal Model of Learning and Policy Diffusion." In: *The American Political Science Review* 102.III, pp. 319–332.

Vos, Ellen (2010). "Responding to Catastrophe: Towards a New Architecture for EU Food Safety Regulation¿' In: *Experimentalist Governance in the European Union. Towards a New Architecture*. Ed. by Charles F. Sabel and Jonathan Zeitlin. Oxford etc.: Oxford University Press, pp. 151–176.

Wadle, Nina (2013). *Privatisierung im Deutschen Strafvollzugs*. Vol. 115. Schriften zum Strafrecht und Strafprozeßzecht. Frankfurt am Main etc.: Peter Lang.

Walter-Rogg, Melanie (2013). "Metropolitan Governance over Federal State Borders in Germany: The Example of the Regions of Hamburg and Central Germany." In: *Governance and Finance of Metropolitan Areas in Federal Systems*. Ed. by Enid Slack and Rupak Chattopadhyay. Ontario: Oxford University Press, pp. 106–154.

Watts, Ronals L. (2008). *Comparing Federal Systems*. 3rd ed. Vol. 50. Queen's Policy Studies Series. Montreal: McGill Queen University Press.

WB (1994). *Development in Practice: Governance. The World Banks Experience*. Washington: The World Bank (WB).

– (2014). *World Development Indicators: Urban Population*. The World Bank (WB), Website: http://data.worldbank.org – rev. November 18th, 2014.

Weaver, R. Kent and Bert A. Rockam (1993). *Do Institutions Matter? Government Capabilities in the United States and Abord*. Washington D.C.: The Brookings Institution.

Weber, Max (1999). *Die Stadt*. Vol. 1. Gesammelte Aufsätze zur Religionssoziologie. Tübingen: J.C.B. Mohr (Paul Siebeck).

– (2006). *Wirtschaft und Gesellschaft*. Parderborn: Voltmedia.

– (2011). *Wirtschaftsgeschichte. Abriss der universalen Sozial- und Wirtschaftsgeschichte*. 6th ed. Berlin: Dunker & Humbolt.

Weßels, Bernhard (2000). "Die Entwicklung des deutschen Korporatismus." In: *Aus Politik und Zeitgeschichte* 2000.B 26-27, pp. 16–30.

Wilde, Mathias (2014a). *Mobilität und Alltag. Einblicke in die Mobilitätspraxis älterer Menschen auf dem Land*. Studien zur Mobilitäts- und Verkehrsforschung. Wiesbaden: Springer VS.

– (2014b). *Sozialwissenschaftliche Mobilitätsforschung – Potentiale für die Planungspraxis?. Lecture at Goethe-Universität Frankfurt am Main, Institut für Humangeographie, Arbeitsgruppe Mobilitätsforschung, 15.05.2014*. Frankfurt am Main: Goethe-Universität Frankfurt am Main.

Williamson, Oliver Eaton (1975). *Markets and Hierarchies. Analysis and Antitrust Implications: A Study in the Economics of Internal Organization*. New York: Free Press.

– (1979). "Transaction-Cost Economics: The Governance of Contractual Relations." In: *Journal of Law and Economics* 22.II, pp. 233–261.

WK (2005a). *Ausbaugegner wollen Akteneinsicht. Ministerium sieht Ablauf in Gefahr, November 2005*. Wiesbadener Kurier, dpa.

– (2005b). *Doch Jumbos auf der Nordwest-Bahn?. Flughafen-Anhörung kommt nicht voran / Mainz beantragt Abbruch des Verfahrens / Wiesbaden hält sich zurück, September 2005*. Wiesbadener Kurier, Harald Kaliwoda.

Womack, James P., Daniel T. Jones, and Daniel Roos (1990). *The Machine that Changed the World*. New York etc.: Rawson Associates etc.

Wonka, Arndt (2008). *Die Europäische Kommission. Supranationale Bürokratie oder Agent der Mitgliedsstaaten?* Vol. 15. Regieren in Europa. Baden-Baden: Nomos.

Wood, Robert C. (1958). "Metropolitan Government, 1975: An Extrapolation of Trends the New Metropolis: Green Belts, Grass Roots or Gargantua¿' In: *The American Political Science Review* 52.I, pp. 108–122.

Woolcock, Michael (2001). "The Place of Social Capital in Understanding Social and Economic Outcomes." In: *Isuma Canadian Journal of Policy Research* 2.I, pp. 11–17.

Wörner, Johann-Dietrich (2004). "Mediation und regionales Dialogforum – Idee und Realität." In: *Elftes Kolloquium Luftverkehr an der Technischen Universität Darmstadt. "Neue Herausforderungen im Luftverkehr"*. Ed. by TUD. Darmstadt: Technische Universität Darmstadt (TUD), pp. 95–106.

– (2010). *Regionales Dialogforum Flughafen Frankfurt. Abschlussdokumentation RDF 2000-2008*. Wiesbaden: Hessische Staatskanzlei.

Wright, Mike et al. (2013). "Introduction." In: *The Oxford Handbook of Corporate Governance*. Ed. by Mike Wright et al. Oxford: Oxford University Press, pp. 1–19.

WT (2003). *Scoping: Fraport mauert. Bürgermeister: "Fliegen, was das Bahnen-system hergibt", 17. April 2002*. Wiesbadener Tagblatt.

WUT (2003). *Pressemitteilung: W.U.T. fordert Einbeziehung des Taunus in die Umweltverträglichkeitsprüfung, 5. April 2003*. Wohnen und Umwelt im Taunus e.V. (WUT).

Zangel, Bernhard and Michael Zürn (2004). "Make Law, not War: Internationale und transnationale Verrechtlichung als Baustein für Global Governance." In: *Verrechtlichung - Baustein für Global Governance?*

Ed. by Bernhard Zangel and Michael Zürn. Vol. 18. EINE Welt. Texte der Stiftung Entwicklung und Frieden. Bonn: J.H.W. Dietz, pp. 12–45.

Zeit, Die (1982). *Das Recht hinter der Scheibe. Prozeßgegen Schubart wird zum Spektakel.* Die Zeit, Claudia Dillmann.

Zeitlin, Jonathan (2005). "Social Europe and Experimentalist Governance: Towards a New Constitutional Compromise¿' In: *EU Law and the Welfare State In Search of Solidarity.* Ed. by Gráinne De Búrca. The Collected Courses of the Academy of European Law. Oxford & New York: Oxford University Press, pp. 213–241.

Zimmermann, Karsten and Huber Heinelt (2012). *Metropolitan Governance in Deutschland. Regieren in Ballungsräumen und neue Formen politischer Steuerung.* Wies-baden: VS Verlag für Sozialwissenschaften.

Zunic, Rocco (2007). *Value-chain governance in der Bio-Pharmazie am Standort Frankfurt am Main.* Frankfurt am Main: Hochschulschrift: Goethe-Universität Frankfurt.

Zürn, Michael (2006). "Global Governance." In: *Governance-Forschung. Vergewisserung über Stand und Entwicklungslinien.* Ed. by Gunnar Folke Schuppert. 2nd ed. Baden-Baden: Nomos, pp. 121–146.

– (2008). "Governance in einer sich wandelnden Welt – eine Zwischenbilanz." In: *Governance in einer sich wandelenden Welt.* Ed. by Gunnar Folke Schuppert and Michael Zürn. PVS – Politische Vierteljahresschrift, Sonderheft 41/2008. Wiesbaden: VS Verlag für Sozialwissenschaften, pp. 553–580.

Kolibri 360 GmbH
Ferdinand-Jühlke-Straße 7
99095 Erfurt, Deutschland
produktsicherheit@kolibri360.de

Druck:
CPI Druckdienstleistungen GmbH
im Auftrag der
Kolibri 360 GmbH
Ferdinand-Jühlke-Str. 7
99095 Erfurt